Biogeochemistry of Environmentally Important Trace Elements

ACS SYMPOSIUM SERIES **835**

Biogeochemistry of Environmentally Important Trace Elements

Yong Cai, Editor
Florida International University

Olin C. Braids, Editor
O.C. Braids & Associates, LLC

American Chemical Society, Washington, DC

Library of Congress Cataloging-in-Publication Data

Biogeochemistry of environmentally important trace elements / Yong Cai, editor, Olin C. Braids, editor.

 p. cm.—(ACS symposium series ; 835)

 "Developed from a symposium sponsored by the Division of Environmental Chemistry, at the 221st National Meeting of the American Chemical Society, San Diego, California, April 1–5, 2001"—P.

 Includes bibliographical references and indexes.

 ISBN 0–8412–3805–7

 1. Arsenic cycle (Biogeochemistry)—Congresses. 2. Mercury cycle (Biogeochemistry)—Congresses. 3. Selenium cycle—Congresses.

 I. Cai, Yong, 1961- II. Braids, O. C. III. American Chemical Society. Meeting. (221st : 2001 : San Diego, Calif.). IV. Series.

QH344 .B5723 2002
628.5′2—dc21   2002026272

The paper used in this publication meets the minimum requirements of American National Standard for Information Sciences—Permanence of Paper for Printed Library Materials, ANSI Z39.48–1984.

Copyright © 2003 American Chemical Society

Distributed by Oxford University Press

PRINTED IN THE UNITED STATES OF AMERICA

Foreword

The ACS Symposium Series was first published in 1974 to provide a mechanism for publishing symposia quickly in book form. The purpose of the series is to publish timely, comprehensive books developed from ACS sponsored symposia based on current scientific research. Occasionally, books are developed from symposia sponsored by other organizations when the topic is of keen interest to the chemistry audience.

Before agreeing to publish a book, the proposed table of contents is reviewed for appropriate and comprehensive coverage and for interest to the audience. Some papers may be excluded to better focus the book; others may be added to provide comprehensiveness. When appropriate, overview or introductory chapters are added. Drafts of chapters are peer-reviewed prior to final acceptance or rejection, and manuscripts are prepared in camera-ready format.

As a rule, only original research papers and original review papers are included in the volumes. Verbatim reproductions of previously published papers are not accepted.

ACS Books Department

Contents

Indexes

Preface

During the past decade great advances have occurred in the under-standing of the significance of trace metals and metalloids (specifically arsenic, mercury, and selenium) in environmental systems. All three elements have been listed as metals of major interest to the U.S. Environmental Protection Agency (USEPA). In order to provide a unique opportunity for the exchange of information and ideas among scientists who work in the biogeochemistry of these elements, we organized a special symposium on the biogeochemistry of trace elements at the 221st American Chemical Society (ACS) National Meeting, which was held in San Diego, California, April 1–5, 2001. The ACS Division of Environmental Chemistry sponsored this symposium. The symposium highlighted the latest developments in the occurrence, transport, and transformation studies of arsenic, mercury, and selenium in the environment. Recent advances in analytical chemistry applicable to biogeochemical studies of these and other trace elements were also presented. Because the toxicity and environmental behavior of trace elements are dependent on their physicochemical forms, studies elucidating their speciation were especially emphasized in this symposium.

Most of the papers in this book were presented in the special symposium of the ACS National Meeting. Others not presented in the symposium are contributions from leading scientists in the field of biogeochemistry of these elements. The symposium and this volume make no attempt to cover every aspect of the biogeochemistry of these elements. Instead, we have attempted to emphasize the areas that are at the center of attention in the scientific community because of their key roles in describing the biogeochemistry of these elements. Three main themes are covered in these chapters: chemical speciation, transport between phases, and transformation and chemical reactions involved in the fate and transport of these trace elements. These main themes are supplemented by some detailed case studies. A few papers are also included to represent the current status on the study of trace element cycling in some localized geographical regions.

We believe that this book addresses some recent accomplishments and trends in the field of biogeochemistry (fate, transport, and transformation in soils, waters, and plants) of the environmentally important trace elements, arsenic, mercury, and selenium. The book is intended to reach a broad audience, including fellow researchers, graduate students, government scientists, and development specialists in this field.

Acknowledgments

We thank all the authors for their contribution to this volume. We also thank the peer reviewers of the chapters for their expertise and efforts. We are grateful to the ACS Division of Environmental Chemistry for the sponsorship of the symposium upon which this book is based. We appreciate the help of Ruth Hathaway and Michael Trehy of the ACS Division of Environmental Chemistry, during the preparation of the symposium and the book. We also acknowledge Stacy VanDerWall and Kelly Dennis in acquisitions and Margaret Brown in editing and production of the ACS Books Department for their help in coordinating the book.

We dedicate this book to Yin Chen; Peter and Eric Cai; and Elaine and Sally Braids.

Yong Cai
Department of Chemistry &Southeast
  Environmental Research Center
Florida International University
University Park
Miami, FL 33199

Olin C. Braids
O.C. Braids & Associates, LLC
Suite 150
9119 Corporate Lake Drive
Tampa, FL 33634

Chapter 1

Biogeochemistry of Environmentally Important Trace Elements: Overview

Yong Cai

Department of Chemistry and Southeast Environmental Research Center,
Florida International University, Miami, FL 33199

The papers included in this book cover three main themes: speciation, transport between phases, and transformation and chemical reactions involved in the biogeochemistry of arsenic, mercury, and selenium. These main themes are supplemented by some detailed case studies. A few papers are also included to represent the current status on the study of trace element cycling in some localized geographical regions.

Trace elements that are of environmental importance include many that are listed in the periodic table as it is these elements that constitute the earth itself. It is, of course, not our intention, for the symposium or this book, to include all these elements. Rather we focus on several elements, arsenic, mercury, and selenium, that have drawn researchers' great attention in the past decade. All three elements have been listed as metals of major interest to the U.S. EPA (*1*).

Arsenic is an element of great concern in the terrestrial as well as aquatic environments because of the high toxicity of certain arsenic species (*2-4*). The natural occurrence of arsenic in the environment is usually associated with sedimentary rocks of marine origin, weathered volcanic rocks, geothermal areas, and fossil fuels. Most of the arsenic derived from anthropogenic sources is released as a by-product of mining, metal refining processes, the burning of fossil fuels, and agricultural use (*4-6*). Recent research suggests that arsenic in

1

drinking water may be more dangerous than previously believed (7). A recent report by the National Academy of Sciences concluded that the previous arsenic standard of 50 µg/L in drinking water does not achieve the U.S.EPA's goal of protecting public health and should be lowered (2,3). EPA has recently decreased its drinking water standard from 50 to 10 µg/L, effective January 1, 2003, to adequately protect public health. The environmental impacts of arsenic contamination and the implementation of new regulations reducing arsenic in drinking water have resulted in a need for detailed studies of the biogeochemical cycling of arsenic and the development of arsenic decontamination technologies.

Mercury is one of the most prevalent and toxic contaminants in the environment (8-10). Mercury is emitted into the environment from a number of natural as well as anthropogenic sources. It has been suggested that anthropogenic emissions are leading to a general increase in mercury on local, regional and global scales (9). Among the metals and metalloids of concern for their potentially harmful effects in the environment, mercury is unique for a number of reasons (10). Mercury and some of its compounds are volatile and can be transported over a long distance, making target populations exposure to mercury potentially serious even in remote areas. Mercury can be efficiently transformed into its most toxic form, methylmercury, under environmental conditions. Mercury is perhaps the only metal which indisputably bioaccumulates and biomagnifies through the food chain, causing harmful effects to animals and humans. These facts have motivated intensive research on mercury as a pollutant of global concern (8,9)

Selenium is one of the most intensively studied inorganic components of diet. Ever since it was recognized in the 1950s that the element, which had until then been known only for its toxic effects, was also an essential nutrient, it has attracted growing interest in both human health and environmental fields of science (11,12). Selenium deficiency diseases and excesses resulting in toxicity in animals and human beings have been reported frequently (11-14, chapter 22 of this volume). There is a rather narrow range between selenium's action as toxicant and as a nutrient to human and animals. Selenium is also a toxicant and possibly a nutrient to plants. Selenium levels exceeding 2 mg kg-1 are found in many seleniferous soils throughout the western U.S., Ireland, Australia, Israel, and other countries (chapter 22 of this volume). These soils are derived from marine parent materials containing high levels of selenium. Anthropogenic sources of selenium in the environment include oil refining, mining, and fossil fuel combustion.

A number of books and reviews have been written on the biogeochemistry of arsenic, mercury, and selenium. To name only a few of them published in the last decade, *Biogeochemistry of Trace Metals* by Adriano, published in 1992 (15), covered a broad aspects on the occurrence, fate, and transport of trace metals in the environment. Stoeppler (16) was editor of a book which dealt with the sampling, sample treatment, chemical speciation, and environmental mobility of heavy metals, including mercury, arsenic, and selenium, in sediment and soils. In 1999, Ebinghaus et al. (9) edited a book which dealt with the characterization, risk assessment and remediation issues in mercury contaminated sites. *Arsenic in the Environment* edited by Nriagu, published in two volumes in 1994, describes in considerable detail the cycling, characterization, and health effects of arsenic (5,17). The Society of Environmental Chemistry and Health has been organizing, biennially, an international conference on arsenic exposure and health effects. Books based on these meetings have been published (18,19). Frankenberger and Benson (12) were editors of a book entitled *Selenium in the Environment* which addresses a number of important issues regarding the fate and transport of selenium in the environment. In 1996 Reilly (11) wrote a detailed monograph on selenium in food and health.

In addition to the books, a number of monograph chapters and journal review articles have described certain phases of the biogeochemistry of these trace elements. One of them is the review paper published by Cullen and Reimer on *Arsenic Speciation in the Environment* (4), which is still a classic and highly cited in its field. Furthermore, reports and documents from various agencies are also important resources for the research in these areas (*e.g. 2,3,8*).

There have been great advances in the past decade in the understanding of the biogeochemical cycling of arsenic, mercury, and selenium in environmental systems. This book is the result of an ACS symposium on trace element biogeochemistry at the 221st ACS National Meeting, which was held in San Diego, April 1-5, 2001. The symposium and the present volume make no attempt to cover every aspect of the biogeochemistry of these elements. Instead we have attempted to emphasize the areas of current interest in the scientific community because of their key roles in elucidating the biogeochemistry of these elements. Three main themes are covered in these chapters: Speciation, transport between phases, and transformation and chemical reactions involved in the fate and transport of these trace elements. These main themes are supplemented by some detailed case studies. A few papers are also included to represent the current status of the study of trace element cycling in some localized geographical regions.

Speciation

Speciation analysis of an element has been defined by Florence (20) as the determination of the concentrations of the individual physico-chemical forms of the element in a sample that together, constitute its total concentration. Speciation analysis involves a complex scheme of operations. Figure 1 illustrates an approach to differentiate speciation analysis into several categories.

4

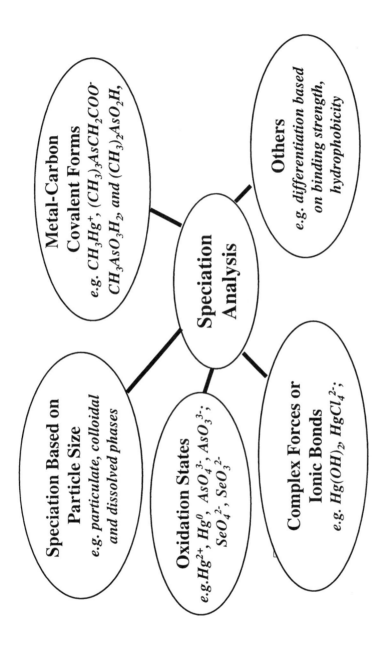

Figure 1 Scheme of an approach showing speciation of trace elements

There has been increasing interest over the past decade in speciation information about elements present in environmental and biological samples, since the toxicological and biogeochemical importance of many metals and metalloids greatly depends upon their chemical forms (21,22). The determination of the total amount of an element is important, but is insufficient to assess its toxicity and overall biogeochemical cycling. Information about concentrations of the individual species of an element, including its organic derivatives, is particularly crucial. Frequently the lack of the speciation information is the major limitation to our understanding of the biogeochemical cycling of the element.

The identification of the chemical forms of an element has become an important and challenging research area in environmental and biomedical studies. Not only should the separation steps (which are required either because of the insufficient sensitivity of the detection technique and/or the problems of interferences caused by concomitant elements) be considered closely in order to maintain the integrity of the species, but the detection techniques should allow both species detection and characterization. Mercury, arsenic, and selenium are three of these elements whose speciation is of particular interest.

Arsenic occurs in a variety of chemical forms in water as a result of many chemical and biological transformations in the aquatic environment. In a review paper (Chapter 2), Watt and Le detail the arsenic species detected in groundwater, surface freshwater, estuarine/coastal, and seawater and the variables that affect the speciation. In Chapter 3, Lee and Nriagu present their work performed to characterize the nature of the interactions between carbonate ions and As(III). Their results from this study, along with data reported previously by the same group (23), seem to support the formation of arsenic carbonate complexes. Arsenic speciation in soils and sediments is much more complicated compared to the speciation in water. Loepper et al. (Chapter 4) describe the assessment of quantity and speciation of arsenic in soils by chemical extraction.

Many different methods have been developed to accurately determine the concentrations of mercury species in various environmental matrices. Only recently has inductively coupled plasma mass spectrometry (ICP/MS) emerged as a competitive technique for mercury species analysis. Hintelmann and Ogrinc (Chapter 21) give a comprehensive description of the various methods in use in their laboratory for mercury species determinations in a variety of matrices, focusing especially on water column measurements. They describe the concept of speciated isotope dilution and stable tracer techniques introducing a novel approach to accurately calculate individual mercury isotope concentrations in multiple isotope addition experiments.

In nature, selenium can be found in a variety of chemical forms at different oxidation states; for example, Se(VI) is usually present as the oxyanion selenate

(SeO_4^{2-}), Se(IV) is present as selenite ($HSeO_3^-$, SeO_3^{2-}), and Se(0) is present as a solid, including both red monoclinic and gray hexagonal forms. Se(-II) is found in a variety of organic compounds, including selenoamino acids and volatile Se forms such as dimethylselenide (DMSe) and dimethyldiselenide (DMDSe). Inorganic selenides, such as hydrogen selenide and metal selenides, are also possible. Fox et al. (Chapter 22) review the recent findings on selenium speciation in soils and plants, with a particular emphasis on reactions and processes important in the field.

Transport between phases

Transport of elements between different phases (water/air; sediment/water; soil to plant, etc.) is clearly one of the key factors that determine the final fate and cycling of these elements in the environment. Several review papers and case studies address this issue in this book. Compared to the intensive studies on the transport of arsenic crossing the water/sediment boundary, arsenic transport from soils to plants has gained little attention. Phytoremediation, an emerging, plant-based technology for the removal of toxic contaminants from soil and water is a potentially attractive approach (*24-26*). This technique has received much attention lately as a cost-effective alternative to the more established treatment methods used at hazardous waste sites. The recent discovery of arsenic hyperaccumulators (*27,28*) will stimulate the research on arsenic uptake by plants. Cai and Ma (Chapter 8) review the mechanisms on the metal tolerance, accumulation, and detoxification in plants with particular emphasis on arsenic in terrestrial plants. In Chapter 9, research results on uptake of arsenic by plants in Southeast England are described by Farago et al.

Many elements are involved in the transfer of gaseous species between earth surface and the atmosphere (Chapters 10-12). While there have been abundant studies and evidence involving the lighter elements (e.g. C, N, S etc), the atmospheric transfer of the heavier elements is much less recognized. The importance of volatile chemical forms of selected trace elements in the environment, including arsenic, mercury, and selenium, have been discussed by several authors. Feldmann (Chapter 10) describes the volatilization of metals from a landfill site in Delta (British Columbia). In Chapter 11, Hirner discusses the chemical and toxicological aspects of the volatile metal and metalloid species associated with waste materials. Tessier et al. (Chapter 12) present their recent results on the biogenic volatilization of trace elements from European estuaries. Despite progress made in the last decade in the research of mercury emission from soils, considerable knowledge gaps still exist. In a comprehensive review, Zhang et al. (Chapter 18) discuss these uncertainties and identify some future priorities for the research of soil mercury emission. This would promote a more extensive and critical assessment of our understanding on soil mercury emission and global mercury cycling. The importance of microbial processes in the regulation of dissolved gaseous mercury (DGM) in deep freshwater has not been previously investigated. Siciliano et al. (Chapter 17) evaluate microbial mercury reductase and oxidase activities in a depth profile from Jack's lake, Canada, and determine the DGM depth profiles for four sampling stations on Lake Ontario, Canada. Their results indicate that microbial processes are an important factor regulating DGM in the hypolimnion. They also discuss the importance of DGM to atmospheric flux rates.

Transformation and Chemical Reactions

As indicated in Figure 1, a variety of forms of these three elements exist in environmental systems. These trace metal species are formed via biotransformation and chemical reactions modulated with redox and pH conditions. A better understanding of these reactions is the major objective to allow a critical assessment of the biogeochemistry of these elements. For example, the complexation and oxidation of arsenite are important pathways in the overall environmental cycling of arsenic. In order to identify soil components responsible for As(III) oxidation and the surface sites which bind the As(V) product, Manning and co-workers (Chapter 5) investigate the complexation and oxidation reactions of As(III) in three soils using standard batch reactions and X-ray absorption spectrometry. The chemistry of arsenic in aquatic environments is complex because of its multiple oxidation states and its association with a variety of minerals through adsorption and precipitation. In Chapter 6, Meng and co-workers provide a detailed review on the redox transformations of inorganic arsenic with the aim of understanding its mobility in aquatic environments. Hydroxyl and superoxide anion radicals are formed naturally and may contribute significantly to redox processes of arsenic in the environment. Studies in this area are scarce. In Chapter 7, Motamedi et al. present the results of their study on the reactions of ultrasonically generated hydroxyl radical with arsenic species in the presence of oxygen and argon over a range of solution pH and arsenic concentrations. This fundamental research on the oxidation of As(III) provides us useful information about the transformation between As(III) and As(V).

While inorganic mercury is the major source of mercury to most aquatic systems, it is methylmercury (CH_3Hg) that bioconcentrates in aquatic food webs and is the source of health advisories worldwide that caution against the consumption of fish containing elevated CH_3Hg (8). Although extensive research has been conducted over the last four decades, further studies are definitely needed for a better understanding of the processes and factors affecting the methylation and bioconcentration of mercury through food webs. A comprehensive review by Benoit and co-workers (Chapter 19) address the geochemical and biological controls over CH_3Hg production and degradation in aquatic ecosystems.

Despite the intensive research efforts on selenium cycling, little is known about the resulting speciation of selenium with mineralization of organic selenium compounds and no information is available for determining the importance of the different decomposition pathways for organic sulfur and selenium present in terrestrial soils. Martens and Suarez (Chapter 23) determine the decomposition of sulfur and selenium pathway intermediates in soil and evolution of volatile sulfur and selenium species indicative of the metabolizing pathways. Speciation of nonvolatilized selenium, following soil incubation with the selenium pathways intermediates, was also determined because the activity of methylation or demethylation pathways in soil may influence the accumulation of organic selenium compounds.

Other Case Studies

Several case studies which represent the current research in the general area of fate and transport of trace elements are included to illustrate actual locations where trace metals have posed significant problems or threats. Yellowknife, Canada has an extensive soil arsenic contaminant problem as a result of 60 years of gold mining activity. In Chapter 13, Reimer and co-workers describe an approach for characterizing arsenic sources and risk at the contaminated sites. Instead of measuring total concentration of arsenic, sequential selective extraction (SSE) and a stimulated gastric fluid extraction (GFE) were used to assess environmentally available and bioavailable fractions in soils. It was found that these techniques can be used to identify actual risks and develop effective remediation strategies. Carbonell-Barrachina et al. (Chapter 14) present a paper on arsenic biogeochemistry in acidified pyrite mine waste from the Aznalcóllar (Southwestern Spain) environmental disaster. Montezuma Well is part of Montezuma Castle National Monument, located in north-central Arizona. It contains elevated concentrations of geogenic arsenic. Compton-O'Brien et al. (Chapter 15) describe the study of occurrence and transport of arsenic in different matrices in this highly arsenic-enriched area. In Chapter 16, Wai and co-workers discuss Blackfoot disease and other related health problems associated with arsenic contamination of groundwater. In Chapter 20, Cossa and co-workers provide a synthesis of current knowledge about mercury dynamics and the state of contamination of the Seine estuary, France. After a study of the sources and level of mercury contamination in the Seine basin and an evaluation of internal inputs into the estuary, this paper considers the distribution of the metal between dissolved and particulate phases, the chemical reactions governing the speciation and bioavailability of mercury, the state of contamination in the biota, and temporal changes in contamination within the last decade. Nguyen and Manning (Chapter 25) describe the spectroscopic and modeling study of lead adsorption and precipitation reactions on a mineral soil. The results from EXAFS analysis revealed a possible lead inner-sphere adsorption mechanism that was used in conjunction with a surface complexation model to quantitatively describe lead binding to soil. Dai and co-workers (Chapter 24) studied the sorption behavior of butyltin compounds in estuarine environments of the Haihe River, China. It is concluded that at acidic condition (pH<4), tributyltin (TBT$^+$) ion binds mainly to the permanent negative mineral charge, while at pH 4-10, ion exchange with =XOH(M) groups in the mineral and complexion with =ROH(M) groups in humic matter with TBT$^+$ ion appear to be the main mechanisms. When pH was higher than 10, hydrophobic partitioning of TBTX (X=OH, Cl etc.) neutral molecules into humic matter seemed to control the sorption and gave a second sorption peak. For dibutyltin (DBT) and monobutyltin (MBT), sorption is most likely controlled by their cationic character. Speciation of divalent free ions in marine waters based on a chelation mechanism has potential applications in metal speciation studies. In Chapter 26, Li and co-workers describe their study using this method to investigate heavy metal concentrations and speciation in Hong Kong coastal waters.

It is the editors' aspiration that this volume will edify and assist researchers in the field and stimulate new avenues of inquiry.

Acknowledgements

I would like to thank Olin Braids for reviewing this paper. I would also thank Anita Holloway for her assistance for the preparation of this manuscript. This is contribution # 170 of the Southeast Environmental Research Center at FIU.

References

1. McKinney, J.; Rogers, R. *Environ. Sci. Technol.* **1992,** *26,* 1298-1299.
2. Agency for Toxic Substances and Disease Registry (ATSDR). *Toxicological Profile for Arsenic*; U.S. Department of Health and Human Services: Atlanta, GA, 1999.
3. National Research Council. *Arsenic in Drinking Water*, National Academy Press: Washington, DC, 1999.
4. Cullen, W. R.; Reimer, K .J. *Chem. Rev.* **1989,** *89,* 713-764.

5. *Arsenic in the Environment, Part 1: Cycling and Characterization;* Nriagu, J. O. Ed.; John Wiley & Sons: New York, 1994; pp 1-430.
6. Welch, A. H; Westjohn, D. B.; Helsel, D. R.; Wanty, R. B. *Ground Water,* **2000,** *38,* 589-604.
7. Kim, M. J.; Nriagu, J. *Sci. Total Environ.* **2000,** *247,* 71-79.
8. U.S. EPA, Office of Air Quality Planning and Standards and Office of Research and Development. *Mercury Study Report to Congress,* EPA-452/R-97-005. 1997, Vol. 1, Executive Summary; Vol. III. Fate and Transport of Mercury in the Environment.
9. *Mercury Contaminated Sites, Characterization, Risk Assessment and Remediation*; Ebinghaus, R.; Turner, R. R.; de Lacerda, L. D.; Vasiliev, O.; Salomons, W. Eds.; Springer: Berlin, 1999; pp 1-538.
10. *Lead, Mercury, Cadmium and Arsenic in the Environment, SCOPE 31;* Hutchinson, T. C.; Meema, K. M. Eds.; John Wiley & Sons: New York, 1987; pp 1-360.
11. Reilley, C. *Selenium in Food and Health;* Blackie Academic & Professional: London, 1996; pp 1-338.
12. *Selenium in the Environment,* Frankenberger Jr., W. T.; Benson, S. Eds.; Marcel Dekker, Inc.: New York, 1994; pp 1-736.
13. Wang, D.; Alfthan, G.; Aro, A, *Environ. Sci. Technol.* **1994,** *28,* 383-387.
14. Sanz Alaejos, M.; Diaz Romero, C. *Chem. Rev.* **1995,** *95,* 227-257.
15. *Biogeochemistry of Trace Metals,* Adriano, D. C. Ed.; Lewis Publishers: Boca Raton, 1992; pp 1-513.
16. *Hazardous Metals in the Environment,* Stoepler, M. Ed.; Elsevier: Amsterdam, 1992; pp 1-541.
17. *Arsenic in the Environment, Part II: Human health and Ecosystem Effects;* Nriagu, J. O. Ed.; John Wiley & Sons: New York, 1994; pp 1-293.
18. *Arsenic Exposure and Health Effects,* Chappell, W. R.; Abernathy, C. O.; Calderon, R. L. Eds.; Elsevier: Oxford, UK, 1999; pp1-416.
19. *Arsenic Exposure and Health Effects,* Chappell, W. R.; Abernathy, C. O.; Calderon, R. L. Eds.; Elsevier: Oxford, UK, **2001**; pp1-416.
20. Florence, T. M. *Talanta,* **1982,** *29,* 354-373.
21. Lund, W. *Fresenius J. Anal. Chem.* **1990,** *337,* 557-564.
22. Van Loon, J. C.; Barefoot, R. R. *Analyst,* **1992,** *117,* 563-570.
23. Kim, M. J.; Nriagu, J.; Haack, S. *Sci. Total Environ.* **2000,** *34,* 3094-3100.
24. U.S. EPA, Office of Research and Development. *Introduction to Phytoremediation,* EPA/600/R-99/107. 2000.
25. Dahmani-Muller, H.; van Oort. F.; Gelie, B.; Balabane, M. *Environ. Pollu.* **2000,** *109,* 231-238.
26. Salt, D. E; Smith, R. D; Raskin, I. *Annu. Rev. Plant Physiol. Plant Mol. Biol.* **1998,** *49,* 643-668.
27. Ma, L. Q.; Komart, K. M.; Tu, C.; Zhang, W.; Cai, Y.; Kennelly, E. D. *Nature,* **2001,** *409,* 579.
28. Francesconi, K.; Visoottiviseth, P.; Sridokchan, W.; Goessler, W. *Sci. Total Enviorn.* **2002,** *284,* 27-35.

Chapter 2

Arsenic Speciation in Natural Waters

C. Watt and X. C. Le

**Department of Public Health Sciences, University of Alberta, 10–102
Clinical Sciences Building, Alberta T6G 2G3, Canada**

Speciation studies are necessary to understand the
biogeochemical cycling of arsenic in aquatic systems. The
species of arsenic present, their behaviour and toxicity will
change depending on the biotic and abiotic conditions in the
water. In groundwater, arsenic is predominantly present as
arsenite (As^{III}) and arsenate (As^V). The major arsenic species
in freshwater are As^{III} and As^V and minor amounts of MMA ,
DMA and methylated As^{III} have also be detected. In seawater,
the arsenic speciation differs in the surface and deep zone. In
addition to the above species, uncharacterized arsenic species
may constitute a significant portion of the total arsenic present
in some water and the identification of these compounds is
necessary to fully understand the arsenic biogeochemistry in
water.

A knowledge of the speciation of As in natural water is important because
the bioavailability and the physiological and toxicologial effects of As depend on
its chemical forms. Arsenic occurs in a variety of chemical forms in water, and
this is a result of many chemical and biological transformations in the aquatic
environment. The mechanism of transport and biological availability of arsenic
depends on the chemical species present.
The main arsenic compounds detected in natural waters are arsenite [As^{III}],

arsenate [As^V], monomethylarsonic acid [MMA^V], monomethylarsonous acid [MMA^{III}], dimethylarsinic acid [DMA^V], dimethylarsinous acid [DMA^{III}], and trimethylarsine oxide [TMAO]. Since there are large differences in chemical behaviour and toxicity among the arsenic species, the determination of speciation is very important in the study of the biogeochemistry of arsenic.

Arsenic speciation and distribution is affected by redox processes. These processes are complicated by adsorption, coprecipitation, desorption and biological mediation. This chapter will examine the species of arsenic detected in groundwater, surface freshwater, estuarine/coastal, and seawater and the variables that affect the speciation. Several reviews have discussed earlier work on these topics (1-4). This paper will emphasize on more recent research.

Groundwater

The major arsenic species in groundwater are the inorganic arsenite (As^{III}) and arsenate (As^V). Extremely high levels of inorganic arsenic in well water have been attributed to devastating arsenic endemic episodes (5-8). In West Bengal, 47 of 47000 tubewell samples contained total arsenic concentrations above 1000 μg/L (7). One hundred and eighty-nine of 9640 tubewell water samples from Bangladesh contained more than 1000 μg/L (7). In Inner Mongolia, approximately 20% of 305 shallow wells tested measured As concentrations between 50 μg/L and 1860 μg/L and 55% of deep wells had concentrations between 50 μg/L and 360 μg/L (8). As^{III}, As^V, (MMA^V), (DMA^V) and trimethylarsenic (TMA including trimethylarsine and trimethylarsine oxide) have been detected in ground water (Table I).

Arsenic is present in groundwater primarily as a result of the dissolution of naturally occurring arsenic-rich minerals and organic matter (9-12). Arsenic sulfides and sulfosalts are one possible source of arsenic in groundwater (10). In the Zimapan Valley, Mexico, the main identified arsenic minerals were arsenopyrite (FeAsS) and the secondary mineral scorodite (FeAsO$_4$.H$_2$O) (12). The oxidation of arsenopyrite releases sulfates and arsenic to the groundwater (11). Arsenopyrite is stable at high pH and low redox potential and becomes less stable as slightly more oxidizing waters move through resulting in arsenic dissolution (12). The reductive dissolution of arsenic-rich iron oxyhydroxides also results in the release of arsenic (13-15). The reduction of arsenic-rich iron oxyhydroxides is influenced by microbial action (15,16). The microbial mediated reduction of As^V sorbed to ferrihydrite in the absence of significant reductive dissolution of the Fe^{III}-oxide solid phase has been studied (17). A *Clostridium sp.* strain was able to reduce aqueous As^V within one day, however, the microbe was not able to cause a significant desorption of As^V from ferrihydrite in 24 days. The sorbed As remained predominantly as As^V and was

Table I: Arsenic Species Detected in Water[a]

Location	As^{III}	As^V	MMA	DMA	Other	Unidentified	Refs.
Groundwater							
Bangladesh (Tubewell for drinking)	30-1200[b]						92
West Bengal,India,Bangladesh (range of means from different districts)	58-164	184-275					7
India	Present in 1 sample	N.D.-133.6					93
Taiwan (Pu-Tai) endemic area	285-683 (462±129)	33-362 (177±109)	<3	<7			94
Taiwan (I-Lan) comparsion area	537-637 (572±42)	24-67 (38±18)	<3	<7			94
Taiwan (Hsin-Chu) control area	<3	<12	<3	<7			94
Taiwan	~24-720[b]		0.5-4.2	2.0-6.9	TMA 3.3-5.1		95
Taiwan (Fushing)	70.2 ±2.6	870±26					96
Taiwan (Chiuying)	51.6 ±1.8	601±22					96
Mexico	Traces-217	4-604	<3	Trace-20			32
Japan	15-70	11-220					31
Germany	N.D.-176.3	N.D.-147.7	<0.001	<0.001			30
Canada (Bowen Island, British Columbia)	0.1-220	0.1-477					97
South West United States	<5	21-120					98
Rivers							
Haya-kawa River, Japan	28[b]		N.D.	N.D.	TMA 2		60
Four rivers in California	0.7-7.4[b]		N.D.	N.D.			37
Hillsborough River, Tampa, Florida	N.D.	0.25	N.D.	N.D.			42
Withlacoochee River, Tampa, Florida	N.D.	0.16	0.06	0.3			42
River Water Reference Material	0.16±0.01	0.18±0.02	<0.02	0.05±0.01	TMA <0.01	0.12±0.02	84
Lakes/Ponds/Reservoirs							
Lake Biwa, Japan	~0.2	~1.9	<0.05	<0.76	MMA^{III} <0.012 DMA^{III} <0.014		44

Continued on next page.

Table 1. *Continued*

Location	As^{III}	As^V	MMA	DMA	Other	Unidentified	Refs.
SubArctic Lakes[f] (pH 1) Yellowknife, NWT	0-22	0.7-520	N.D.-0.5	N.D.-0.7	TMA N.D.-0.2	N.D.-0.6	58
SubArctic Lakes[f] (pH 6) Yellowknife, NWT	17250[h]		N.D.-0.04	N.D.	TMA N.D.	N.D.-1.2	58
Mono Lake, Calif. (Alkaline)	97.5[h]		N.D.	N.D.			37
Pyramid Lake, Calif.	0.3-11.2[h]		N.D.	1.1			37
Six lakes in California	1.1[h]	0.01-1.5	N.D.-0.6	0.03-2.4			37
Elkhorn Slough, Calif.			0.07	0.2			37
Davis Creek Reservoir/natural (filtered samples)	Trace-1.2		0.2-0.3	0.03-0.9			37
Davis Creek Reservoir/natural (unfiltered samples)	0.8-2.6[h]		0.2-0.3	N.D.-0.8			37
Lake Echols, Tampa, Florida	2.7	0.4	0.1	0.3			42
Lake Magdalene, Tampa, Florida	0.9	0.5	0.2	0.1			42
Remote Pond, Withlacoochee Forest, Tampa, Florida		0.3	0.1	0.6			42
Research Pond, Tampa, Florida	0.8	1.0	0.05	0.1			42
Pavin Lake, France	0-8.2	0.28-4.4	Not quantified	Not quantified			46
Seawater							
Pacific Ocean	1.1-1.6[h]		0.009-0.02	0.02-0.2			77,78
Antarctic Ocean (average)	0.003	1.0	0.007	0.02			64
East Indian Ocean (average)	0.2	0.4	0.03	0.05			64
North Indian Ocean (average)	0.2	0.5	0.03	0.2			64
North Sea (over several months)	0.1-1.5[h]		Trace-0.05	0.09-0.3			99
China Sea (average)	0.2	0.3	0.02	0.08			64
Indonesian Archipelago (average)	0.2	0.4	0.03	0.09			64
Estuaries							
Seine Estuary, France (over several months)	N.D.-1.2	0.4-2.1	N.D.-0.05	N.D.-0.3			68

Location							Ref
Continental Shelf off the Gironde Estuary, France	0.01-0.3	0.8-1.5	N.D.	0-0.1			86
Tamar Estuary, South-west England[c]	2.7[b]		0.2-0.5	0.02-1.3			74
Tamar Estuary (Porewaters) South-west, England[c]	29.6[b]		0.4	0.5			74
Tamar Estuary (Porewaters), South-west, England[c]	5-60[b]		0.04-0.7	0.1-0.4			71
Southampton Water, U.K.	0.03-0.1	Not analyzed	0.03-0.08	0.05-0.5			67
Humbar Estuary, U.K. (over four seasons)	0.7-2.3[b]		N.D.-0.02	N.D.-0.2			69
Tagus Estuary, Atlantic coast of Europe[c]	2.8-14.7[b]		0.01-0.06	0.07-0.2	TMA 0.01-0.04	28-39% in 4 subsamples	100
Patuxent River Estuary, Maryland (over two years) (For DMA, only maximum values included in range)	0.1-0.2	0.1-1.1	<0.3	0.2-0.6			75
Suisun Bay, Calif.	1.5[b]		0.1	0.07			37
Bay, Causeway, Tampa, Florida	0.1	1.4	N.D.	0.2			42
Tidal Flat, Tampa, Florida	0.6	1.3	0.08	0.3			42
McKay Bay, Tampa, Florida	0.06	0.3	0.07	1.0			42

[a]All concentrations in μg/L. N.D. = species non-detected
[b]The sum of $As^{III} + As^V$
[c]Anthropogenically influenced.

not solubilized during microbial growth. It has also been suggested that As is leached into groundwater because of an interaction between carbonate ions and As minerals in rocks (*10*). In laboratory experiments, bicarbonate (HCO_3^-) was able to leach As from sulfide minerals under anaerobic conditions (10).

Species Stability, Preservation and On-site Speciation

As^V has commonly been reported as the predominant water-soluble species in groundwater. However, studies now show that As^{III} may be present in higher concentrations. As^{III} was determined to be the predominant species in most groundwater samples measured from tanks and wells in Hanford, Michigan (*18*). In thirteen of sixteen wells and three of four tanks, As^{III} was present as $86 \pm 6\%$ of the total arsenic (*18*). The procedures used for sample handling and analysis and the sample matrix may result in the oxidation of As^{III} to As^V(*19-22*). As the methods of sampling, preservation and analysis improve, there is increasing evidence that As^{III} might be more prevalent (*19-23*).

The instability of arsenic species in water samples is very important. As^{III} is more difficult than As^V to remove from water (1). Therefore, it is important to determine the correct concentrations of the two species. Many preservation methods have been used to maintain the original arsenic species distribution in water samples. Edwards et al. used ascorbic acid and HCl as preservatives in different types of water (24). In spiked synthetic water both preservatives were able to maintain the concentration (4 μg/L) of As^{III} and As^V for 28 days. In spiked natural waters with ascorbic acid, As^{III} slowly transformed to As^V during 8 days of observation. To prevent iron precipitation, HCl (25) and EDTA (26) have been used to preserve arsenic species in iron-rich (Fe^{III}) drinking waters. Both HNO_3 and HCl caused As^{III} oxidation to As^V in spiked deionized water and Ottawa River samples (27) Lindemann et al. tested the storage of water samples at temperatures of $-20°$, 3°C and 20°C and found that 3°C to be the most effective temperature to preserve arsenic species (28).

Methods of on-site separation of As^{III} and As^V species immediately after water sample collection are particularly useful. The portability of the methods allow for complete separation and no preservation of the samples are needed. Le et al. used disposable solid-phase cartridges for the speciation of particulate and soluble arsenic (*19*). A measured volume of a water sample is passed through a 0.45 μm membrane filter and a silica-based strong anion exchange cartridge connected in serial. Particulate arsenic is captured on the filter and the anion exchange cartridge retains As^V. Arsenite is not retained and is detected in the effluent. The anion-exchange cartridge is subsequently eluted with 1M hydrochloric acid (HCl) and the eluent is analyzed for As^V concentration. Kim also used a portable ion-exchange method with a strong anion-exchange resin in the field and found that neither As^V nor As^{III} changed its oxidation state during the experiment (20). However, upon examination of samples from private wells in Genesee County, Michigan, the oxidation of inorganic arsenite to arsenate occurred within three days of storage in a refrigerator (20). Immediately after sampling, As^{III} represented 82-95% of the total arsenic and after three days, 25-30% of the original As^{III} was oxidized to As^V (20). In another study, As^{III} dropped to approximately 70% of its original value after 6 months of storage (21).

Predicted and Observed Soluble Arsenic Speciation

Thermodynamic considerations alone predict that As^V should be the predominant arsenic species stable in oxygenated waters and As^{III} should be the only arsenic species stable in reductive environments (1,29). However, As^{III} and As^V have been detected in both oxidizing and reducing environments. The presence of dissolved iron (Fe^{II}) and manganese and little or no dissolved oxygen were considered indicative of a reducing environment (13). The oxidation state of arsenic depends on the redox potential, the pH and biological mediation (1).

As^V is favoured thermodynamically in water with a positive redox potential and high ionic strength (1,30,31). However, As^{III} has also been detected as the predominant species in groundwater under oxidative conditions (13,30). Kondo et al. found a linear correlation (-0.76) between the $\log[As^{III}/As^V]$ ratio and the redox potential (31). At redox potentials typical of natural waters, Eh-pH diagrams suggest that decreases in pH should increase the amount of As^{III} present compared to As^V (1). However, this is rarely the case. In well waters, Del Razo et al. could not find a relationship between the As^{III}/As^V ratio and the pH (6.3-8.8) of the samples (32). In this study, the extraction process from the aquifer could have resulted in changes in the oxidation state (32).

Dissolved organic carbon concentrations were found to be considerably higher in wells during times when As^{III} was present in contaminated wells in Kulheim, Germany (30). In one well with a positive redox potential, As^{III} was the dominant species. The authors suggest that this may be due to the stabilization of As^{III} by organic matter (30). The oxidation of arsenite by oxygen occurs very slowly (33). In one study where pure oxygen and air were used for oxidation, only 57% and 54% of As^{III} respectively, were oxidized after five days (33). Planar-Friedrich et al. examined 19 wells in the Rioverde basin, Mexico and six of them showed higher concentrations of As^{III} than As^V (34). These wells likely favoured reducing conditions. The authors hypothesize that the lacustrine sediments probably contain organic rich strata resulting in biological degradation and subsequent oxygen consumption.

Carbonate ions aid in the release of arsenic from minerals (10). Kim et al. suggest that complexes of As^{III} and carbonate ions $(As(CO_3)_2^-$ and $As(CO_3)(OH)_2^-$ may be incorrectly identified as As^V with the anion exchange resins used in many studies. Also, As^{III} and $AsCO_3^+$ may be mistaken for each other on the basis of anion exchange alone (*10*). Phosphate from fertilizer and decayed organic matter promoted the growth of sediment biota and may have enhanced the release of arsenic (*16*). Chemolithotrophic bacteria has been attributed to the reduction of arsenic-rich hematite and maghemite in the current tailings impoundment at Campbell Mine in Balmertown, ON, Canada (*36*).

Particulate Matter

In oxidizing conditions As^V becomes associated with particulate material and is released in reducing conditions. Suspended matter in natural water may occur as undissolved mineral and organic species that bind heavy metals. Arsenic will adsorb to particulate and this fraction can be substantial (*11,19,35*). As^V readily adsorbs onto iron(III) oxyhydroxide particles. As^V and As^{III} can react with sulfide ions to form insoluble arsenic sulfide precipitates (*10*). In groundwater samples in the United States, particulate arsenic accounted for more than 50% of the total arsenic in 30% of the samples collected (*35*). In one groundwater sample, particulate arsenic accounted for more than 96% of the total arsenic (*35*). In a site contaminated by mine wastes, the particulate material contained more than 220 times the amounts of As than in solution (11).

In sediments in parts of the Bengal basin in India and Bangladesh, As adsorbs to Fe-hydroxide coated sand grains and clay minerals (16). It is trapped in sediments and released due to reducing conditions and microbial action (*16*).

Fresh Surface Water

The geochemical cycle of arsenic in fresh surface water and the environmental biochemistry and speciation of arsenic have been examined in detail (*1,2,37*). The concentration of arsenic in freshwater systems varies considerably with the geological composition of the drainage area and the extent of anthropogenic input. Geothermal inputs are the source of elevated arsenic concentrations in some lakes and rivers (*38-40*). Chemical weathering and anthropogenic activities may also result in the contamination of surface water.

Typical concentrations of arsenic in freshwater have been reported as 0.1 – 80 µg/L (*1,41*). The major arsenic species in freshwater are inorganic As^{III} and As^V (Table I). Minor amounts of MMA^V, DMA^V and methylated As^{III} species can also be present in natural waters (*1,42-45*).

Fresh Surface Water Arsenic Speciation

In Lake Pavin, France, As^V represented 88-95% of total dissolved arsenic in the oxic compartment and As^{III} was the dominant species in the anoxic zone (50-90%) *(46)*. However, As^{III} has also been detected throughout the water column of the lake *(46)*. As^V was the dominant species present in the hypolimnion of Lake Biwa (northern basin) through all seasons *(45)*. During summer stratification in Lake Griefen, arsenite was dominate in the epilimnion, but in the anoxic hypolimnion, arsenate remained the dominant species *(47)*. Seyler & Martin suggest that a slow reduction rate from As^V to As^{III} may be responsible for the presence of As^V in the anoxic layers *(46)*.

The observed disequilibrium speciation in the anoxic layers of Lake Pavin is attributed to a kinetic control of the arsenate reduction reaction *(46)*. The theoretical As^{III}/As^V ratio was calculated to be 10^{10} in the redoxcline (50-65m), however, the observed ratio was measured at 1.05 *(46)*. In Upper Mystic Lake in the Aberjona Watershed, arsenic speciation also appeared kinetically controlled *(48)*. As^{III} represented 30% of the total arsenic for the colder months and $48\pm7\%$ in the summer and fall. Arsenate decreased from 62+/-28% of the total arsenic in the winter and early spring to $28\pm13\%$ in the summer months. The presence of As^{III} and methylarsenic in the surface water suggests biological transformation *(48)*. Laboratory studies have shown that the oxidation of geothermal As^{III} in streamwaters of the Eastern Sierra Nevada are likely microbially mediated *(49)*. Sohrin et al. found that the As^{III}/As^V ratio was lower during more extensive anoxia and suggested that the rates of reduction and oxidation may vary depending on physicochemical conditions and biotic composition *(45)*. It may be difficult to predict the arsenic species present based exclusively on redox status because dissimilatory reduction of As^V may also occur under aerobic conditions when facilitated by microorganisms *(50)*. Microbial populations from *Sphingomonas, Caulobacter, Rhizobium,* and *Pseudomonas* genera were isolated from limed reprocessed mine tailings, and all were capable of rapidly reducing As^V in aerated serum bottles *(50)*.

Mobility of Arsenic

The distribution and mobilization of the arsenic species in the sediments is controlled by redox potentials, pH, microbially mediated transformation and adsorption processes *(51 and refs therein)*. Reducing conditions beneath the sediment water interface results in the dissolution of iron oxyhydroxides and the release of As^V to the porewater and overlying water. Dissolved Fe^{II}

concentrations up to 41 mg/L in Lake Coeur D'Alene at a depth of 10 to 15 cm (below the sediment-water interface) indicates that oxygen is absent and iron is being reduced in this region (*52*). Reductive dissolution of ferric hydroxides is a microbially mediated process and two strains of dissimilatory Fe^{III}-reducing bacteria were isolated from surface sediments of Lake Coeur D'Alene, Idaho contaminated with mine tailings. (*52*).

The mobility of arsenic may increase as the sediment environment becomes more reducing. The As^V is either reduced to As^{III}, which is more mobile or reductive dissolution of iron occurs (*51*). In rivers, arsenic is mainly bound to sediments (*51*). Organic rich sediments create reducing conditions that reduce As^V to As^{III} and cause possible desorption of the less strongly held As^{III}. As^V is not necessarily reduced to As^{III} under reductive condtions. In laboratory experiments, the bacterium *Shewanella alga* strain BrY promoted As mobilization from a crystalline ferric arsenate as well as from sorption sites within whole sediments (*53*). As^V sorbed to sediments from Lake Coeur d'Alene, ID, a mining impacted environment enriched in both Fe and As, was solubilized by the activity of the bacterium *Shewanella alga* BrY and As^{III} wasn't detected (*53*). Therefore, the authors concluded that arsenic mobility can be enhanced by bacterial mediation in the absence of arsenic reduction.

Seasonal Variation

Arsenic concentrations and species change with seasons (*45,48,63*). In the summer, hypolimnion waters spiked with DMA^V decreased less than 20% whereas in the winter in oxic deep waters the decrease was 100% (*37*). The dominant redox reactions of inorganic As in Lake Greifen, Switzerland in the epilimnion were biologically mediated reduction by phytoplankton in summer and oxidation of As^{III} by Mn oxides in fall (*47*). In Lake Biwa, As^V was the dominant species during the winter mixing and decreased during summer stagnation (*45*). As^{III} increased in spring and fall and DMA^V was the dominant species in summer (*45*). MMA^{III} and DMA^{III} were present in small amounts in the epilimnion and hypolimnion of Lake Biwa and did not show significant seasonal change (*45*). DMA^{III} may be thermodynamically unstable in oxic aquatic solutions (*57*). The concentrations of methylarsenicals may also vary year by year. Hasewaga et al. reported that MMA^{III} was uniformly distributed throughout the water column of Lake Biwa in 1993 but increased throughout the water column in April and October 1994 (*44*).

Effects of Sulfide on Arsenic Speciation

The presence of sulfide will complicate As speciation (1). Sulfide may affect the species of Fe, Mn, and As present and their solubilities *(1,2,29,46,51)*. In the presence of sulfide and under low oxygen conditions, iron sulfide minerals may form and arsenic may coprecipitate with iron sulfides as arsenopyrite (FeAsS) or form arsenic sulfides *(1,29)*. At pH values below approximately 5.5 and Eh values ~0V, realgar (AsS) and orpiment (As_2S_3) will form *(29)*. Reduced As species existed in the sediment of a contaminated wetland collected from the deeper anoxic parts of the wetland in the spring and summer *(54)*. In the spring, there was 22% sorbed arsenite and 78% arsenic sulfide species in the solid phase *(54)*. In a study using municipal sewage sludge, Carbonell-Barrachina et al. found that at a redox potential of -250mV, arsenic solubility was controlled by the formation of insoluble sulfides and as a result soluble As contents significantly decreased as compared to levels measured at 0 mV *(55)*.

The increases of both As^V and As^{III} in the water column of Upper Mystic Lake were also correlated with that of sulfide *(56)*. Dissolved As^{III} is thermodynamically favored in sulfate-reducing waters and is not as strongly adsorbed as As^V *(1,56)*. As the release of sulfide diminishes, As^{III} is oxidized to As^V and As^V is scavenged by particles *(1,51,56)*.

Effects of Phytoplankton

The uptake of As^V by phytoplankton can lead to arsenic transformation in aerobic aquatic environments. It is not clear whether As^{III} and methylarsenicals are produced directly by phytoplankton or indirectly by degradation due to biotic composition differences and excretion and degradation pathways. The freshwater algae *Closterium aciculare* was able to produce As^{III} and methylarsenicals in the presence of As^V in culture mediums *(57)*. Arsenite and DMA^V were the major species present in the culture medium and the minor species included MMA^V, DMA^{III}, and MMA^{III} *(57)*. In the epilimnion of Lake Biwa, the As^{III} appeared to be produced by phytoplankton (45). In the hypolimnion sources may include sediment release and the reduction of As^V throughout the water column *(45)*.

With the exception of two alkaline lakes (Mono and Pyramid), the methylated forms of arsenic comprised an average of 24% of the total dissolved As in the surface waters of lakes examined by Anderson & Bruland *(37)*. The average total methylarsenic concentration was 0.7% of total dissolved arsenic in sediment porewaters contamined by gold mines *(58)*. MMA^{III} and DMA^{III} have been found in the aerobic surface waters of Lake Biwa *(44)*. MMA^{III} and DMA^{III} are intermediates in the biosynthesis of organoarsenicals *(1)*. Recently, MMA^{III} has been detected in various strains of freshwater phytoplankton *(57)*. In Lake

Biwa, the dominant species present was DMA^V and was produced in the photic zone and also in the hypolimnion (44). DMA^V and As^{III} have been observed in the oxic euphotic zone in the northern basin of Lake Biwa and in anoxic conditions (45).

Chlorophyll a is often used as a surrogate measure of the amount of phytoplankton in a water sample. The relationship between Chlorophyll a and methylarsenical concentration is sometimes used to determine if As^{III} and organoarsenicals are being directly produced by phytoplankton. However, not all studies have found a significant correlation (37,45). When single sample days were examined, there was some linear correlation suggesting that DMA^V and As^{III} in Lake Biwa do originate from phytoplankton and are dependent on sampling time (the species and amounts of phytoplankton present may vary throughout the year) (45). Abdel-Moati found a correlation between the methylation of As and freshwater diatoms (chlorophyll a)(r=0.8827) (59). In Lake Biwa, methylarsenicals in the hypolimnion were not directly produced by phytoplankton (44). The transparency of Lake Biwa during this study was at 1.0-2.4m and chlorophyll a was practically absent below the depth of 6m (44).

Effects of Freshwater Animals

Freshwater animals are able to convert dimethylarsenic to trimethylarsenic compounds (60,61). In laboratory studies, freshwater fish were able to accumulate and transform arsenic when exposed to water containing 10 mg/dm^3 arsenate for a duration of 7 days (61). As^{III}, As^V, mono-, di-, and tri- methyl arsenic compounds were detected in eight tissues of the freshwater fish $T.$ $mossambica$. A large proportion trimethylated arsenic (22-67%) was present in most tissues. In another laboratory study, some freshwater fish and marsh snails also contained higher amounts of trimethylarsenic than dimethylarsenic compounds(60).

Microorganisms in Sediments

Hasewaga et al. have suggested that methylated arsenic species detected in the sediment interstitial water of Lake Biwa could be the result of detritus degradation and/or the results of in situ bacterial methylation (44). Manganese-, iron-, and sulfate-reducing bacteria from anaerobic lake sediments can methylate and demethylate arsenic (62). Bright et al. found that the extent of methylation by sediment microorganisms and the proportion of different methylated species varied for different culture conditions (62). The biota between the epilimnion and hypolimnions of lakes may differ resulting in different proportions of methylarsenicals present (*37*).

Effects of pH and Temperature on Biomethylation

The pH of water may be a factor in the biomethylation of arsenic. In municipal sewage sludge, arsenic biomethylation was measured at pH 6.5 and 8.0 but was significantly reduced at pH 5.0 (*55*). Temperature may also have an impact on methylation. Significant increases in DMA and MMA occurred as the surface water temperature of Upper Mystic Lake warmed from 5°C to 19°C (*48*). In the hypolimnion of Lake Biwa, DMA^V concentration and temperature did not correlate linearly but the DMA^V concentration in surface water was proportional with temperature (*44*).

Effects of Phosphate

Phosphate concentrations may affect the methylation process (*1,57*). Arsenate is chemically similar to phosphate and may take its place in biotic pathways. In laboratory experiments the phosphate concentration affected the DMA^V production by *C. aciculare* (*57*). DMA^V production was enhanced when the phosphate/arsenate ratio decreased in the culture medium at the beginning of stationary growth. Other studies have also found that there is no clear relationship between the arsenate/phosphate ratio and methylated forms of arsenic (*37,45*). Anderson and Bruland found no correlation between arsenate/phosphate ratios and increased DMA^V concentrations in the epilimnion of the Davis Creek Reservoir *(37)*. The authors conclude that because DMA^V does not react significantly with particles increases in DMA^V and decreases in phosphate may be due to the removal of phosphate by organic matter *(37)*. Crowley Lake contains high phosphorus concentrations due to inputs from geothermal water (38). Methylated arsenic did not contribute significantly to the total arsenic concentration in any Crowley Lake water samples (*38*). Bright et al. also did not find any relationship between methyarsenic and phosphate in sediment cores of subarctic lakes (*58*). In Lake Biwa, the southern basin was more productive, however, the northern basin contained similar concentrations of methylarsenicals (*45*). The high concentrations of phosphorus in the southern basin may have decreased the arsenic metabolism efficiency of phytoplankton *(45)*. In the summer months, more than 95% of the arsenic in the epilimnion of the Crowley Lake reservoir was arsenate even though the waters were very productive during this time (*38*). Some species of phytoplankton may be able to discriminate between phosphate and arsenate better than others (*64*).

Arsenothiols

It is generally believed that the arsenic species detected by hydride generation are the oxy species $(CH_3)_xAsO(OH)_{3-x}$ (x=0-3) (44,45). However, arsenic/sulfur compounds might be expected to dominate in the reducing environments (1). The presence of the thio analogue [i.e., oxythioarsenate, $H_3As^VO_3S$] in water from an arsenic-rich, reducing environment has been demonstrated (65). Bright et al. detected MMA, DMA and TMA in the reduced environment of lake sediment porewater indicating the possible presence of arsenothiols in freshwater (58). The authors suggest that dimethylarsinothiols in sediment porewaters could be a result of exocellular production from DMA^V in reducing conditions or DMA^V could be produced during upward diffusion of porewater containing dimethylarsenothiols into a more oxic environment (58).

Estuaries

Observed Arsenic Speciation

The behaviour of arsenic in estuaries is unique due to the mixture of seawater and freshwater. In the Saguenay Fjord Estuary, Canada, dissolved arsenic showed very low concentrations in the freshwaters of the Saguenay River (~0.03µg/L) and higher concentrations in the marine waters (~1µg/L) (66). The proportions of arsenic species vary with the extent of anthropogenic input and biological activities. For example, As^V was the only detectable arsenic species in the Itchen estuary in southern England most of the year (67). However, 30% of the total arsenic was in the form of methylated species when the temperature was above 12°C and when productivity was at a maximum. In the most riverine site, DMA^V was found only in May, early June and early August. In the Seine Estuary, mean MMA^V concentrations measured in October 1993 (0.025µg/L) and September 1994 (0.022 µg/L) were comparable, however, in February 1995 the MMA^V concentration was below detection limits (68). It should be noted that the detection limit of the technique used was 0.022µg/L. Except in the winter, DMA related to phytoplankton activity represents a large fraction of dissolved total arsenic (68). As^{III} concentrations increased with temperature but were less than 0.07 µg/L everywhere in February (68). Methylated species were not detected in the winter and midsummer in the upper Humbar Estuary (69). Methylarsenicals detected in surface waters comprised 2-25% and 10-82% of the total dissolved arsenic in Tosa Bay and Uranouchi Inlet respectively (70).

During an estuarine algal bloom in the Patuxent River (subestuary), arsenate was rapidly reduced to arsenite and methylated species (72). The rates of arsenate reduction and the species produced varied depending on the phytoplankton species composition and the season. The phytoplankton species composition may vary depending on the location in the estuary (72). Phosphate depletion cannot be a necessary prerequisite for arsenate assimilation in saline water because in Southampton Water phosphate concentration remained high throughout 1988, yet arsenic reduction and methylation occurred (73).

In the Tamar Estuary, dissolved inorganic arsenic was elevated in the higher salinity region (74). Riedel observed that arsenate rises to maximum concentrations before salinity reaches a maximum, and falls off during the period of highest salinities (75). Inorganic arsenic was linearly and inversely related to salinity in the upper Humbar Estuary but in the lower estuary dissolved inorganic arsenic behaved conservatively but was directly related to salinity (69).

Sediment and Sediment Porewater

Inorganic arsenic, MMA and DMA were observed in samples of estuarine sediment porewater (0.04-0.7 µg/L) collected from the Tamar Estuary (71). Arsenic is released to porewaters of the Saguenay Fjord Estuary from the degradation of organic matter (66 and refs therein). Reduced As^{III} was the predominant species in the suboxic/anoxic porewaters of the Saguenay Fjord Estuary. Although both iron and manganese oxides can oxidise As^{III} to As^{V} it is thought that manganese oxides are primarily responsible for the oxidation (66).

The methylation of arsenic may prevent the adsorption to sediment and result in less retention in the estuary (72). The ratio of dissolved to particulate arsenic distribution is controlled mainly by the iron content of particles (68). The adsorption of arsenic on particulate matter occurs as a function of turbidity, salinity, pH, redox potential and biological activity (68). Arsenic in suspended particulate matter in the Humbar Estuary ranged from 7.4-76.4 µg/g (69).

Seawater

Arsenic concentrations in seawater are fairly stable at about 0.5-2 µg/L. (1,2,41). The major species in seawater are As^{V} and As^{III} and minor species are MMA and DMA (Table I). Although arsenobetaine, arsenosugars, tetramethylarsonium, arsenocholine, and trimethylarsine oxide (TMAO) are found in marine life, they are not detected in surface seawater (76). Uncharacterized arsenic can average 25% of total arsenic in seawater (76). In marine sediments/porewater the major species are As^{V}, As^{III}, minor species are

MMA^V, DMA^V and trace species are TMAO and arsenobetaine (76). Arsenosugars, TeMA, and arsenocholine have not been detected (76). As^V was the dominant species in both the East Indian and Antarctic Oceans surface water representing 52% and 97% respectively (64). As^{III} accounts for 27% and 0.2% in those regions. Fractions of MMA and DMA were 4% and 17% in east Indian and 0.8% and 2% in Antarctic waters (64). The surface waters of the Antarctic were more productive than the surface waters of the East Indian Ocean, but the MMA and DMA concentrations in the Antarctic surface were less than the east Indian Ocean (64). At sampling stations in the Pacific Ocean, inorganic arsenic concentrations were low at the surface and deep/bottom zones and MMA and DMA were observed throughout the water column in the north western and equatiorial regions (77). Both MMA^V and DMA^V displayed maximum concentrations in surface water and abruptly dropped with depth from 0-200m (78).

In the coastal waters of Tosa Bay (Pacific Ocean), As^V was the dominant arsenic species all year. The maximum concentrations of As^{III} and DMA^{III+V} were observed in surface waters (70). In the Uranouchi inlet, the concentrations of the trivalent methylarsenic species MMA^{III}and DMA^{III} were low compared with those of the pentavalent species in most waters. During one period in the spring, MMA^{III} was more abundant than MMA^V (increased from 0.004-0.04μg/L) (70). The total of MMA^{III} and DMA^{III} was approximately 18% of the total methylarsenical concentrations (70). The concentrations of inorganic arsenic increased with depth to maximum concentrations at a depth of 2000m and then slowly decreased to minimum concentrations at a depth of 5000m (78).

Microbial Transformation of Arsenic in SeaWater

In marine systems, arsenic can be transformed by bacteria, algae and other marine organisms (76). Arsenosugars in algae and arsenobetaine in animals, may be released to the sea at some point, as a consequence of grazing or predation by animals or the aging of the organism (79). Though these arsenic species have not been detected in seawater, there is increasing evidence that they may exist. *Fucus gardneri* was able to convert arsenate to As^{III} and DMA^V after exposure in water (80). MMA was not detected. There was an increase in one arsenosugar but it was accompanied by a decrease in another arsenosugar so there was no conclusive evidence that arsenosugars were produced by *Fucus gardneri* (80). In laboratory experiments, arsenobetaine and arsenocholine were completely degraded by microorganisms in seawater (79). Arsenobetaine was transformed within hours initially to dimethylarsinoylacetate and then to DMA. Arsenocholine behaved similarly but at a slower rate (79). The speed of the degradation processes may offer an explanation for the absence of arsenobetaine in natural waters (79).

Organisms in marine sediments can also degrade arsenobetaine to inorganic arsenic via TMAO under aerobic conditions (81). In laboratory experiments, arsenobetaine-containing growth media were incubated with suspended substances as a source of marine microorganisms (82). The mixture was able to degrade arsenobetaine to inorganic arsenic, TMAO and DMA^V (82).

Uncharacterized Arsenic Species

In addition to the well-characterized arsenic species discussed above in estuaries and seawater, there are also reports of unidentified arsenic species in water (58,76,83-86). After UV-irradiation of surface water from Uranouchi Inlet, the inorganic and dimethylarsenic concentrations detected with hydride generation rapidly increased (83). The UV-labile arsenic fractions comprised 15-45% and 4-26% of the total arsenic in Uranouchi Inlet and Lake Biwa, respectively. UV-labile fraction is the increment in measurable arsenic concentration before and after the 2.5h of UV irradiation. Between filtered (0.45μm) and unfiltered samples, the difference in the uncharacterized fraction was significant compared with that of inorganic and methylarsenicals (83). The authors suggest that the uncharacterized arsenic captured on the 0.45 μm filter was derived from the organoarsenicals in biological organic detritus (83). In sediment porewater, there was an increase of 18-420% in total dissolved As concentration observed after irradiation (58). Michel et al. reported a difference between the sum of known arsenic species and total arsenic of 13% in the euphotic layer of an estuary (86). These studies suggest that substantial amounts of arsenic remain to be characterized.

Concluding Remarks

High levels of arsenic in aquatic systems can pose a serious hazard to the health of humans and the environment. In order to predict the potential toxicity to an aquatic system, the biogeochemical changes that will affect arsenic speciation must be considered. The arsenic species that are present in aquatic systems determine its toxicity and mobility. The measured levels of a particular chemical species in a water column represents the complex balance between production, chemical and microbial transformation and physical dispersal. It is necessary to not only understand the nature of the arsenic species that are formed but also how they may be released into the water. The new refractory arsenic found in water samples should be a focal point of further research on the biogeochemical processes of arsenic cycling in the aqautic environment. They may offer important details regarding the arsenic speciation in natural waters and biological production in organisms.

Treatment options will depend on many factors, including the oxidation state of arsenic and the concentration and species in source water (*26,35*). As^V is removed more easily than As^{III} (*35,87*). There are many different treatment options for the removal of arsenic from groundwater. Most recently, zerovalent iron filings, iron-sulphide minerals (pyrite and pyrrhotite), and aluminum-based coagulation have been used to remove arsenite and arsenate from groundwater (*87-89*). Another possibility is to oxidize arsenite to arsenate before removal (*33,90*). Kim & Nriagu have attempted to oxidize arsenite in natural waters using ozone, air and oxygen (*33*). The presence of iron and/or manganese will accelerate the oxidative removal of arsenic. Clay minerals contain iron oxides and are commonly present in the subsurface environment and are very effective at adsorbing As^V than As^{III} (*91*). Hug et al. found that dissolved oxygen and micromolar hydrogen peroxide did not oxidize As^{III} on a time scale of hours (*90*). In solutions containing 0.06-5mg/L $Fe^{II,III}$, over 90% of As^{III} could be oxidized photochemically within 2-3 h by illumination with 90 W/m^2. Fe^{III} citrate complexes accelerated As^{III} oxidation. Understanding arsenic speciation in water will contribute to more effective approaches to the removal of arsenic.

Acknowledgments

This work was supported by the Natural Sciences and Engineering Research Council of Canada, the Canada Research Chairs Program, and the Canadian Water Networks NCE.

References

1. Cullen, W. R.; Reimer, K. J. *Chem. Rev.* **1989**, *89*, 713-764.
2. Tamaki, S.; Frankenberger, W. T. *Rev. Environ. Cont. Toxicol.* **1992**, *124*, 79-110.
3. Welch, A. H. *Ground Water* **2000**, *38*, 589-604.
4. National Research Council. Chemistry and Analysis of Arsenic Secies in Water, Food, Urine, Blood, Hair, and Nails. In *Arsenic in Drinking Water*; National Academy Press: Washington, D.C., 1999; pp. 27-81.
5. Chowdhury, U. K.; Biswas, B. K.; Chowdhury, T. R.; Samanta, G.; Mandal, B. K.; Basu, G. C.; Chanda, C. R.; Lodh, D.; Saha, K. C.; Mukherjee, S. K.; Roy, S.; Kabir, S.; Quamruzzaman, Q.; Chakraborti, D. *Environ. Health Perspect.* **2000**, *108*, 393-397.

6. Chowdhury, U. K.; Biswas, B. K.; Dhar, G.; Samanta, G.; Mandal, B. K.; Chowdhury, T. R.; Chakraborti, D.; Kabir, S.; Roy, S. Groundwater Arsenic Contamination & Suffering of People in Bangladesh. In *Arsenic Exposure and Health Effects;* Chappell, W. R.; Abernathy, C. O.; Calderon, R. L. Eds.; Elsevier: Amsterdam, 1999; pp. 165-182.
7. Samanta, G.; Chowdhury, T. R.; Mandal, B. K.; Biswas, B. K.; Chowdhury, U. K.; Basu, G. C.; Chanda, C. R.; Lodh, D.; Chakraborti, D. *Microchem. J.* **1999**, *62*, 174-191.
8. Luo, Z. D.; Zhang, Y. M.; Ma, L.; Zhang, G. Y.; He, X.; Wilson, R.; Byrd, D. M.; Griffiths, J. G.; Lai, S.; He, L.; Grumski, K.; Lamm, S. H. Chronic Arsenicism and Cancer in Inner Mongolia-Consequences of Well-water Arsenic Levels greater than 50ug/L. In *Arsenic Exposure and Health Effects;* Abernathy, C. O.; Calderon, R. L.; Chappell, W. R. Eds.; Chapman & Hall: London, 1999; pp. 55-68.
9. Woo, N. C.; Choi, M. J. *Environ. Geol.* **2001**, *40*, 305-311.
10. Kim, M. J.; Nriagu, J.; Haack, S. *Environ. Sci. Technol.* **2000**, *34*, 3094-3100.
11. Roussel, C.; Bril, H.; Fernandez, A. *J. Environ. Qual.* **2000**, *29*, 182-188.
12. Armienta, M. A.; Villasenor, G.; Rodriguez, R.; Ongley, L. K.; Mango, H. *Environ. Geol.* **2001**, *40*, 571-580.
13. Korte, N. E. *Environ. Geol. Wat. Sci.* **1991**, *18*, 137-141.
14. Nickson, R.; MCarthur, J.; Burgess, W.; Ahmed, K. M.; Ravenscroft, P.; Rahmann, M. *Nature* **1998**, *395*, 338-339.
15. Nickson, R. T.; McArthur, J. M.; Ravenscroft, P.; Burgess, W. G.; Ahmed, K. M. *Appl. Geochem.* **2000**, *15*, 403-413.
16. Acharyya, S. K.; Lahiri, S.; Raymahashay, B. C.; Bhowmik, A. *Environ. Geol.* **2000**, *39*, 1127-1137.
17. Langner, H. W.; Inskeep, W. P. *Environ. Sci. Technol.* **2000**, *34*, 3131-3136.
18. Hering, J. G.; Chiu, V. Q. *J. Environ. Eng.* **2000**, 471-474.
19. Le, X. C.; Yalcin, S.; Ma, M. *Environ. Sci. Technol.* **2000**, *34*, 2342-2347.
20. Kim, M. J. *Bull. Environ. Contam. Toxicol.* **2001**, *67*, 46-51.
21. Raessler, M.; Michalke, B.; Schramel, P.; Schulte-Hostede, S.; Kettrup, A. *Intern. J. Environ. Anal. Chem.* **1998**, *72*, 195-203.
22. Roig-Navarro, A. F.; Martinez-Bravo, Y.; Lopez, F. J.; Hernandez, F. *J. Chromatogr. A* **2001**, *912*, 319-327.
23. Korte, N. E.; Fernando, Q. *Crit. Rev. Environ. Control* **1991**, *21*, 1-40.
24. Edwards, L.; Patel, M.; McNeill, Chim, L.; Frey, M.; Eaton, A.D.; Antweller, R.C.; Taylor, H.E. *J. Am. Wat. Works Assoc.* **1998**, *90*, 103-113.
25. Borho, M.; Wilderer, P. *SRT Aqua* **1997**, *46*, 138-143.
26. Gallagher, P.; Schwegel, C. A.; Wei, X.; Creed, J. T. *J. Environ. Monit.* **2001**, *3*, 371-376.

30

27. Hall, G.E.M.; Pelchat, J.G.; Gauthier, G. *J. Anal. At. Spectrom.* **1999**, *14*, 205-213.
28. Lindemann, T.; Prange, A.; Dannecker, W.; Neidhart, B. *Fresenius J. Anal. Chem.* **2000**, *368*, 214-220.
29. Ferguson, J. F.; Gavis, J. *Wat. Res.* **1972**, *6*, 1259-1274.
30. Raessler, M.; Michalke, B.; Schulte-Hostede, S.; Kettrup, A. *Sci. Tot. Environ.* **2000**, *258*, 171-181.
31. Kondo, H.; Ishiguro, Y.; Ohno, K.; Nagase, M.; Toba, M.; Takagi, M. *Wat. Res.* **1999**, *33*, 1967-1972.
32. Del Razo, L. M.; Arellano, M. A.; Cebrian, M. E. *Environ. Poll.* **1990**, *64*, 143-153.
33. Kim, M. J.; Nriagu, J. *Sci. Tot. Environ.* **2000**, *247*, 71-79.
34. Planer-Friedrich, B.; Armienta, M. A.; Merkel, B. J. *Environ. Geol.* **2001**, *40*, 1290-1298.
35. Chen, H. W.; Frey, M. M.; Clifford, D.; McNeill, L. S.; Edwards, M. *J. Am. Wat. Works Assoc.* **1999**, *91*, 74-85.
36. Mccreadie, H.; Blowes, D. W.; Ptacek, C. J.; Jambor, J. L. *Environ. Sci. Technol.* **2000**, *34*, 3159-3166.
37. Anderson, L. C. D.; Bruland, K. W. *Environ. Sci. Technol.* **1991**, *25*, 420-427.
38. Kneebone, P. E.; Hering, J. G. *Environ. Sci. Technol.* **2000**, *34*, 4307-4312.
39. Gihring, T. M.; Druschel, G. K.; Mccleskey, R. B.; Hamers, R. J.; Banfield, J. F. *Environ. Sci. Technol.* **2001**, *35*, 3857-3862.
40. Buyuktuncel, E.; Bektas, S.; Salih, B.; Evirgen, M. M.; Genc, O. *Fres. Environ. Bull.* **1997**, *6*, 494-501.
41. Andreae, M. O. *Deep-Sea Res.* **1978**, *25*, 391-402.
42. Braman, R. S.; Foreback, C. C. *Science* **1973**, *182*, 1247-1249.
43. Howard, A. G.; Comber, S. D. W. *Appl. Organomet. Chem.* **1989**, *3*, 509-514.
44. Hasegawa, H. *Appl. Organomet. Chem.* **1997**, *11*, 305-311.
45. Sohrin, Y.; Matsui, M.; Kawashima, M.; Hojo, M.; Hasegawa, H. *Environ. Sci. Technol.* **1997**, *31*, 2712-2720.
46. Seyler, P.; Martin, J. M. *Environ. Sci. Technol.* **1989**, *23*, 1258-1263.
47. Kuhn, A.; Sigg, L. *Limnol. Oceanogr.* **1993**, *38*, 1052-1059.
48. Aurillo, A. C.; Mason, R. P.; Hemond, H. F. *Environ. Sci. Technol.* **1994**, *28*, 577-585.
49. Wilkie, J. A.; Hering, J. G. *Environ. Sci. Technol.* **1998**, *32*, 657-662.
50. Macur, R. E.; Wheeler, J. T.; McDermott, T. R.; Inskeep, W. P. *Environ. Sci. Technol.* **2001**, *35*, 3676-3682.
51. Mok, W. M.; Wai, C. M. Mobilization of arsenic in contaminanted river waters. In *Arsenic in the Environment, Part I:Cycling and Characterization;* Nriagu, J. O. Ed.; John Wiley & Sons, Inc., 1994.

52. Cummings, D. E.; March, A. W.; Bostick, B.; Spring, S.; Caccavo, F.; Fendorf, S.; Rosenzweig, R. F. *Appl. Environ. Microbiol.* **2000**, *66*, 154-162.
53. Cummings, D. E.; Caccavo, F.; Fendorf, S.; Rosenzweig, R. F. *Environ. Sci. Technol.* **1999**, *33*, 723-729.
54. La Force, M. J.; Hansel, C. M.; Fendorf, S. *Environ. Sci. Technol.* **2000**, *34*, 3937-3943.
55. Carbonell-Barrachina, A. A.; Jugsujinda, A.; Burlo, F.; Delaune, R. D.; Patrick Jr., W. H. *Wat. Res.* **2000**, *34*, 216-224.
56. Spliethoff, H. M.; Mason, R. P.; Hemond, H. F. *Environ. Sci. Technol.* **1995**, *29*, 2157-2161.
57. Hasegawa, H.; Sohrin, Y.; Seki, K.; Sato, M.; Norisuye, K.; Naito, K.; Matsui, M. *Chemosphere* **2001**, *43*, 265-272.
58. Bright, D. A.; Dodd, M.; Reimer, K. J. *Sci. Tot. Environ.* **1996**, *180*, 165-182.
59. Abdel-Moati, A. R. *Wat. Air Soil Pollut.* **1990**, *51*, 117-132.
60. Kaise, T.; Ogura, M.; Nozaki, T.; Saitoh, K.; Sakurai, T.; Matsubara, C.; Watanabe, C.; Hanaoka, K. *Appl. Organomet. Chem.* **1997**, *11*, 297-304.
61. Suhendrayatna, A. O.; Tsunenori, N.; Shigeru, M. *Appl. Organomet. Chem.* **2001**, *15*, 566-571.
62. Bright, D. A.; Brock, S.; Cullen, W. R.; Hewitt, G. M.; Jafaar, J.; Reimer, K. J. *Appl. Organomet. Chem.* **1994**, *8*, 415-422.
63. Chen, C. Y.; Folt, C. L. *Environ. Sci. Technol.* **2000**, *34*, 3878-3884.
64. Santosa, S. J.; Wada, S.; Tanaka, S. *Appl. Organomet. Chem.* **1994**, *8*, 273-283.
65. Schwedt, G.; Rieckhoff, M. *J. Chromatogr. A* **1996**, *736*, 341-350.
66. Mucci, A.; Richard, L. F.; Lucotte, M.; Guignard, C. *Aqua. Geochem.* **2000**, *6*, 293-324.
67. Howard, A. G.; Apte, S. C. *Appl. Organomet. Chem.* **1989**, *3*, 499-507.
68. Michel, P.; Chiffoleau, J. F.; Averty, B.; Auger, D.; Chartier, E. *Cont. Shelf Res.* **1999**, *19*, 2041-2061.
69. Kitts, H. J.; Millward, G. E.; Morris, A. W.; Ebdon, L. *Estuar. Coast. Shelf Sci.* **1994**, *39*, 157-172.
70. Hasegawa, H. *Appl. Organomet. Chem.* **1996**, *10*, 733-740.
71. Ebdon, L.; Walton, A. P.; Millward, G. E.; Whitfield, M. *Appl. Organomet. Chem.* **1987**, *1*, 427-433.
72. Sanders, J. G.; Riedel, G. F. *Estuaries* **1993**, *16*, 521-532.
73. Howard, A. G.; Comber, S. D. W.; Kifle, D.; Antai, E. E.; Purdie, D. A. *Estuar. Coast. Shelf Sci.* **1995**, *40*, 435-440.
74. Howard, A. G.; Apte, S. C.; Comber, S. D. W.; Morris, R. J. *Estuar. Coast. Mar. Sci.* **1988**, *27*, 427-443.
75. Riedel, G. F. *Estuaries* **1993**, *16*, 533-540.

76. Francesconi, K. A.; Edmonds, J. S. *Adv. Inorg. Chem.* **1997**, *44*, 147-189.
77. Santosa, S. J.; Mokudai, H.; Takahashi, M.; Tanaka, S. *Appl. Organomet. Chem.* **1996**, *10*, 697-705.
78. Santosa, S. J.; Wada, S.; Mokudai, H.; Tanaka, S. *Appl. Organomet. Chem.* **1997**, *11*, 403-414.
79. Khokiattiwong, S.; Goessler, W.; Pedersen, S. N.; Cox, R.; Francesconi, K. A. *Appl. Organomet. Chem.* **2001**, *15*, 481-489.
80. Granchinho, S. C. R.; Polishchuk, E.; Cullen, W. R.; Reimer, K. J. *Appl. Organomet. Chem.* **2001**, *15*, 553-560.
81. Hanaoka, K.; Tagawa, S.; Kaise, T. *Hydrobiol.* **1992**, *235/236*, 623-628.
82. Hanaoka, K.; Koga, H.; Tagawa, S.; Kaise, T. *Comp. Biochem. Physiol.* **1992**, *101B*, 595-599.
83. Hasegawa, H.; Matsui, M.; Okamura, S.; Hojo, M.; Iwasaki, N.; Sohrin, Y. *Appl. Organomet. Chem.* **1999**, *13*, 113-119.
84. Sturgeon, R. E.; Siu, K. W. M.; Willie, S. N.; Berman, S. S. *Analyst* **1989**, *114*, 1393-1396.
85. de Bettencourt, A. M.; Andreae, M. O. *Appl. Organomet. Chem.* **1991**, *4*, 111-116.
86. Michel, P.; Boutier, B.; Herbland, A.; Averty, B.; Artigas, L. F.; Auger, D.; Chartier, E. *Oceanol. Acta* **1997**, *21*, 325-333.
87. Jingtai, H.; Fyfe, W. S. *Chin. Sci. Bull.* **2000**, *45*, 1430-1434.
88. Su, C.; Puls, R. W. *Environ. Sci. Technol.* **2001**, *35*, 1487-1492.
89. Gregor, J. *Wat. Res.* **2001**, *35*, 1659-1664.
90. Hug, S. J.; Canonica, L.; Wegelin, M.; Gechter, D.; Gunten, U. V. *Environ. Sci. Technol.* **2001**, *35*, 2114-2121.
91. Lin, Z.; Puls, R. W. *Environ. Geol.* **2000**, *39*, 753-759.
92. Tanabe, K.; Yokota, H.; Hironaka, H.; Tsushima, S.; Kubota, Y. *Appl. Organomet. Chem.* **2001**, *15*, 241-251.
93. Nag, J. K.; Balaram, V.; Rubio, R.; Alberti, J.; Das, A. K. *J. Trace Elem. Med. Biol.* **1996**, *10*, 20-24.
94. Chen, S. L.; Yeh, S. J.; Yang, M. H.; Lin, T. H. *Biol. Trace Elem. Res.* **1995**, *48*, 263-274.
95. Lin, T. H.; Huang, Y. L.; Wang, M. Y. *J. Toxicol. Environ. Health* **1998**, *53*, 85-93.
96. Hung, T. C.; Liao, S. M. *Toxicol. Environ. Chem.* **1996**, *56*, 63-73.
97. Boyle, D.R.; Turner, R.J.W.; Hall, G.E.M. *Env. Geochem. and Health* **1998**, *20*, 199-212.
98. Robertson, F. N. *Env. Geochem. Health* **1989**, *11*, 171-176.
99. Millward, G. E.; Kitts, H. J.; Comber, S. D. W.; Ebdon, L.; Howard, A. G. *Estuar. Coast. Shelf Sci. (1996)* **1996**, *43*, 1-18.
100. de Bettencourt, A. M. *Neth. J. Sea Res.* **1988**, *22*, 205-212.

Chapter 3

Arsenic Carbonate Complexes in Aqueous Systems

Janice S. Lee and Jerome O. Nriagu

Department of Environmental Health Sciences, School of Public Health,
University of Michigan, Ann Arbor, MI 48109

Exploratory research involving new arsenic carbonate complexes has been conducted, and results strongly point to their existence. Attempt was made to form the complexes by bubbling CO_2 gas in arsenic trioxide solution for 24 hours. Ion chromatography of the solution showed shifts and peak broadening among the As(III) peak and the emergence of new peaks not found in solutions containing only CO_2 or As(III) in Milli-Q water. Corresponding peaks obtained from atomic absorption spectroscopy showed similar elution times and increased peak areas. Results from a previous study suggest that the carbonation of sulfide minerals is an important process in leaching arsenic into groundwater under anaerobic conditions, which provides further evidence for the formation of arsenic carbonate complexes. Likely arsenic carbonate complexes are believed to be $As(CO_3)_2^-$, $As(CO_3)(OH)_2^-$, $As(CO_3)_2(OH)^{2-}$, and $As(CO_3)^+$.

Introduction

Many communities around the world are exposed to elevated levels of arsenic in their drinking water. Exposure to arsenic is associated with cancer of the skin, bladder, liver, lung, and kidney, and hyperpigmentation and hyperkeratosis. Natural contamination of groundwaters with arsenic have been found in many parts of the world. Groundwaters that are mostly reducing and contain high concentrations of dissolved iron and manganese are associated with elevated levels of arsenic. Anoxic wells in Bangladesh show arsenic concentrations that correlate with concentrations of dissolved iron and bicarbonate, as well as increasing arsenic concentrations with increasing depth of wells (1). Groundwater in alluvial aquifers from Delta Plains, Eastern India contains calcium, magnesium and bicarbonate with elevated levels of iron, phosphate and arsenic (2).

Mining wastes that are contaminated with arsenic and metals are accompanied by high bicarbonate levels. Woo and Choi (3) found significant correlations of arsenic with Ca, Mg, and HCO_3 in water near a gold-mine waste site in Korea, implying that arsenic concentrations in water could also be increased by desorption following dissolution of carbonate mineral such as calcite ($CaCO_3$), magnesite ($MgCO_3$), cerrusite ($PbCO_3$), rhodochrosite ($MnCO_3$), and siderite ($FeCO_3$).

It has been reported that arsenic in groundwater evolves from the oxidation of arsenic-rich pyrite during the water extraction process (4, 5) or from the reductive dissolution of arsenic-rich iron oxyhydroxides (6). The latter process solubilizes iron and its absorbed load and increases bicarbonate concentration indirectly. However, no known abiotic mechanism exists for the abstraction of arsenic from the aquifer rocks under reducing conditions. Results from a recent study suggest the carbonation of arsenite minerals to be an important process in leaching arsenic into groundwater under anaerobic conditions (7). Kim et al. (7) suggests that the leaching of As into groundwater is caused by the direct interaction between HCO_3^- and As minerals in the aquifer rocks, and that the reaction of ferrous ion with the thioarsenite from carbonation process can result in the formation of arsenopyrite, a common mineral in arsenic-rich aquifers. It has been postulated that the formation of stable arseno-carbonate complexes is responsible for leaching arsenic from host rocks under anaerobic conditions. Using orpiment, As_2S_3, as an example, the following reactions can be used to depict dissolution by carbonate:

$$As_2S_3 + HCO_3^- \rightarrow As(CO_3)^+ + HAsS_2 + S^{2-}$$
$$As_2S_3 + 2HCO_3^- \rightarrow HAs(CO_3)_2 + HAsS_2 + S^{2-}$$
$$HAsS_2 + H_2O \rightarrow HAsS_2(OH)^- + H^+$$
$$HAsS_2 + HCO_3^- \rightarrow 2HS^- + As(CO_3)^+$$
$$HAsS_2 + HCO_3^- + 2H_2O \rightarrow As(CO_3)(OH)_2^- + 2HS^- + 2H^+$$
$$HAsS_2 + HCO_3^- + H_2O \rightarrow As(CO_3)(OH) + 2HS^- + H^+$$
$$S^{2-} + 2H^+ \rightarrow 2HS^-$$

The notion that carbonate forms some kind of complex with arsenite was the basis of the old pharmaceutical curative known as Fowler's solution, which is an aqueous solution of As_2O_3 in K_2CO_3 (8). Pentavalent arsenic can also form complexes with carbonate, as can be seen when CO_2 increases the solubility of hydrometallurgical arsenate precipitates. (9). The influence of CO_2 on the solubility of metal arsenates dates back to 1948 when Mass (10) observed an increase in the solubility of copper arsenate in the presence of CO_2. Table I shows the effect of CO_2 on the stability of copper, magnesium, cadmium, barium, and lead arsenates after 126 days. Arsenic and total metal ion concentrations for samples without CO_2 were determined at pH values similar to the pH of the CO_2 containing solutions, in order to control possible pH effects. The stability of these arsenates is greatly compromised in the presence of CO_2. It can be seen that CO_2 significantly increases arsenic concentration, with a greater than 1000 fold increase for cadmium arsenate in the presence of CO_2.

Since carbonate exists naturally in the environment and is commonly found in groundwater, it is possible for arsenic carbonate complexes to form in groundwaters that contain arsenic. However, as Schaufelberger points out (8), apart from a note in Ephraim in 1920 on the rhodanide-arsenite complexes, the existence of intermediate arsenic carbonate complexes has never been thoroughly investigated or reported.

This chapter presents work performed to characterize the nature of the interactions between carbonate ions and As(III). The results from this study, along with Kim et al's study (7), support the formation of arsenic carbonate complexes.

Table I. Effect of Carbon Dioxide on Dissolution of Metal Arsenates

Metal Arsenate	With 1% CO_2		Without CO_2	
	As	Me	As	Me
Copper	0.090	0.006	0.030	0.009
Magnesium	0.020	0.009	0.005	0.006
Cadmium	0.020	5.34E-6	1.47E-5	5.25E-5
Barium	0.001	7.21E-4	7.88E-4	0.0017
Lead	8.25E-5	0.001	1.33E-7	0.002

NOTE: Metal(Me)/Arsenic(As) molar ratio is 1.5:1; Concentrations expressed in mol/L; Total Me ion concentrations= (Fe+Cu+Cd+Zn).

SOURCE: Data are from Reference 11.

Materials and Methods

Arsenic carbonate complexes were prepared by continuous bubbling of 1.90E-4mol/L (285mg/L) arsenic trioxide (As_2O_3) solution with CO_2 for 24 hours. As(III) solution was prepared using 1.79E-4mol/L (268mg/L) As_2O_3. As(V) solution was prepared by dissolving 3.75E-5mol/L (93mg/L) sodium arsenate (Na_2HAsO_4) in deoxygenated water, and used to check the quality of As(III) containing solutions. The initial pH of the solutions before bubbling was ~6. The solutions became slightly acidic (pH ~4-5) after purging with CO_2. Arsenic trioxide and sodium arsenate were obtained from Aldrich and all solutions were prepared in Milli-Q water.

Solutions were analyzed by Alltech's Odyssey High Performance Ion Chromatography (IC) System using a Waters IC-PAC Anion Column (4.6 x 150mm) and a Wescan Cation Column (4.6 x 50mm) to detect negatively and positively charged species respectively. The IC System is equipped with the Alltech 550 Conductivity Detector, which uses a high frequency (10KHz) sinusoidal applied potential and synchronous detection to measure minute changes in conductance of the column effluent. The detector utilizes a temperature controlled cell housing to enhance baseline stability by reducing thermal noise and drift. The mobile phase and the sample are heated before reaching the detector cell. Table II lists the conditions under which anion and cation analysis was conducted. Sodium carbonate and bicarbonate were obtained from Fisher Scientific, and the ethylenediamine and tartaric acid were from Aldrich. Total arsenic was analyzed using a graphite furnace atomic absorption spectrophotometer (GFAAS, Perkin-Elmer 4100ZL) equipped with an electrodeless discharge lamp. A matrix modifier consisting of 5 µg of Pd and 3 µg of $Mg(NO_3)_2$ was used for each 20 µL sample for the GFAAS analysis.

Table II. Ion Chromatography Conditions for Anion and Cation Analysis

	Anion Analysis	*Cation Analysis*
Mobile phase	0.85mM $NaHCO_3$	1.25mM Ethylenediamine
	0.90mM Na_2CO_3	1.67mM Tartaric acid
Flow rate	1.40 ml/min	0.40 ml/min
Column temperature	25°C	25°C
Injection volume	0.50 ml	0.50 ml

Results and Discussion

Figure 1 shows the chromatogram obtained with an anion column for (a) arsenic trioxide solution saturated with CO_2, (b) arsenic trioxide solution with no CO_2, and (c) As(V) solution. Milli-Q water saturated with CO_2 is not shown; the IC trace contains some noise and no significant peaks. Figure 1 shows the arsenic trioxide solution saturated with CO_2 to contain distinct peaks not found in the other solutions. The two solutions containing As(III) show three identical peaks with elution times of 4-10 minutes(min). The presence of impurities found in Milli-Q water and/or As_2O_3 could account for these minor peaks. For example, when the IC traces for these As(III)-containing solutions were compared to an IC trace for the anion standard, two of the small peaks corresponded to fluoride (5 min) and chloride (8 min). It is common to find trace amounts of chlorine in highly purified water and chemicals. The same reasoning can be applied to explain the occurrence of three similar peaks also found in the As(V) solution. In addition to fluoride and chloride, nitrate (17 min) and sulfate (37 min) are also present in the IC trace for As(V) solution. As(V) was eluted at 40 minutes. The two peaks with elution time of 11 and 12 minutes found in the As(III) solutions can be attributed to As(III) since there is no equivalent peak in the As(V) solution. The slight shift in peak elution time for As(III) in solutions with and without CO_2 is most likely due to the slight change in pH of the two solutions.

The three distinct peaks with elution times of 16-25 minutes are only found in the CO_2-saturated As(III) solution, and indicate the presence of dissolved species not found in the other solutions. We are tentatively attributing these peaks to contain $As(CO_3)_2^-$, $As(CO_3)(OH)_2^-$, and $As(CO_3)_2(OH)^{2-}$. Eluent from IC was collected every minute and analyzed for arsenic using GFAAS. Results showed corresponding arsenic peaks at similar elution times. The two peaks with elution times greater than 30 minutes, can be attributed to sulfate and the presence of As(V). Again, the change in retention times for sulfate and As(V) is caused by minor pH differences between the As(V) solution and arsenic trioxide solution saturated with CO_2.

Since the cation column was shorter in length compared to the anion column, arsenic was eluted in less than five minutes and isolation of arsenic peaks from other peaks was difficult. Therefore, eluents from the IC were collected sequentially and analyzed for total arsenic using GFAAS. Figure 2 shows the arsenic concentrations obtained by GFAAS after sequential collection from the IC analysis of positively charged species in As(III) solutions with and without CO_2. The As(III) solution saturated with CO_2 shows one peak with a short retention time and another peak with an elution time of about 3 minutes (Figure 2). The first peak is most likely H_3AsO_3 because this species should elute out rapidly since it is neutral and therefore not retained by the column, and the second peak is interpreted as $As(CO_3)^+$.

We also observed that the presence of CO_2 significantly increases the concentration of arsenic in solution. After bubbling the arsenic trioxide solution

38

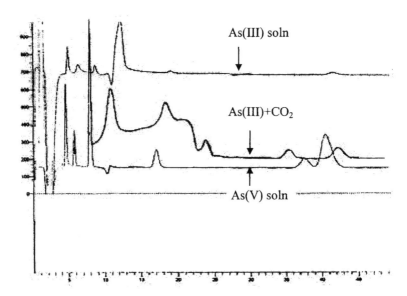

Figure 1. Chromatogram from IC showing peaks for As(V) solution (93mg/L, pH~6), As$_2$O$_3$ solution saturated with CO$_2$ (285mg/L, pH~4-5), and As$_2$O$_3$ solution without CO$_2$ (268mg/L, pH~4-5).

Analysis of Positively Charged Species in Solutions of As(III) + CO$_2$

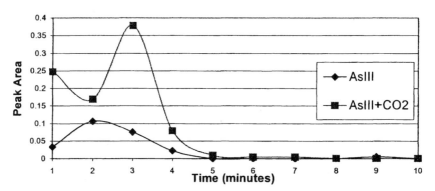

Figure 2. GFAAS analysis of concentrations of arsenic in sequentially collected samples from an IC equipped with a cation column. Graphs for As$_2$O$_3$ solutions saturated with CO$_2$ and without CO$_2$ are compared.

for 24 hours with CO_2, eluent was collected sequentially and the peak area obtained with the anion column for As(III) increased from 0.917 to 2.896 (Figure 3). Difference in elution times for As(III) species in Figure 2 and 3 is due to the use of different columns and IC operating conditions, as well as method of detection. Enhanced solubility of As_2O_3 can be explained by the formation of stable complexes between dissolved As(III) and the carbonate ions.

The preliminary studies point to the formation of several stable arseno-carbonate complexes, including negatively and positively charged species in solutions containing As(III) and carbonate ions. Current speciation methods are based on ion exchange methods that separate arsenic species into a neutral fraction (As(III)) and a negatively charged fraction (As(V)). This method would not be adequate for groundwaters that have high carbonate concentrations, and may account for the general misconception that H_3AsO_3 and $H_xAsO_4^{3-x}$ are the only arsenic species in groundwater. As Kim et al. showed (7), formation of arsenic carbonate complexes provides an effective mechanism for dissolution of arsenic minerals in host rocks by groundwater under anaerobic conditions.

In this study, prepared solutions were neutral to slightly acidic. The main arsenic carbonate complexes are believed to be $As(CO_3)_2^-$, $As(CO_3)(OH)_2^-$, $As(CO_3)_2(OH)^{2-}$, and $As(CO_3)^+$. In our solutions, H_3AsO_3 is the dominant arsenic species (the pK_a values for As(III) are 9.2, 12.1, and 13.4). H_2CO_3 is the major carbonate species, while small amounts of HCO_3^- may also exist (the pK_a values for carbonate system are 6.3 and 10.3). Therefore, the following reactions are postulated:

$$H_3AsO_3 + 2HCO_3^- + H^+ \rightarrow As(CO_3)_2^- + 3H_2O$$
$$H_3AsO_3 + HCO_3^- \rightarrow As(CO_3)(OH)_2^- + H_2O$$
$$H_3AsO_3 + 2HCO_3^- \rightarrow As(CO_3)_2(OH)^{2-} + 2H_2O$$
$$H_3AsO_3 + H_2CO_3 + H^+ \rightarrow As(CO_3)^+ + 3H_2O$$

Kim et al. (7) also estimated the stability constants for these arsenic carbonate complexes since none have been reported in literature. Stability constants were derived by extrapolating from a plot of ionic radii against the literature values for trivalent element carbonate complexes using the lanthanide and actinide series elements. The stability constants are estimated to be about 10^7 for B_1 and $10^{12.5}$ for B_2, suggesting that the carbonate complexes to be stable inorganic arsenic species. Since these arseno-carbonate complexes are believed to be fairly stable, they may actually contribute significantly to the amounts of arsenic found in well water. The following scheme for the conversion of carbonate ion pairs to the oxyanions of arsenic in groundwater was suggested (7):

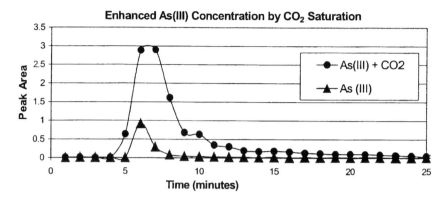

Figure 3. GFAAS analysis of As(III) concentrations sequentially collected with an anion column for As_2O_3 solutions saturated with and without CO_2 for 24hours.

$$As(CO_3)_2^- + 3H_2O \rightarrow H_3AsO_3 + 2HCO_3^- + H^+$$
$$As(CO_3)^+ + 3H_2O \rightarrow H_3AsO_3 + HCO_3^- + 2H^+$$
$$As(CO_3)_2^- + 4H_2O \rightarrow HAsO_4^{2-} + 2HCO_3^- + 5H^+ + 2e^-$$
$$As(CO_3)(OH)_2^- + H_2O \rightarrow H_3AsO_3 + HCO_3^-$$

The presence of charged arsenic carbonate complexes may explain some of the discrepancies reported regarding the geochemistry of arsenic in groundwater. It also challenges the current belief that arsenate and arsenite are the only stable arsenic species in natural waters.

Acknowledgements

The authors thank Myoung-Jin Kim for initial works and stimulating discussions that contributed to this study. The authors also thank Xiao-Qin Wang for her technical assistance and insightful comments on this work.

References

1. Nickson, R. Thesis, University college, Londin, 1997.
2. Bhattacharya, P.; Chatterjee, D.; Jacks, G. *Water Resources Develop.* **1997,** 13, 79-92.
3. Woo, N.C.; Choi, M.J. *Environ. Geol.* **2001,** 40, 305-311.

4. Das, D.; Basu, G.; Chowdhury, T.R.; Chakraborty, D. in *Proc. Int. Conf. Arsenic in Groundwater*, **1995,** Calcutta.

5. Saha, A.K.; Chakrabarti, C. in *Proc. Int. Conf. Arsenic in Groundwater*, **1995,** Calcutta.

6. Nickson, R.; McArthur, J.; Burgess, W.; Ahmed, K.M.; Ravenskroft, P.; Rahman, M. *Nature* **1998,** 395, 338.

7. Kim, M.J.; Nriagu, J.; Haack, S. *Environ. Sci. Technol.* **2000,** 34, 3094-3100.

8. Schaufelberger, F.A. In *Arsenic in the Environment Part I: Cycling and Characterization;* Nriagu, J.O., Ed.; John Wiley & Sons, Inc.: NY, 1994; pp 403-415.

9. Gonzalez, V.L.E.; Monhemius, A.J. In *Arsenic Metallurgy, Fundamentals and Applications;* Reddy, R.G.; Hendrix, J.L.; Queneau, P.B., Eds.; Metallurgical Society: PA, 1988; pp 405-418.

10. Mass, R. *Ann. Chim.* **1949,** 4, 459-504.

11. Harris, G.B.; Monette, S. In *Arsenic Metallurgy, Fundamentals and Applications;* Reddy, R.G.; Hendrix, J.L.; Queneau, P.B., Eds.; Metallurgical Society: PA, 1988; pp 469-488.

Chapter 4

Quantity and Speciation of Arsenic in Soils by Chemical Extraction

R. H. Loeppert[1], A. Jain[2], M. A. Abd El-Haleem[3], and
B. K. Biswas[1]

[1]Soil and Crop Sciences Department, Texas A&M University,
College Station, TX 77843
[2]Center for Water Quality, Florida A&M University, Tallahassee, FL 32307
[3]College of Agriculture, Zagazig University, Benha, Egypt

Chemical extraction represents an important tool for understanding arsenic bonding and speciation in soils, as well as for assessment and management of arsenic contaminated sites. The object of this paper is to review extraction mechanisms and discuss pitfalls to the successful application of extraction methodologies.

Introduction

Arsenic, which is toxic to animals and plants and an environmental hazard, can be present in soils as a result of natural processes or anthropogenic activities. The toxicity and environmental hazard of arsenic in a specific situation are strongly influenced by its solubility, speciation, and retention/release characteristics. A prerequisite for the successful assessment and management of an arsenic-contaminated soil is a knowledge of the soil arsenic, including arsenic concentration, solubility, speciation, mode of retention (bonding environment), and mobilization potential. Chemical extraction represents an indispensable tool for understanding these characteristics. The geoscientist, analyst, regulator, and engineer are faced with the challenge of obtaining the best possible information, while avoiding the pitfalls of improper extraction methodologies and incorrect interpretation of results. The objective of the current paper is to discuss the chemical principles of extraction of arsenic from soils, as a tool for environmental assessment and management. The emphasis in the current paper is with inorganic As(III) and As(V). The discussion of extraction is preceded by a brief discussion of arsenic species and predominant bonding mechanisms, since the prerequisite of any consideration of extraction procedure is a consideration of the probable mode(s) of bonding of soil arsenic.

Arsenic in Soil

Forms of Dissolved Soil Arsenic

Arsenic can exist in both inorganic and organic forms. Inorganic arsenic exists predominantly as As(III) and As(V) in aqueous systems. Although As(III) is more thermodynamically stable under reduced conditions (e.g., flooded soils) and As(V) is more stable under well oxidized conditions, both oxidation states will often exist concurrently, because of the relatively slow redox transformations between arsenic species (1). Also, the redox processes involving arsenic are often biological and intracellular, where the redox conditions might be very different than that of the bulk soil. The result is that As(III) and As(V) are both observed (sometimes transiently) under both oxidized and reduced soil conditions.

Inorganic As(III) (arsenite) in solution exists predominantly as $H_3AsO_3^\circ$ ($pK_a = 9.22$) and its conjugate bases $H_2AsO_3^-$ ($pK_a = 12.13$), $HAsO_3^{2-}$ ($pK_a = 13.4$), and AsO_3^{3-} (2). Under most Earth surface conditions (approximately pH 4 to 8.5) the neutral species, $H_3AsO_3^\circ$, is dominant. Under very reduced conditions, As(III) can be microbially transformed to arsine AsH_3, but this species is only a minor component of the soil solution because of its volatility.

Inorganic As(V) (arsenate) can exist in solution as $H_3AsO_4^\circ$ ($pK_a = 2.20$) and its conjugate bases $H_2AsO_4^-$ ($pK_a = 6.97$), $HAsO_4^{2-}$ ($pK_a = 11.53$), and AsO_4^{3-} (2). Under Earth surface conditions, inorganic As(V) exists predominantly as the anionic species $H_2AsO_4^-$ and $HAsO_4^{2-}$.

Bonding of Soil Arsenic

The predominant mode of bonding of inorganic As(III) and As(V) in oxidized soils involves complexation at the surfaces of Fe, Al, and Mn oxides. Bonding is usually stronger to Fe oxides than to Al and Mn oxides. In the case of Fe oxides, bonding of both As(V) and As(III) is predominantly by inner sphere rather that outer sphere complexation (Figure 1) (3-6). Spectroscopic studies have indicated that As(V) and As(III) each form predominantly bidentate, binuclear bridging surface complexes (3, 5, 6); i.e., two oxygens from a given As(V) or As(III) ligand bond to adjacent surface structural Fe atoms (Figure 1).

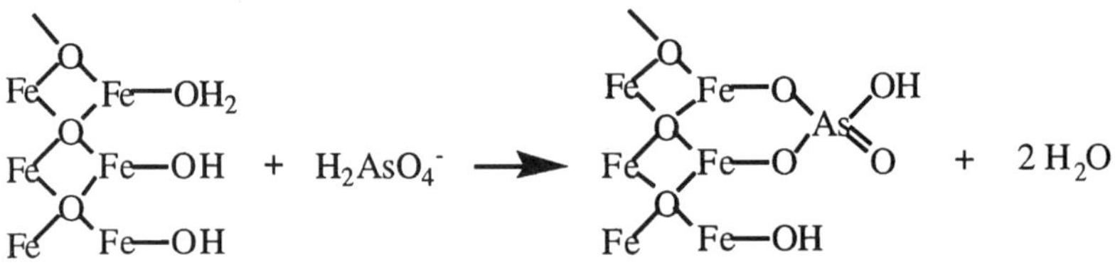

Figure 1. Bidentate, binuclear complex of arsenate on Fe oxide surface.

With inner sphere complexation (chemisorption), the arsenic competes with surface structural H_2O and OH^- groups for coordination of surface structural Fe. Thus the arsenic is chemically bonded to the surface and is considered to be a chemical component of the mineral surface. Arsenic is strongly bound to both poorly crystalline Fe oxides (e.g., ferrihydrite) and well crystalline Fe oxides (e.g., goethite, hematite).

Bonding of inorganic As(III) and As(V) by soil Fe oxide can occur at pH values either above or below the zero point of charge (zpc) of the oxide, i.e., to either negatively or positively charged surfaces (Figure 2) (7); however, As(III) and As(V) follow different trends in the influence of pH on adsorption behavior. As(V) is more strongly retained by soil Fe oxide at low pH, compared to the stronger retention of As(III) at high pH. The adsorption maxima for retention of arsenic by ferrihydrite are at approximately pH 4-5 for As(V) and pH 9 for As(III). Above pH 8, a considerable decrease in retention of As(V) is usually observed, due to the increased negative surface potential of the oxide mineral and the increased competition of OH^- for surface adsorption sites with increasing pH. These effects of pH on adsorption have considerable significance to the solubility and extraction of arsenic.

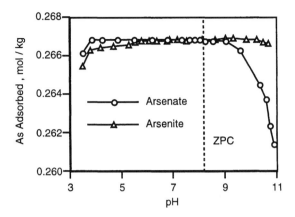

Figure 2. Adsorption of As(III) and As(V) on ferrihydrite. (Adapted with permission from reference 7. Copyright 1998.)

In reduced sulfidic systems, the association of As(III) with metal sulfides (8), as surface complexes, solid solutions, or separate solid phases, also plays an important role in arsenic retention. As(III) can be strongly retained by both the amorphous (reactive) sulfide minerals and the well crystalline sulfide minerals such as pyrite.

Solubility of Soil Inorganic Arsenic

Because of the relative ease and strength of bonding of inorganic arsenic to soil Fe oxides, dissolved arsenic concentrations are relatively low in most soil situations. Only under conditions in which the concentration of dissolved inorganic arsenic exceeds the concentration of available Fe oxide ligand-binding sites is the concentration of dissolved arsenic expected to be appreciable. Such a condition can exist where a large amount of soluble arsenic is added to a soil as a result of anthropogenic activities or where the soil is altered to result in a decrease in concentration of available surface sites. The latter condition could occur when the lowering of soil redox potential has resulted in the dissolution of soil Fe oxide, with a resulting decreased concentration of Fe-oxide surface sites. Under flooded (reduced) conditions arsenic is generally more soluble than under oxidized conditions.

Under most soil conditions the solubility of arsenic is controlled by adsorption processes, i.e., by oxide minerals (under oxidizing conditions) and sulfide minerals (under reduced conditions). The Ca, Mn, Fe(II), and Al salts of

As(V) are generally too soluble to control the concentration of soil pore-water arsenic (9). Only under conditions where the quantity of arsenic exceeds the availability of surface ligand-binding sites and sufficient dissolved metal cation is present is the solubility of arsenic likely to be controlled by arsenic-containing salts, such as Ca arsenate and Ba arsenate.

Principles of Arsenic Extraction

Arsenic in soil can exist as both soluble species and bound species. The concentration and speciation of soluble arsenic is of interest because of the impact of dissolved arsenic on water quality and because these species directly impact plants and microbes.

Extraction of Soluble Arsenic

Two of the more commonly used extractants to assess soluble arsenic in soils are deionized water and 0.01 M $CaCl_2$ (10). The extraction of soluble arsenic must be made under conditions where neither the quantity of dissolved arsenic nor the concentration of available surface adsorption sites is altered as a result of the extraction process. This requires the use of a mild neutral extracting agent, such as water or NaCl, that will not result in appreciable change in pH of the soil suspension or react specifically with soil colloid surfaces. Also, care must be taken that the redox condition of the soil is not altered. For example, the oxidation of a reduced soil could result in precipitation of dissolved Fe^{2+} as Fe(III) oxide, with a resulting change in concentration of surface sites available for adsorption of arsenic. Also, oxidation of reactive sulfides upon exposure to air can result in a considerable change in arsenic solubility and adsorption behavior. Soils must be sampled and maintained under conditions which ensure that the redox condition of the sample reflects the exact redox condition of the in situ soil. Exposure to atmospheric oxygen must be prevented. These essential precautions require creativity and advanced planning on the part of the scientist.

Extraction of Bound Arsenic

The first decision of the scientist, regulator, or engineer is whether the primary interest is with total arsenic or with understanding the various reservoirs of arsenic in the soil. In the former case, the soil matrix must be totally dissolved to obtain complete release of arsenic from organic and inorganic soil components, prior to analysis. Total dissolution is usually accomplished by digestion with concentrated mineral acids. If the primary interest is to understand the form, species, mobilization potential, or bioavailability of arsenic, then a selective extraction procedure might be more appropriate. In the selective extraction procedures, the extraction of bound arsenic requires either dissolution or partial dissolution of the arsenic-bonding solid phase (and the resulting release of arsenic) or ligand exchange of adsorbed arsenic by a competing ligand. The mechanisms of dissolution most commonly utilized in soil-extraction procedures are ligand-enhanced dissolution, H^+-enhanced dissolution, OH^--enhanced dissolution, and reductive dissolution. These individual processes are summarized below.

Mechanisms of Extraction of Bound Arsenic

Ligand Exchange

Extraction of arsenic by ligand exchange involves the desorption of arsenic by a competing ligand, e.g., phosphate or OH^- (equation 1). Phosphate has been widely used as an extractant for soil arsenic.

$$Fe\text{-oxide-}AsO_4H + HPO_4^{2-} \rightarrow Fe\text{ oxide-}PO_4H + HAsO_4^{2-} \qquad (1)$$

The ligand exchange process is influenced by the relative affinity of the competing ligands for the oxide surface (thermodynamic factor) and the rate of ligand exchange (kinetic factor).

As(V) and phosphate are similar in chemistry, including ion size and acid dissociation constants of the protonated species. Although As(V) and phosphate retention by Fe oxide decreases with increasing pH and there is usually a slight preference for As(V) versus phosphate adsorption at any given pH, these ligands show very similar adsorption behavior across the entire pH range (Figure 3) (11). On the other hand, As(III) and phosphate exhibit very different adsorption behavior, i.e., phosphate retention is greatest at low pH and As(III) retention is greatest at high pH (Figure 4) (11). In this experiment, a significant amount of As(III) was retained in the presence of a large excess of phosphate across the entire pH range, but especially at pH 8 to 10. This retention behavior of As(III) indicates that certain sites at the Fe oxide surface exhibit a strong preference for As(III) versus phosphate (11). Even in the presence of 0.1 M Na phosphate, appreciable As(III) is retained by ferrihydrite (11) and goethite (12).

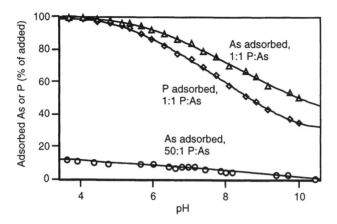

Figure 3. Competitive adsorption of As(V) and phosphate by ferrihydrite.
Adapted with permission from reference 11. Copyright 2000.

The rates of As(III) and As(V) desorption by phosphate follow very different trends (Figure 5) (*13*). For example, in the case of desorption from goethite, maximum As(III) desorption was achieved within approximately 4 h, whereas, the quantity of As(V) desorbed continued to increase for up to 100 h (Figure 5) (*13-14*). In both the As(III) and As(V) systems, it was difficult to achieve quantitative desorption of arsenic upon reaction with phosphate at pH 5.

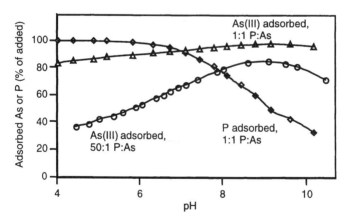

Figure 4. Competitive adsorption of As(III) and phosphate by ferrihydrite.
(Adapted from reference 11).

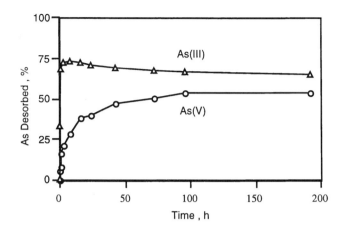

Figure 5. Influence of time on desorption of As(III) and As(V) by Na Phosphate (pH 5.0) at a 200:1 P:As molar ratio.

Previous studies have indicated that the ligand exchange of phosphate on goethite is both acid catalyzed and base catalyzed (*15*), and the rate of ligand exchange is accelerated under conditions of higher acidity or basicity, compared to near neutral conditions. Similar pH-dependent behavior might be expected for ligand exchange reactions involving As(V) and phosphate.

There are several general conclusions that can be made regarding the use of ligand exchange by phosphate for the assessment of bound arsenic in soils: (i) desorption of As(V) is strongly influenced by extraction time and pH; (ii) As(III) is more readily desorbed at low pH and As(V) at high pH; and (iii) desorption of As(III) is not complete at any pH, indicating that there are As(III) adsorption sites on Fe oxides for which phosphate is not highly competitive.

Ligand-enhanced Dissolution

Ligand-enhanced dissolution involves complexation of surface structural cation, e.g., Fe^{3+}, by an organic complexing agent (e.g., oxalate; citrate; DTPA, diethylenetriaminepentaacetic acid) and dissolution of the mineral surface (equation 2). Arsenic is subsequently released as a result of dissolution of the metal oxide ligand-binding sites.

$$\text{Fe-oxide-AsO}_4 + \text{L}^- \rightarrow \text{Fe}^{3+}\text{-L} + \text{As}_{aq} \tag{2}$$

The overall reaction occurs in two steps: (i) rapid adsorption of ligand at the Fe oxide surface, and (ii) subsequent slow dissolution of Fe^{3+} (16).

Among the reagents that have been used most extensively for the extraction of arsenic by a ligand-enhanced dissolution are sodium oxalate and ammonium oxalate. At pH 3, ammonium oxalate (in the dark) will quantitatively dissolve poorly crystalline Fe oxide, e.g., ferrihydrite; whereas, the reactions with well crystalline Fe oxides, e.g., goethite or hematite, are much slower (17-18). Thus, ammonium oxalate extraction in the dark allows the selective dissolution of poorly crystalline soil Fe oxides and possibly the selective extraction of arsenic from these phases. This reaction must be performed in the dark, since in the light, goethite and hematite are also appreciably dissolved due to the photoreduction of Fe(III) (18).

When the oxalate reaction is performed in the light, As(III) is substantially transformed to As(V), thus preventing the quantitative speciation of arsenic. In the dark, the likelihood of oxidation of As(III) or reduction of As(V) is decreased; therefore, the quantitative speciation of arsenic is feasible.

During the extraction of a near neutral arsenic-contaminated soil by pH 3 NH_4 oxalate in the dark, the maximum desorption of arsenic was observed at short reaction times (< 30 min), followed by a gradual decrease in desorbed arsenic. The rapid desorption of arsenic occurred during the period of rapid dissolution of the reactive Fe-oxide phase, followed by gradual readsorption of arsenic to the remaining Fe oxide as the Fe-oxide dissolution rate decreased. The apparent readsorption of arsenic in the presence of oxalate indicates that oxalate was not highly competitive with As(V) for the ligand-binding sites. These results also indicate that the desorption of As by oxalate from ferrihydrite and goethite is predominantly attributable to ligand-enhanced dissolution of the Fe oxide rather than ligand exchange. In previous studies it was observed that addition of phosphate to the oxalate prevented the readsorption of As(V) to goethite (12). This latter procedure could not be used to selectively desorb arsenic from amorphous Fe oxide, since the addition of phosphate would also result in release of phosphate from well crystalline phases.

H$^+$-enhanced Dissolution

During reaction of arsenic-contaminated soil with mineral acid, e.g., HCl or HNO_3, the predominant mechanism of arsenic release is H$^+$-enhanced dissolution (equation 3).

$$\text{Fe-oxide-AsO}_4 + \text{H}^+ \rightarrow \text{Fe}^{3+} + \text{H}_2\text{O} + \text{As}_{aq} \tag{3}$$

The short range order oxides, e.g., ferrihydrite, are more readily dissolved than the crystalline oxides (goethite and hematite) (*17*). Hydrochloric acid (0.1 or 1.0 M) is also used for selective dissolution of the reactive sulfide minerals; however, the crystalline sulfides, e.g., pyrite, are not readily dissolved by this treatment (equation 4) (*8*).

$$FeS + 2 H^+ \rightarrow Fe^{2+} + H_2S \tag{4}$$

We have observed that soils differ considerably in the rate of dissolution of arsenic upon reaction with 0.1 M HCl (Figure 6). There is usually an initial rapid release of arsenic during the first 30 min of reaction, followed by a considerable slowing in reaction rate. There is sometimes an eventual decrease in desorbed arsenic, which is sometimes negligible (as with soil A, Figure 6) or substantial (as with soil C). This phenomenon is attributable to the rapid desorption of arsenic during the initial period of rapid Fe oxide dissolution, followed by a gradual readsorption of arsenic as the Fe oxide approaches equilibrium with respect to dissolved Fe. The overall desorption/readsorption pattern is determined by the initial bonding environment of the arsenic and the availability of mineral sites for readsorption. Acid concentration, reaction time, and soil-to-solution ratio can strongly influence arsenic extraction patterns (*19*), especially under conditions of mineral-acid treatment in which the Fe-oxide phase is not totally dissolved. Methodological variables can have an overriding influence on the results of chemical extractions, which illustrates the importance of uniform methodologies and well defined analytical protocol. In several recent studies, we observed that As(III) was partially oxidized to As(V) during mineral acid extractions under conditions of both light and dark. But oxidation of As(III) does not always occur, and the conditions that promote oxidation are not yet totally clear. Therefore, any analysis of arsenic species following mineral acid extraction should be interpreted with caution, even when extractions are performed in the dark.

OH⁻-enhanced Dissolution

Arsenic extraction at high pH is attributable to OH⁻-enhanced dissolution of Fe oxide (equation 5), but also to competitive adsorption of OH⁻ at the oxide surface and the increasingly negative surface potential with increasing pH which makes the system less favorable for adsorption of negatively charged arsenic species.

$$Fe\text{-oxide-}AsO_4 + OH^- \rightarrow Fe(OH)_4^- + As_{aq} \tag{5}$$

The mobilization of arsenic is highly dependent on the pH of the extracting solution. With higher pH extractants, higher amounts or arsenic are often extracted from oxidized soils (e.g., reference 20). This phenomenon is attributable to the relative ease of extraction of As(V) at high pH. Soils dominated by As(III) would likely exhibit different behavior. NaOH (0.1 M) is a relatively effective extractant of As(V) from goethite and ferrihydrite, but it is much less effective for As(III) (12).

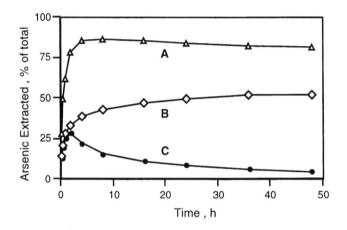

Figure 6. Influence of reaction time on extraction of arsenic from three soils (A, B, and C) with 0.1 M HCl.

Reductive Dissolution

Several reducing reagents, e.g., 0.1 M hydroxylamine hydrochloride (pH 2.0), 0.25 M hydroxylamine hydrochloride in 25 % acetic acid, and citrate dithionite, have been used to extract free Mn and Fe oxides, and arsenic associated with these phases, from soil (equation 6).

$$Fe\text{-}oxide\text{-}AsO_4 + e^- + L\text{-} \rightarrow Fe^{2+}\text{-}L + As_{aq} \qquad (6)$$

The citrate dithionite method can be used to determine the total free Fe-oxide content of a soil (18). Under some conditions hydroxylamine hydrochloride extraction will underestimate the Fe associated with crystalline Fe oxides (21).

Sequential Extraction

The objective of an arsenic sequential extraction procedure is to quantify arsenic in the various bonding environments in the soil, e.g., soluble, readily exchangeable, organic arsenic, MnO_2 adsorbed, amorphous Fe-oxide adsorbed, crystalline Fe-oxide adsorbed, $CaCO_3$ bound, Ca arsenate, Fe(III) arsenate, reactive sulfide bound, crystalline sulfide bound, and residual. As well as providing information regarding bonding environment, sequential extractions can give important clues concerning bioavailability, bioaccessibility, and arsenic release potential. Sequential extraction procedures are described in several recent studies (*21-31*). Simplified extraction schemes that group arsenic into broad general categories (e.g., the easily extractable fraction, arsenic solubilized under reducing conditions, and strongly bound arsenic) have been effectively utilized in some cases (e. g., reference 32).

All sequential extraction procedures are subject to potential errors, since the individual extraction steps are often nonspecific for the intended phase. Also, the individual extraction steps can sometimes result in phase transformations of arsenic that will influence the extractability of arsenic in subsequent extraction steps. For this latter reason, scientists often prefer single extractions to sequential extractions. The results of any sequential extraction should be interpreted with caution; yet if used properly and critically, sequential extraction represents a useful tool for assessing soil-arsenic status.

Assessment of Bioavailable and Bioaccessible Arsenic

Numerous extractants have been utilized for the assessment of "bioavailable" arsenic, e.g., deionized water, acetic acid, 0.5 M $NaHCO_3$, 0.01 M $CaCl_2$, 0.01 M $NaNO_3$, and dilute HCl/HF. Many of these same extractants have been used to assess "bioavailable" phosphate. Several studies have indicated that 0.01 M $CaCl_2$ provides the best assessment of "bioavailable" arsenic (*10,33*). In this discussion, the term "bioavailable", when used in the context of chemical extraction, is placed in quotation marks to infer that chemical extraction is only providing a chemical approximation of bioavailability. The only true way to assess arsenic bioavailability is to determine the actual acquisition of arsenic by the target organism.

Two general philosophies / approaches have been used to determine the "bioavailable" arsenic fraction by means of chemical extraction. The first approach is based on the assumption that soils should be compared under uniform conditions. Therefore, the soils are extracted with the selected extracting agent following a uniform soil pretreatment (usually air drying). Of course, the actual solubility and bioavailability of arsenic in the soil is strongly influenced by in situ soil variables, such as redox potential and localized microbial and plant processes. The second approach to assessment of "bioavailable" arsenic is to extract arsenic under conditions as close as possible to the in situ soil conditions, with careful attention to maintaining the actual soil redox condition. This approach allows the extraction of "bioavailable" arsenic under the actual soil conditions, but does not allow the comparison of different soils under uniform conditions. The researcher's approach is dictated by the experimental objective.

Under some conditions bound arsenic can be mobilized as a result of soil biotic and abiotic processes. Therefore, the term bioaccessible has been used to represent the fraction that is potentially mobilizable and bioavailable. Several reagents have been used to assess bioaccessible arsenic, including phosphate (arsenic bioaccessible as a result of ligand exchange), hydroxylamine hydrochloride (arsenic bioaccessible under mildly reducing conditions), oxalate (arsenic bioaccessible as a result of dissolution of poorly crystalline Fe oxides), and 0.1 M HCl (arsenic bioaccessible as a result of dissolution of reactive sulfides and oxides).

EPA Toxic Characteristic Leaching Procedure (TCLP)

The TCLP extract is buffered at pH 4.9 by Na acetate (*34*). At this pH, the adsorption of As(V) by Fe oxide is close to its approximate maximum (*7*). Therefore, this extractant could underestimate the arsenic leaching potential of neutral and alkaline soils. Also, care should be taken to maintain the oxidation state of reduced soils. The oxidation of reduced soils and the precipitation of Fe oxide could result in an underestimation of arsenic leaching potential by the TCLP procedure (*35*).

Conclusions

Chemical extraction of arsenic represents a powerful tool for understanding concentration, speciation, bonding environment, and release potential of arsenic in soils. It also represents an indispensable tool for the assessment and management of arsenic contaminated sites. Yet there are many pitfalls to the successful application of extraction methodologies. The successful application depends on the analyst's understanding of the extraction process and mechanism, and the critical evaluation of each individual system in which chemical extraction is to be utilized.

Literature Cited

1. Inskeep, W. P.; McDermott, T. R. In Environmental Chemistry of Arsenic; Frankenberger, W. T., Ed.; Marcel Dekker: New York, 2002 ; pp 183-215.
2. Wagman, D. D.; Evans, H. H.; Parker, V. B.; Schumm, R. H.; Harlow, I.; Bailey, S. M.; Churney, K. L.; Butall, R. L. J. *Phys. Chem. Ref. Data II Suppl.* **1982**, *2*, 392.
3. Waychunas, G. A.; Rea, B. A.; Fuller, C. C.; Davis, J. A. *Geochem. Cosmochim. Acta* **1993**, *57*, 2251-2269.
4. Manceau, A. *Geochim. Cosmochim. Acta* **1995**. *59*, 3647-3653.
5. Fendorf, S.; Elck, M. J.; Grossl, P.; Sparks, D. L. *Environ. Sci. Technol.* **1997**, *31*, 315-320.
6. Manning, B. A.; Fendorf, S. E.; Goldberg, S. *Environ. Sci. Technol.* **1998**, *32*, 2383-2388.
7. Raven, K. P.; Jain, A.; Loeppert, R. H. *Environ. Sci. Technol.* **1998**, *32*, 344-349.
8. Morse, J. W. *Mar. Chem.* **1994**, *46*, 1-6.
9. Sadiq, M. *Water Air Soil Pollut.* **1997**, *93*, 117-136.
10. Houba, V. J. G.; Temminghoff, E. J. M.; Gailkhorst, G. A. *Commun. Soil Sci. Plant Anal.* **2000**, *31*, 1299-1396.
11. Jain, A; Loeppert, R. H. *J. Environ. Qual.* **2000**, *29*, 1422-1430.
12. Jackson, B. P.; Miller, W. P. *Soil Sci. Soc. Am. J.* **2000**, *64*, 1616-1622.
13. Abd El-Haleem, A. M.; Loeppert, R. H.; Hossner, L. R. *Agron. Abstr.* **2000**, 239.
14. Liu, F.; De Cristofaro, A.; Violante, A. *Soil Sci.* **2001**, *166*, 197-208.
15. Mott, C. J. B. In *The Chemistry of Soil Processes*; Greenland, D. J.; Hayes, M. H. B., Ed.; Wiley: Chichester, UK; pp 179-219.
16. Stumm, W.; Furrer, G. In *Aquatic Surface Chemistry*; Stumm, W., Ed.; Wiley: New York, 1987; pp 197-219.
17. Schwertmann, U. In *Iron Nutrition and Interactions in Plants*; Chen, Y.; Hadar, Y., Ed.; Kluwer: Dordrecht, Netherlands, 1991; pp 3-27.
18. Loeppert, R. H.; Inskeep, W. L. In *Methods of Soil Analysis, Part 3, Chemical Methods*; Sparks, D. L., Ed.; SSSA Book Series, 5; Soil Science Society of America: Madison, WI, 1996; pp 639-664.
19. Chappell, J.; Chismell, B.; Olszowy. H. *Talanta.* **1995**, *44*, 323-329.
20. Bissen, M.; Frimmel, F. H. *Fresen. J. Anal. Chem.* **2000**, *367*, 51-55.
21. La Force, M. J.; Fendorf, S. *Soil Sci. Soc. Am. J.* **2000**, *64*, 1608-1615.
22. Keon, N. E; Swartz, D. J.; Brabander, C.; Harvey, C.; Hemond, H. F. *Environ. Sci. Technol.* **2001**, *35*, 2778-2784.
23. Lombi, E.; Sletten, R. S.; Wenzel, W. W. *Water Air Soil Pollut.* **2000**, *124*, 319-332.
24. Onken, B. M.; Adriano, D. C. *Soil Sci. Soc. Am. J.* **1997**, *61*, 746-752.

25. Szakova, J.; Tlustos, P.; Balik, J.; Pavlikova, D.; Vanek, V. *Fresen. J. Anal. Chem.* **1999**, *363*, 594-595.
26. Terashima, S.; Taniguchi, M. *Bunseki Kagaku* **1996**, *45*, 1051-1058.
27. Wenzel, W. W.; Kirchbaumer, N.; Prochaska, T.; Stingeder, G.; Lombi, E.; Adriano, D. C. *Anal. Chim. Acta* **2001**, *436*, 309-323.
28. Ariza, J. L. G.; Giraldez, I.; Sanchez-Rodas, D.; Morales, E. *Talanta* **2000**, *52*, 545-554.
29. Gomez-Ariza, J. L.; Giraldez, I.; Sanchez-Rodas, D.; Morales, E. *Int. J. Environ. Anal. Chem.* **1999**, *75*, 3-18.
30. McLaren, R. G.; Naidu, R.; Smith, J.; Tiller, K. G. *J. Environ. Qual.* **1998**, *27*, 348-354.
31. Zhang, T. H.; Shan, X. Q.; Li, F. L. *Commun. Soil Sci. Plant Anal.* **1998**, *29*, 1023-1034.
32. Gleyzes, C.; Tellier, S.; Sabrier, R.; Astruc, M. *Environ. Technol.* **2001**, 27-38.
33. Szakova, J.; Tlustos, P.; Balik, J.; Pavlikova, D.; Balikova, M. *Chem. Listy* **2001**, *95*, 179-183.
34. U.S. Environmental Protection Agency. *Test Methods for Evaluating Solid Waste, Physical/Chemical Methods*, 3rd edition; SW-846, Method 1311; U. S. Government Printing Office: Washington, DC, 1992.
35. Meng, X.; Korfiatis, G. P.; Jing, C.; Christodoulatos, C. *Environ.Sci. Technol.* **2001**, *35*, 3476-3481.

Chapter 5

Arsenic(III) Complexation and Oxidation Reactions on Soil

Bruce A. Manning[1], Scott E. Fendorf[2], and Donald L. Suarez[3]

[1]Department of Chemistry and Biochemistry, San Francisco State University, 1600 Holloway Avenue, San Francisco, CA 94132
[2]Department of Geological and Environmental Sciences, Stanford University, Palo Alto, CA 94305
[3]U.S. Salinity Laboratory, Agricultural Research Service, U.S. Department of Agriculture, 450 Big Springs Road, Riverside, CA 92507–4617

The complexation and oxidation reactions of arsenic(III) were investigated in three soils using standard batch reactions, a stirred reactor, and X-ray absorption spectroscopy. The objective was to identity soil components responsible for As(III) oxidation and the surface sites which bind the As(V) product. Speciation of As(III) and As(V) was determined using HPLC coupled with hydride generation atomic absorption spectrometry. Certain soils and soil minerals such as manganese oxides caused rapid oxidation of As(III) and formation of strongly adsorbed As(V) surface complexes. The coordination environment of As(V) in soil was determined to be predominantly an Fe oxide surface complex. Based on results from extended X-ray absorption fine structure (EXAFS) spectroscopy, the As-Fe inter-atomic distance of 3.38 angstroms was indicative of an inner-sphere, bidentate surface complex. This paper discusses the kinetics and mechanism of As(III) oxidation in soil and the use of EXAFS as a probe of the As(V) coordination environment.

Introduction

The complexation and oxidation of arsenite (As(III)) are important pathways in the overall environmental cycling of As. The environmental and human health impacts of elevated concentrations of As in groundwater have received increased attention due to this element's toxicity (*1-3*). The predominant forms of As in soil and water are inorganic arsenate (As(V)) and arsenite (As(III)). The As(III) species exists as undissociated arsenious acid ($H_3AsO_3^0$) below pH 9.2 (*4*) and predominates under mildly reducing conditions (*4-7*). Because As(III) is more weakly bound to most soil minerals than As(V) (*8, 9*) it is more mobile and bioavailable. In addition, As(III) is substantially more toxic than As(V) (*10, 11*).

An important reaction in the environmental fate of As(III) is heterogeneous oxidation on soil mineral surfaces (*8, 12-16*). Manganese oxides are extremely important naturally occurring minerals because they readily oxidize many reduced species such as As(III) (*14-16*). The total manganese content of soils has been shown to be an important factor in determining the rate of As(III) oxidation (*12*). The partial oxidation of As(III) has been observed on phyllosilicate clay minerals (*8*). Other surfaces which can potentially oxidize As(III) include titanium dioxide (*13*) and zero-valent iron (Fe(0)) which is being developed for groundwater remediation (*17, 18*). The objectives of this study were to investigate both the complexation of As(III) and the oxidation of As(III) to As(V) in soil using a liquid chromatographic technique for As(III)/(V) speciation in combination with X-ray absorption spectroscopy (EXAFS and XANES) for analysis of As oxidation state in the solid phase.

Materials and Methods

The <2-mm size fractions of three arid-region soils from California were collected and included an Aiken clay, a Fallbrook mixed fine-sandy loam, and Wyo loam. These materials have been characterized previously (*9, 12*) and the soil properties of interest in the present study include soil surface area, total extractable Fe and Mn, and pH. Surface areas measured by BET N_2 were 44.8, 19.4, and 78.2 m^2 g^{-1} for Aiken, Fallbrook, and Wyo, respectively. A citrate-dithionite extraction was performed by shaking 4 g soil in 120 mL of 0.57 *M* sodium citrate ($Na_3C_6H_5O_7 \cdot 5H_2O$) and 0.1 *M* sodium dithionite ($Na_2S_2O_4$) for 16 h (*19*). Following shaking, the suspensions were centrifuged (4100 *g*, 10 min) and 15 mL aliquots were removed, filtered (0.1 μm), and acidified with high purity HNO_3. Samples were then analyzed for Al, Fe, and Mn by inductively coupled plasma atomic emission spectrometry (ICP-AES). Determination of pH was made in 1:10 soil:Deionized (DI) water suspensions.

The oxidation of As(III) by soil was investigated by reacting 2 g soil with 20 mL of 67 µM As(III) in a background electrolyte solution of 0.5 mM $CaCl_2$. The soil-As(III) reactions were stopped at time points between 0.1 and 48 hours and the tubes were centrifuged (14,500 g, 5 min) and the solutions were filtered (0.1 µm). The soil solids remaining in the reaction tubes were then resuspended and shaken for 1 h in 20 mL of 10 mM KH_2PO_4:K_2HPO_4 buffer at pH 7 (10 mM PO_4) to recover exchangeable As(III) and As(V) from the soil. In addition to centrifuge tube work, a closed system propeller-stirred reactor experimental approach was used allowing gas and pH control. The Fallbrook soil, which was found to completely oxidize As(III), was investigated to determine the mechanism of As(III) oxidation. Aerated and N_2-purged Fallbrook stirred suspensions (100 g soil in 1 L 0.5 mM $CaCl_2$) were equilibrated 4 d prior to As(III) addition (270 µM As(III)). After As(III) addition, the reactions were monitored by removal of 5 mL suspension aliquots followed by analysis of dissolved Fe and Mn by ICP-AES as well as As(III)/(V) speciation.

Arsenic speciation analysis was performed using high performance liquid chromatography coupled with hydride generation atomic absorption spectrophotometry (HPLC-HGAAS) (20). A Dionex AS11 Ionpac anion exchange column was used for As(III)/(V) separation using a 30 mM NaOH/1% methanol mobile phase (flow rate 1 mL min^{-1}). The column was in-line with flow-through hydride generator (Varian VGA 76) and a Perkin Elmer 3030B AA spectrometer monitoring the 193.7 nm wavelength with peak areas for As(III) and As(V) measured using a Hewlett Packard 3393A integrator.

Selected samples from this study were investigated using X-ray absorption spectroscopy. The XAS technique was used to determine the oxidation state of As and coordination environment of the resulting surface complexes. Because detection limits with XAS are much higher than HGAAS, a higher As(III) starting concentration was used (1 mM As(III)). The soil-As(III) mixtures were analyzed by HPLC-HGAAS and the solids were collected and rinsed with DI water on ashless filter paper for XAS. Samples were stored as wet pastes on the filter paper prior to XAS analysis in sealed N_2-purged polycarbonate test tubes. Sodium salts ($NaAsO_2$ and Na_2HAsO_4) and solutions of 1 mM As(III) and As(V) were also analyzed. The XAS spectra were collected at the Stanford Synchrotron Radiation Laboratory (SSRL) on beamline 4-3 using a Si(220) monochromator and a 13 element germanium semiconductor detector. The K edge of As (11867 eV) was examined using an energy range of 11667 to 12867 eV and at least 5 individual scans were collected for each sample. The extended X-ray absorption fine structure (EXAFS) data were analyzed using EXAFSPAK (21) and FEFF 7.0 (22) software. Individual scans were averaged and the background X-ray absorbance was removed by fitting a linear polynomial equation through the pre-edge region. This was followed by fitting a spline function through the extended X-ray absorption region of the spectra (11920-12867 eV) to isolate the contributions from backscattering of outgoing photoelectrons by coherent shells of atoms (nearest neighbor atoms) around the central As atom. Fitting of the theoretical EXAFS expression involved using the interatomic distances (R_{As-O} and R_{As-Fe}) and the debeye-waller parameters (σ^2, a measure of disorder in the shell) as adjustable parameters while fixing the coordination numbers (N). Coordination numbers were then adjusted in a second analysis while holding R and σ^2 constant.

Results and Discussion

A rapid uptake of soluble As(III) was observed with the Aiken soil (Figure 1) with 98% removal by adsorption at 18 h. No substantial dissolved As(V) was observed in the As(III)-Aiken suspensions when analyzed by HPLC-HGAAS. The Wyo soil had similar experimental results as Aiken (data not shown), possibly because both soils had high surface areas and citrate-dithionite extractable Fe (12.8 and 8.1 g kg^{-1} for Aiken and Wyo, respectively). The Fallbrook soil also displayed an initially rapid As(III) uptake reaction, but was significantly slower than Aiken and Wyo (Figure 2). The primary difference was the formation of measurable As(V) in As(III)-Fallbrook suspensions resulting from As(III) oxidation. Solution As(V) reached a maximum at 18 h followed by a steady decline between 18 and 48 h. The surface-bound As species were partly recovered in a phosphate buffer (PO$_4$) exchange reaction with the soil solids and the results are shown in Figures 3 and 4. Between zero and 1 h the As(III)/(V) speciation in the PO$_4$ extract of the Aiken soil revealed that As(III) was initially the predominant species recovered (Figure 3). Detectable As(V) was also observed, reaching a maximum at 8 h followed by a steady decline. This phenomenon of a decline in recoverable As(III) and As(V) was observed for all soils indicative of a long-term aging reaction where As becomes increasing insoluble and more strongly bound. The Fallbrook soil substantially oxidized As(III) and both As(III) and As(V) were recoverable in the PO$_4$ extract (Figure 4). Approximately 23% of added As(III) was recovered exchangeable As(V) in the Fallbrook soil at 18 h which declined slowly between 18-48 h.

The long term As(III) oxidation reaction on Fallbrook soil was investigated using a continuously-stirred reactor and a higher As(III) concentration (270 µM). The results for the N$_2$-purged suspension showed a rapid As(III) uptake reaction

6161

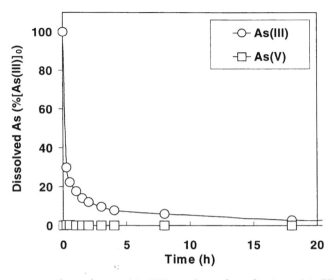

Figure 1. Time-dependence of As(III) uptake and production of As(V) in Aiken soil suspension.

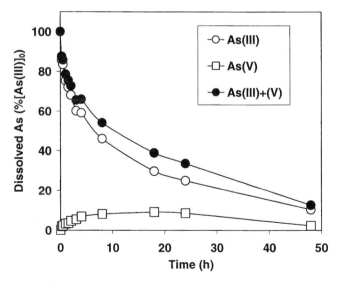

Figure 2. Time-dependence of As(III) uptake and production of As(V) in Fallbrook soil suspension.

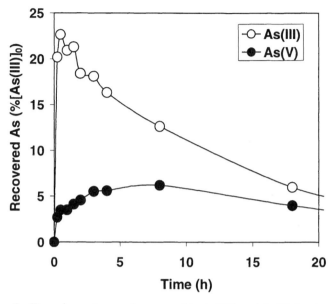

Figure 3. Time-dependence of recoverable As(III) and As(V) in phosphate
extraction of As(III)-treated Aiken soil suspension.

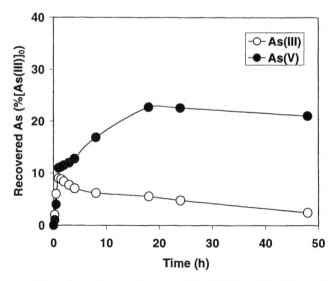

Figure 4. Time-dependence of recoverable As(III) and As(V) in phosphate
extraction of As(III)-treated Fallbrook soil suspension.

occurred between zero and 8 h followed by a slower reaction between 8 and 150 h (Figure 5a). The higher initial As(III) concentration may have saturated the available sites for complexation and oxidation. The production of dissolved As(V) occurred in the early time period which reached a maximum at 8 h followed by a gradual decline. The corresponding results for dissolved Fe and Mn (Figure 5b) revealed that production of Mn in the As(III)-treated soil steadily increased from 0-96 h and then leveled off at 4.8 mg/L (8.6 μM) at 120 h. Dissolved Fe was not detected throughout the reaction. A control reactor (zero As(III)) was also tested and a significantly higher concentration of Mn was found in the As(III)-treated system than in the control. This suggested that dissolved Mn release may have occurred as a result of oxidation of As(III) by soil Mn oxides. The Mn release behavior has been investigated and shown with model synthetic Mn oxide reacted with As(III) (15).

X-ray absorption spectroscopy was used for analysis of the solid phase As speciation. The solid phase As concentrations for As(III)- treated soils were between 0.010 and 0.012 mmol g^{-1}. The X-ray absorption near edge structure (XANES) spectra of As(III)-treated soils and As(III)/As(V) standards are shown in Figure 6. The As(III) and As(V) standards are given as a reference with absorption maxima occurring at 11872 and 11876 eV, respectively. The oxidation state of As is determined by the energy of the X-ray edge. Oxidation of As(III) to As(V) causes a shift in the X-ray absorption edge to a higher energy (approximately 1.5-2 eV per unit oxidation state change) allowing solid phase oxidation state of As to be evaluated. The Fallbrook soil XANES spectrum shows clear evidence of complete oxidation of As(III) to As(V), whereas the Aiken and Wyo soils display a distinct shoulder in the XANES spectrum corresponding to the As(III) energy at 11872 eV. This is evidence that these soils partially oxidized As(III) and contain mixtures of As(III) and As(V). The Aiken and Wyo soils may partially stabilize As(III) by adsorption on mineral surfaces which do not oxidize As(III) such as Fe oxyhydroxides (23). Batch experiments with the Aiken and Wyo soils revealed a more rapid adsorption reaction than the Fallbrook soil, possibly reflecting a greater quantities of reactive surface sites on Fe oxides.

The structure and nearest-neighbor environment of As(III) and As(V) bound on soils was analyzed by EXAFS spectroscopy. The EXAFS function ($\chi(k) \times k^3$) for As(III)-treated soils (Figure 7) shows the best fit of the theoretical EXAFS expression (dashed lines) to the experimental EXAFS data (points). The fit (dashed line) represents a sine wave with contributions from a shell of oxygen atoms surrounding the As atom (As-O shell) and a second shell of metal atoms (As-Fe shell) further away from the As atom. The results from fitting adjustable parameters of the theoretical EXAFS function to the experimental data are given in Table I. Partial As(III) oxidation by the Aiken and Wyo soils is evident by a both the coordination number (N) and interatomic distance (R) of the As-O shell.

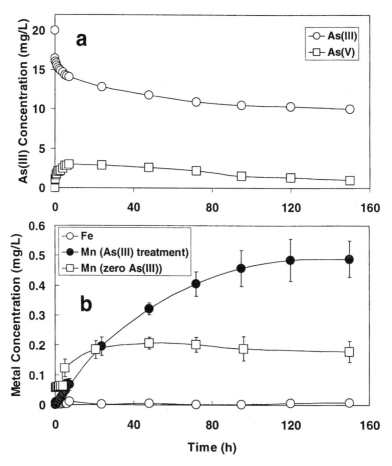

Figure 5. Dissolved As(III) and As(V) (a) and metals (Mn and Fe) (b) in continuously stirred reactor (N$_2$-purged) of As(III)-treated Fallbrook soil suspension and a control (zero As(III)).

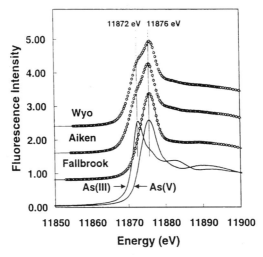

Figure 6. XANES spectra of As(III) treated Wyo, Aiken, and Fallbrook soils. Also shown are As(III) and As(V) referencestandards.

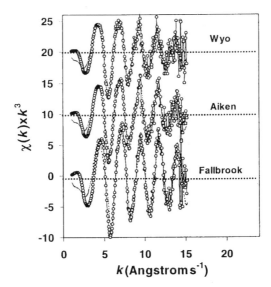

Figure 7. Experimental EXAFS spectra (points) and theoretical fits (dashed lines) of As(III) treated Wyo, Aiken, and Fallbrook soils.

The N value of 3.54 at a distance of 1.71-1.72 Å is indicative of a mixed As(III)+As(V) environment. Previous EXAFS work on As has given R_{As-O} values of 1.79 Å for As(III) (*23*) and 1.69 Å for As(V) (*13, 24*). In contrast, the Fallbrook soil was best fit using a $N=4$ and the R_{As-O} values of 1.69 Å.

Table I. EXAFS Parameters Optimized for As(III)-Treated Soils

Sample	*Shell*	N [a]	R(Å)	σ^2	E_0
Aiken	As-O	3.54	1.71	2.87×10^{-3}	-2.55
	As-Fe$_1$	1.18	3.09	8.98×10^{-3}	-2.55
	As-Fe$_2$	2.36	3.38	1.25×10^{-2}	-2.55
Wyo	As-O	3.54	1.72	2.90×10^{-3}	-3.12
	As-Fe$_1$	1.18	3.01	1.12×10^{-2}	-3.12
	As-Fe$_2$	2.36	3.38	1.37×10^{-2}	-3.12
Fallbrook	As-O	4.00	1.69	1.40×10^{-3}	-2.27
	As-Fe$_1$	1.18	2.85	2.10×10^{-2}	-2.27
	As-Fe$_2$	2.36	3.38	1.60×10^{-2}	-2.27

[a] N = coordination number, R = interatomic distance (error = ±0.02 Å), σ^2 = debeye waller parameter, E_0 = energy offset (threshold E_0 shift in eV, used for all shells).

Figure 8 shows the Fourier transform of the experimental and fitted EXAFS data. The fit line (dashed) gives an excellent description of the experimental data. The Fourier transform data in Figure 8 are not phase-corrected and therefore are shifted slightly (-0.4 angstroms) from the true interatomic distance. The contribution of a second shell of atoms was investigated and the best fit involved a composite shell of Fe atoms at two distances. Previous EXAFS work on As adsorbed on model Fe oxide compounds has been most successful when a multiple Fe shell is used in the fitting routine. The second peak (As-Fe$_1$ + As-Fe$_2$) in the Fourier transform plot is from photoelectron backscattering from next nearest neighbor Fe atoms and is evidence of "inner-sphere" covalent surface complexes of As on mineral surfaces. The first contribution to this shell (As-Fe$_1$) is from approximately 1.18 Fe atoms at 3.02 ± 0.06 Å (Aiken) and at 3.00 ± 0.06 Å (Wyo). A stronger contribution to the peak (As-Fe$_2$) comes from a shell of 2.36 Fe atoms at 3.38 ± 0.06 Å (all soils). The Fallbrook soil was best fit using a shorter As-Fe$_1$ distance at 2.85 ± 0.06 Å. The error estimate in the distance of ± 0.06 Å is derived from the sum of squares of differences between

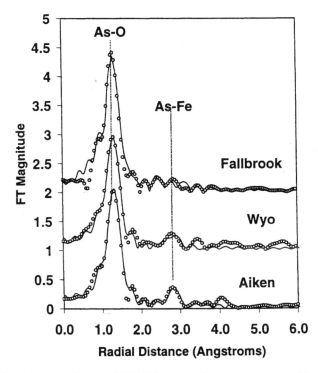

Figure 8. Fourier transforms of EXAFS spectra (corresponding to Figure 7.) of As(III) treated Wyo, Aiken, and Fallbrook soils.

68

the fit and experimental data. The As-Fe$_2$ distance of 3.38 Å is similar to an As-Fe shell of 3.30 Å previously reported for As(V)-treated goethite (α-FeOOH) which has been ascribed to a bidentate, binuclear As(V) surface complex where the AsO$_4$ tetrahedron is attached to adjacent apices of edge-sharing Fe octahedra (*25, 26*).

Based on XAS data (EXAFS and XANES) and stirred-reactor experiments, we postulate that As(III) oxidation by soil is the result of oxidation on mineral surfaces (probably MnO$_2$) followed by formation of both dissolved As(V) and adsorbed As(V) on mineral surfaces. A release of dissolved Mn was measured in the Fallbrook soil which is evidence of a reductive dissolution of soil Mn oxide by As(III). This natural process may be advantageous from an environmental standpoint given the greater mobility and toxicity of As(III). The removal of anoxic groundwater containing dissolved As(III) may be effectively remediated by reaction with simple soil-sand mixtures that contain sufficient Mn oxide and which adsorb the As(V) reaction product.

References

1. Research Plan for Arsenic in Drinking Water; U.S. Environmental Protection Agency, Office of Research and Development: Washington, D.C., 1998; EPA/600/R-98/042.
2. Dhar, R. K.; Biswas, B. K.; Samanta, G.; Mandal, B. K.; Chakraborti, D.; Roy, S.; Jafar, A.; Islam, A.; Ara, G.; Kabir, S.; Khan, A. W.; Ahmed, S. A.; Hadi, S. A. *Current Sci.* **1997**, *73*, 48-59.
3. Bhattacharya, P.; Chatterjee, D.; Jacks, G. *Int. J. Water Res. Manag.* **1997**, *13*, 79-92.
4. Korte, N. E.; Fernando, Q. *Crit. Rev. Environ. Control* **1991**, *21*, 1-39.
5. Masscheleyn, P. H.; Delaune, R. D.; Patrick, W. H., Jr. *J. Environ. Qual.* **1991**, *20*, 522-527.
6. Tye, C. T.; Haswell, S. J.; O'Neil, P.; Bancroft, K. C. C. *Anal. Chim. Acta* **1985**, *169*, 195-200.
7. Sadiq, M.; Zaida, T. H.; Mian, A. A. *Water Air Soil Pollut.* **1983**, *20*, 369-377.
8. Manning, B. A.; Goldberg, S. *Environ. Sci. Technol.* **1997a**, *31*, 2005-2011.
9. Manning, B. A.; Goldberg, S. *Soil Sci.* **1997b**, *162*, 886-895.
10. Knowles, F. C.; Benson, A. A. *Trends in Biochem. Sci.* **1983**, *8*, 178-180.
11. Coddington, K. *Tox. Environ. Chem.* **1986**, *11*, 281-290.
12. Manning, B. A.; Suarez, D. L. *Soil Sci. Soc. Am. J.* **2000**, *64*, 128-137.
13. Foster, A. L.; Brown, G. E., Jr.; Parks, G. A. *Environ. Sci. Technol.* **1998**, *32*, 1444-1452.

14. Nesbitt, H. W.; Canning, G. W.; Bancroft, G. M. *Geochim. Cosmochim. Acta* **1999**, *62*, 2097-2110.
15. Scott, M. J.; Morgan, J. J. *Environ. Sci. Technol.* **1995**, *29*, 1898-1905.
16. Oscarson, D. W.; Huang, P. M.; Liaw, W. K.; Hammer, U .T. *Soil Sci. Soc. Am. J.* **1983**, *47*, 644-648.
17. Su., C.; Puls., R. W. *Environ. Sci. Technol.* **2001**, *35*, 1487-1492.
18. Farrell, J.; Wang, J.; O'Day, P.; Conklin, M. *Environ. Sci. Technol.* **2001**, *35*, 20026-2032.
19. Holmgren, G. G. S. *Soil Sci. Soc. Am. Proc.* **1967**, *31*, 210-211.
20. Manning, B. A.; Martens, D. A. *Environ. Sci. Technol.* **1997**, *31*, 171-177.
21. George, G. N.; Pickering , I. J. *EXAFSPAK: A Suite of Computer Programs for Analysis of X-ray Absorption Spectra.* Stanford Synchrotron Radiation Laboratory: Stanford, CA; **1993**.
22. Rehr, J. J.; Mustre de Leon, J.; Zabinsky, S. I.; Albers, R. C. *J. Am. Chem. Soc.* **1991**, *113*, 5135-5140.
23. Manning, B. A.; Fendorf, S. E.; Goldberg, S. *Environ. Sci. Technol.* **1998**, *32*, 2383-2388.
24. Fendorf, S.; Eick, M. J.; Grossl, P.; Sparks, D. L. *Environ. Sci. Technol.* **1997**, *31*, 315-320.
25. Foster, A. L.; Brown, G. E., Jr.; Tingle, T. N.; Parks, G. A. *Amer. Mineral.* **1998**, *83*, 553-568.
26. Waychunas, G. A.; Rea, B. A.; Fuller, C. C.; Davis, J. A. *Geochim. Cosmochim. Acta.* **1993**, *57*, 2251-2269.

Chapter 6

A Review of Redox Transformation of Arsenic in Aquatic Environments

Xiaoguang Meng, Chuanyong Jing, and George P. Korfiatis

Center for Environmental Engineering, Stevens Institute of Technology, Hoboken, NJ 07030

Arsenic is a redox sensitive element which can exist in As(V), As(III), As(0), and As(-III) oxidation states under redox conditions in the natural environment. The mobility and toxicity of arsenic are determined by its oxidation states. Thermodynamic data in the literature are summarized and used to construct a pe-pH diagram for an arsenic-iron-sulfur system. The chemical species of arsenic in aqueous systems usually are not at equilibrium status because of slow rates of redox reactions. Microbially mediated reductions and oxidations of arsenic play an important role in the transformation of arsenic in sediments, soil, geothermal water, surface water, and water treatment sludge. Heterogeneous oxidation of As(III) species takes place in suspensions containing manganese dioxides, titanium dioxides, and clay minerals. The fate and transport of arsenic is closely related to the redox reactions of sulfur and iron. Oxidized arsenic species such as As(V) are typically adsorbed on iron oxides in soil and sediments under oxic conditions. At low redox potentials, reduced arsenic species such as As(III) are associated with sulfide and pyrite minerals. The occurrence of arsenic in groundwater is mainly attributed to reductive dissolution of iron oxides and oxidative dissolutions of arsenic-rich pyrite and sulfide minerals under moderate reducing conditions (i.e., arsenic mobilization zone).

Introduction

Naturally occurring arsenic is commonly found in groundwater in many countries such as Bangladesh, India, Ghana, Chile, Argentina, Philippines, China, Mexico, Poland, Hungry, Japan, and the USA (*1*). The release of arsenic from the arsenic-bearing aquifer sediments may have polluted more than 3 million of the approximately 5 million existing wells in Bangladesh, affecting up to 70 million people. Chronic ingestion of arsenic contaminated drinking water can cause skin lesions, skin, bladder, and lung cancers, anemia, and circulatory problems (*2*). People who consume waters with 3 μg/L arsenic daily have about a 1 in 1,000 risk of developing bladder or lung cancer during their lifetime. At 10 μg/L, the risk is more than 3 in 1,000. The U.S. Environmental Protection Agency (EPA) has reduced the maximum contaminant level (MCL) of arsenic in drinking water from 50 to 10 μg/L in October 2001 (*3*).

The chemistry of arsenic in aquatic environments is complex because of its multi-oxidation states and its association with a variety of minerals through adsorption and precipitation. The occurrence of elevated arsenic in groundwater has been attributed to various reactions such as reduction, oxidation, and competitive adsorption. The complexity of geochemical formations of the aquifer sediments and water chemistry often makes the identification of the mechanisms controlling arsenic release difficult. The redox reactions of arsenic species are characterized by slow kinetics and are usually mediated by the activity of microorganisms. Inorganic arsenic species are the predominant forms although arsenic can form organic compounds such as methylated arsenic in aquatic systems (*4,5*). This review summarizes the redox transformations of inorganic arsenic with the aim of understanding its mobility in aquatic environments.

Thermodynamic Predictions of Arsenic Species

Various pe-pH or Eh-pH diagrams have been constructed for different arsenic systems. These diagrams provide useful information on the predominant forms of arsenic under different chemical conditions. A simple Eh-pH diagram with only arsenic species predicts the oxidation states of arsenic in water (*6*). Since arsenic can form numerous minerals with sulfide under anoxic conditions, some Eh-pH diagrams include arsenic and sulfide (*7,8*). The more comprehensive Eh-pH diagrams comprise arsenic, sulfide, and ferrous because arsenic is usually associated with pyrite minerals under reducing conditions in natural environments (9).

A pe-pH diagram (Figure 1) for an arsenic-sulfur-iron system was constructed based on the thermodynamic data in Table I. Similar concentrations of the element as the levels observed in natural waters were used in the thermodynamic calculations so that the solids formed would be representative of natural precipitates. The total arsenic concentration was 10^{-5} M (550 μg/L). Naturally occurring arsenic in groundwater is usually below 100 μg/L in the U.S. (4) while a concentration range of 50 to 500 μg/L is common in contaminated well water in Bangladesh (10). Total concentrations of sulfur and iron in the system were 10^{-5} M each.

As(V) is the stable oxidation state under oxic conditions or in oxygenated waters (Figure 1). In a neutral pH range, As(V) exists in oxyanionic forms of $H_2AsO_4^-$ and $HAsO_4^{2-}$. At pH less than 1.5 arsenate can form scorodite (FeAsO$_4 \cdot$H$_2$O) precipitates with Fe(III). At high total Fe(III) concentration, scorodite precipitate can be formed in near neutral pH. Scorodite is commonly found in association with As-bearing ore deposits as a weathering product of arsenopyrite (FeAsS) (11). If calcium were included, calcium arsenate precipitates would have a stable region at pH higher than 10. The formation of calcium arsenate precipitates is responsible for the removal of As(V) during lime treatment of water. Lime treatment is not very efficient for the removal of As(V) because of relatively high solubility of the precipitates (12). Lime and cement have also been used to immobilize arsenic in contaminated solids and sludge (13,14,15).

In a moderately reducing environment As(III) becomes stable. It is present predominantly as H_3AsO_3 when pH is less than 9. As(III) is usually present in anoxic systems such as groundwater, sediment porewater, and geothermal water (16,17,18). In aquatic environments with neutral pH, arsenite has higher mobility than arsenate because of its lower sorptive affinity to sediment and soil particles. As(III) is considered more toxic than As(V) (19).

When the pe is low enough for the formation of sulfide, As(III) can form sulfide minerals such as orpiment (As_2S_3) at relatively low pH. Pyrite is included in the diagram because arsenic is usually associated with the mineral. Arsenopyrite is stable under more reducing conditions. The dominant oxidation state in FeAsS is As(-1) and it is part of the AsS^{2-} unit (22). Arsenopyrite and arsenic-rich pyrite (arsenian pyrite) are commonly found in arsenic-rich aquifers and natural deposits (23, 24). Arsenic content in arsenian pyrite ranges from less than 0.5 to 10 wt% (24). Most of the arsenic in the arsenian pyrite is probably in a structural position similar to that in arsenopyrite, where it substitutes for sulfur in the S_2^{2-} units yielding AsS^{2-}. The release of arsenic from the arsenian pyrite may be responsible for the contamination of groundwater in some regions of the USA (23).

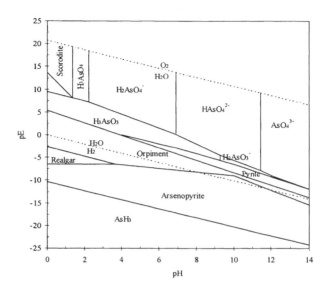

Figure 1. pE-pH diagram for the system of As-S-Fe-H$_2$O with total As, sulfur and iron 10^{-5} M.

Under low reducing conditions As(III) is reduced to its zero valent form. Although the solubility of arsenic metal in water is very low, it rarely occurs in natural environments (7). At relatively low pH, arsenic forms the covalent compound, AsS (realgar), which has low solubility in water. At extremely low pe values arsenic is reduced to As(-III) and forms arsine (AsH$_3$) gas. Arsine is only slightly soluble in water and is highly toxic.

The pe-pH diagrams show the domains of arsenic species at equilibrium and predict the direction a non-equilibrium system will proceed. Arsenic species in aquatic environment are seldom at equilibrium status because of the redox reactions slow rate. No oxidation of As(III) by dissolved oxygen in distilled demineralized water was detected after 37 days (25). As(III) can be present in oxic environments as a metastable species due to incomplete reduction of As(V) (26). As(V) and As(III) usually coexist in anoxic groundwater and porewaters (6,16,27).

Table I. Chemical Reactions of Arsenic in As-S-Fe-H$_2$O system.

No	Reactions	Log K	Ref
1	$H_3AsO_4 + 2H^+ + 2e^- = H_3AsO_3 + H_2O$	18.898	20
2	$H_2AsO_4^- + 3H^+ + 2e^- = H_3AsO_3 + H_2O$	21.130	20
3	$HAsO_4^{2-} + 4H^+ + 2e^- = H_3AsO_3 + H_2O$	28.065	20
4	$HAsO_4^{2-} + 3H^+ + 2e^- = H_2AsO_3^- + H_2O$	18.783	20
5	$AsO_4^{3-} + 4H^+ + 2e^- = H_2AsO_3^- + H_2O$	30.241	20
6	$AsO_4^{3-} + 3H^+ + 2e^- = HAsO_3^{2-} + H_2O$	18.226	20
7	$H_3AsO_4 = H_2AsO_4^- + H^+$	-2.232	20
8	$H_3AsO_4 = HAsO_4^{2-} + 2H^+$	-9.167	20
9	$H_3AsO_4 = AsO_4^{3-} + 3H^+$	-20.625	20
10	$H_3AsO_3 = H_2AsO_3^- + H^+$	-9.282	20
11	$H_3AsO_3 = HAsO_3^{2-} + 2H^+$	-21.297	20
12	$H_3AsO_3 = AsO_3^- + 3H^+$	-34.669	20
13	$Fe^{3+} + H_3AsO_4 + 2H_2O = FeAsO_4 \cdot 2H_2O$ (scorodite) $+ 3H^+$	10.92	9
14	$2H_3AsO_3 + HS^- + 3H^+ = As_2S_3$ (orpiment) $+ 6H_2O$	61.066	20
15	$H_3AsO_3 + HS^- + 2H^+ + 2e^- = AsS$ (realgar) $+ 6H_2O$	19.747	20
16	$Fe^{2+} + AsS + 2e^- = FeAsS$ (arsenopyrite)	12.782	9
17	$Fe^{2+} + 2HS^- = FeS_2$ (pyrite) $+ 2H^+ + 2e^-$	18.508	20
18	$H_3AsO_3 + 6H^+ + 6e^- = AsH_{3(g)} + 3H_2O$	-8.86	21
19	$3Ca^{2+} + 2H_3AsO_4 + 4H_2O = Ca_3(AsO_4)_2 \cdot 4H_2O(s) + 6H^+$	-22.30	20
20	$3Pb^{2+} + 2H_3AsO_4 = Pb_3(AsO_4)_2 \cdot 4H_2O(s) + 6H^+$	-5.8	20
21	$Al^{3+} + H_3AsO_4 + 2H_2O = AlAsO_4 \cdot 2H_2O(s) + 6H^+$	-4.8	20
22	$3Zn^{2+} + 2H_3AsO_4 + 2.5H_2O = Zn_3(AsO_4)_2 \cdot 2.5H_2O(s) + 6H^+$	-13.65	20
23	$3Cu^{2+} + 2H_3AsO_4 + 2H_2O = Cu_3(AsO_4)_2 \cdot 2H_2O(s) + 6H^+$	-6.10	20

Biotic Redox Reactions

Microorganisms are able to reduce and oxidize arsenic species in aquatic systems. Various species of anaerobic bacteria can achieve growth by dissimilatory (i.e., respiratory) reduction of arsenate to arsenite (28,29). Biological processes can mobilize arsenic in sediments by either direct or indirect mechanisms.

Several bacterial, algal, and fungal species in aquatic and enteric environments can directly reduce arsenate (*30,31,32*). The glucose fermenting microorganism (CN8) can reduce soluble arsenate to arsenite rapidly (*33*). However, no obvious reduction of sorbed arsenate on ferrihydrite was detected during approximately 25 days of incubation. The microorganism does not reduce the Fe(III) in ferrihydrite to Fe(II). Therefore, the release of arsenate from the solid phase was limited.

Reductive dissolution of iron hydroxides in sediments is an important mechanism of arsenic release. Cummings et al. (*34*) demonstrated that dissimilatory iron-reducing bacteria *Shewanella alga* strain BrY reduced Fe(III) in synthetic scorodite and did not respire As(V). As a result of the dissimilatory reduction, As(V) and Fe(II) were released into solution. The iron-reducing bacteria solubilized arsenate sorbed to sediments by reductively dissolving iron hydroxides. The rate of reductive dissolution for high surface area ferrihydrite was much higher than for crystalline goethite or hematite, resulting in different rates of arsenic release (*35*).

Sediments contain diverse species of microorganisms such as sulfate-reducing bacteria, strain MIT-13, strain SES-3, and *Desulfotomaculum auripigmentum* which can mobilize arsenic through respiratory reduction of As(V) and Fe(III) (*36,37,38*). *Sulfurospirillum barnesii* is capable of anaerobic growth using Fe(III) or As(V) as electron acceptors (*39*). The bacterium can reduce both soluble and sorbed As(V) on the surfaces of ferrihydrite and aluminum hydroxide to As(III). The rates of As(V) reduction by microorganisms have been measured at 2×10^{-7} to 5×10^{-7} µmol of As cell^{-1} day^{-1} in both strain MIT-13 and in strain SES-3.

Sulfate-reducing bacteria may also promote As(V) reduction by producing hydrogen sulfide (*40,41*). As(V) reduction by hydrogen sulfide is rapid at low pH and follow a second-order kinetic model (*42*). The rate of reduction was 300 times greater at pH 4 than at pH 7.

In addition to reductive release of arsenic from sediments, biotic reductions can also convert arsenic from soluble to solid forms. Under very low reducing conditions, As(III) and sulfide produced by the activities of microorganisms can form sulfide minerals and arsenian pyrite (*43*). Rochette et al. (*42*) reported the formation of orpiment at high sulfide to arsenic ratios (20:1). The low solubility of these minerals limits the release of arsenic into water.

While the anaerobic microorganisms discussed above can reduce As(V), some aerobes are able to oxidize As(III) under oxic conditions (*44,45*). The rapid oxidation of As(III) in streamwaters of the Eastern Sierra Nevada was attributed to activities of bacteria attached to submerged macrophytes (*46*). When antibiotics were added into the streamwater samples, the oxidation was stopped, indicating that biotic activity was responsible for As(III) oxidation. However, the bacterial species were not identified. The oxidation of As(III) followed the pseudo-first-order with a half-life of about 0.3 h. As(III) is often the predominant species at the point of discharge of geothermal springs. As(III) was converted to As(V) by microbially mediated oxidation within a few minutes (47). Gihring et al. (48) found that *Thermus aquaticus* and *Thermus thermophilus* caused the rapid oxidation of As(III). The *Thermus* species are common in hot springs and thermally polluted domestic and industrial waters.

Enhanced Abiotic Oxidation

Minerals in sediments and soils can directly and indirectly oxidize As(III) at high rates. Manganese dioxides directly oxidize As(III) to As(V) in a pH range 5 to 10 (49,50). As the results of the redox reactions, Mn(IV) in the oxides is reduced to soluble Mn(II). The rate of As(III) oxidation by manganese dioxides is not affected by the concentration of dissolved oxygen because Mn(IV) acts as the electron acceptor (49).

Oscarson et al. (51) studied the oxidation of As(III) in suspensions prepared by adding lake sediment samples to deionized water. The addition of $HgCl_2$ into the sediments as an inhibitor of biological activity did not affect the rate of As(III) oxidation, indicating an abiotic oxidation process. Flushing N_2 or air through the suspensions also had no effect on As(III) oxidation. Selective removal of manganese oxides from the sediments decreased the extent of As(III) oxidation (52).

TiO_2 is able to catalyze photooxidation of As(III) to As(V) in oxygenated water under ultraviolet and solar irradiation (53,54). In solutions containing 0.05 to 10 mg/L As(III) and 1 to 50 mg/L TiO_2 total oxidation of As(III) occurred within minutes. Variation of pH from 5 to 9 did not have significant effect on the rate of the oxidation (53). The oxidation was attributed to the photogenerated hydroxyl radicals (54). It has been reported that soluble Fe(II) and Fe(III) also significantly increased the photooxidation rate of As(III) (55,56).

Clay minerals have high surface area and are important components of soils and sediments. Several natural clays such as kaolinite and illite have been found to enhance the oxidation of As(III) (57,58). Pure kaolinite and illite (i.e., aluminosilicates) contain Al(III) and Si(IV) which do not participate in electron transfers under ambient conditions. Therefore, the heterogeneous oxidation of As(III) may be caused by small amounts of other minerals in natural clays. Further studies by Foster et al. (59) demonstrated that As(III) oxidation in slurries of Georgia kaolinite was strongly dependent on both light and concentration of oxygen. They concluded that the oxidation was caused by a Ti-containing phase in the clay. The content of Ti in the Georgia kaolinite was approximately 1.8%.

Experimental results obtained with synthetic solids showed that aluminum hydroxides, gibbsite, and amorphous silica did not enhance the oxidation of As(III) by oxygen (57,59,60). Thermodynamic considerations suggest that soluble Fe(III) should oxidize arsenite only under acidic conditions. Numerous As(III) adsorption experiments have been conducted using ferric hydroxides and oxides. The distinct adsorption behaviors of As(III) and As(V) (16) indicate that no obvious As(III) oxidation should occur in the ferric hydroxide and oxide suspensions. The oxidation states of arsenic on the solid surfaces can be determined using x-ray absorption spectroscopy and with chemical extraction methods. However, errors may be introduced by some chemical extraction processes because the oxidation rate of As(III) by oxygen can be increased in acid and base solutions (56,57).

Under oxic conditions, both biotic and abiotic oxidations of As(III) would take place in soil and sediments. Depending on the contents of microorganisms, organic matter, and Mn and Ti oxides either biotic or abiotic mechanisms can be rate control processes for the oxidation of As(III).

Effects of Redox Transformation on Arsenic Mobility

The mobility of arsenic in aquatic environments is controlled by arsenic oxidation states, stability of solid minerals, and the chemical composition of waters. Groundwaters containing high arsenic concentrations usually have low dissolved oxygen and high soluble iron. Arsenic can be released into water as a result of either reduction of As(V) and iron hydroxides or oxidation of arsenic sulfide minerals. Coexisting solutes such as phosphates and silicates can compete with arsenic for the surface sites, thus increasing the mobility of As(V) and As(III) (16,61). It was also assumed that the formation of arseno-carbonate complexes could increase the release of arsenic from arsenic sulfide minerals (23).

The occurrence of high arsenic concentration in groundwater in Bangladesh has been hypothetically attributed to microbially mediated reduction of arsenic-rich iron hydroxides (62,63). The groundwater sediments in Bangladesh have a relatively high content of organic matter. It is assumed that the reductive dissolution of iron hydroxides is driven by microbial metabolism of organic matter (CH_3COO^-):

$$8FeOOH + CH_3COO^- + 15H_2CO_3 \rightarrow 8Fe^{2+} + 17HCO_3^- + 12H_2O \quad (1)$$

The reductive release of arsenic was supported by laboratory incubation experiments. Incubation of lake sediment samples under controlled redox potentials has shown that under oxidizing conditions (pe = 3.4 to 8.4) and at pH

5, soluble arsenic was approximately 25 µg/L and 65-98% of the soluble arsenic was present as As(V) (27). The redox potential was controlled by purging the sediment suspensions with argon gas and air. Upon reduction of pe to –3.4, the soluble arsenic concentration increased to approximately 250 µg/L. Under the same reducing conditions, sulfide and soluble iron concentration increased to 2.8 and 152 mg/L, respectively. When contaminated estuarine sediment was incubated at pH 7, soluble arsenic concentration increased dramatically with pe decreasing from approximately –1.4 to –2.2 (64). However, no data at very low Eh values were available for assessing the formation of arsenic sulfides and arsenian pyrite in those studies.

The transformations of As(V) and sulfate in an iron hydroxide sludge were studied over a wide range of redox potentials by incubating the sludge generated in a surface water treatment plant (65). The biological activity in the sludge containing high content of organic matter reduced the pe values from 6.3 (i.e. Eh=373 mV) to –3.9 (Eh=-231 mV) during 22 days of incubation in a closed container and in the dark. Thermodynamic considerations predicted that at a neutral pH the reduction of dissolved oxygen, As(V), ferric hydroxide [FeOOH(am)], and sulfate would occur at pe values of approximately 13.1, 1, -1, and –2, respectively.

The incubation results in Figure 2 showed that at a pe of 6.3, soluble As(III) and As(V) concentrations were less than 5 µg/L. When the pe value was reduced to less than zero, soluble As(III) and As(V) concentrations increased abruptly. As(III) concentration reached a maximum of approximately 95 µg/L at pe of about –3. The soluble As(III) and As(V) concentrations diminished when pe was lower than –4. The decrease of As(III) was concurrent with the reduction of sulfate concentration at pe of approximately –3. On the basis of the observed mobility of arsenic, the pe range was divided into three redox zones: defined as adsorption (pe>0), mobilization (-4<pe<0), and reductive fixation (pe<-4) zones.

The distribution of arsenic and iron in the incubation system was calculated using the thermodynamic constants and the adsorption constants for ferric hydroxide (65). The model calculations indicted that As(V) was adsorbed on ferric hydroxide at pe>0 (Figure 3). The reduction of As(V) to As(III) began at pe value of approximately zero. The reductive dissolution of ferric hydroxide was observed at pe of less than -0.5. When pe was reduced to approximately – 2.7, pyrite was formed. The concurrent formation of pyrite with the decrease in As(III) concentration (Figure 2) at pe of about –3 suggested that As(III) formed precipitates with pyrite. Thermodynamic calculations indicated that arsenic would form orpiment with sulfide at very low pe values (pe < –5.5). The modeling results demonstrate that the arsenic mobilization zone was caused by dynamic transitions between ferric hydroxides and the pyrite minerals. As(III) and As(V) were released into water because neither reduced nor oxidized minerals were stable in the mobilization or transition zone.

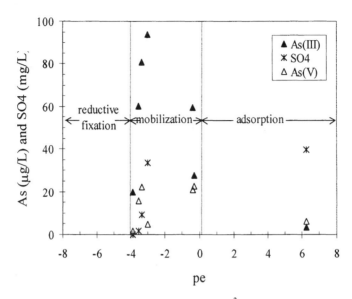

Figure 2. Soluble concentrations of As(III), SO_4^{2-}, and As(V) as a function of
pe during incubation of ferric hydroxide sludge sample in a closed
system. pH range 6.98-7.76. (Modified from Ref 65).

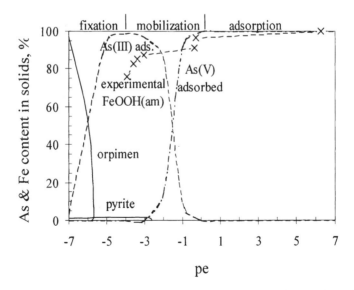

Figure 3. Model predictions of the transformations of As and Fe in the solid-
phase. "experimental FeOOH(am)" denotes the observed
FeOOH(am) content used in the model calculations. (from Ref 65)

The three arsenic zones were observed along 6 m of depth profile in the Campbell Mine gold-tailings impoundment (66). Total arsenic concentration in the porewater was below 1.5 mg/L in the upper 4.5 m of the profile because of adsorption of As(V) on iron oxides. At depth of 5.2 m, total arsenic increased to a maximum concentration of 107 mg/L due to microbially mediated reduction of iron oxides and As(V). The arsenic concentration decreased abruptly to less than 0.5 mg/L at a depth of 6 m because of the formation of arsenic sulfide minerals. Sulfate concentration decreased from 2300 mg/L at 3.7 m depth to 22 mg/L to 5.8 m.

The results in Figures 2 and 3 show that high Fe(II) and As(III) concentrations are associated with reducing environments. However, the chemical conditions in the waters alone are not sufficient for identifying the mechanisms of arsenic release. Elevated As(III) and Fe(II) concentrations can be caused by reductive dissolution of ferric hydroxide or oxidation of sulfide minerals. Arsenian pyrite and various arsenic sulfides are common arsenic-bearing minerals in groundwater aquifers (4,67). Arsenic can be released from these phases by oxidative processes (11,68). The pe boundaries of the mobilization zone may vary for oxidative and reductive release of arsenic because of slow rates of the redox reactions. The redox potentials for oxidation release of arsenic from different minerals such as arsenian pyrite and orpiment are not the same. Information such as chemical forms of the arsenic-rich minerals in the aquifers is required for determining the reactions controlling arsenic release into groundwater.

References

1. Jain, C. K.; Ali, I. *Water Res.* **2000**, *34*, 4304-4312.
2. *Arsenic in Drinking Water: 2001 update,* National Research Council, pp21-62, National Academy Press, 2001, Washington, DC.
3. EPA announces arsenic standard for drinking water of 10 parts per billion, *EPA headquarters press release*, Washington, DC, 10/31/2001.
4. Welch, A. H.; Westjohn, D. B; Helsel, D. R.; Wanty, R. B. *Ground Water.* **2000**, *38*, 589-604.
5. Sohrin, Y.; Matsur, M. *Environ. Sci. Technol.* **1997**, *31*, 2712-2720.
6. Masscheleyn, P; Delaune, R.; Patrick, J. W. *Environ. Sci. Technol.* **1991**, *25*, 1414-1419.
7. Ferguson, J. F.; Gavis, J. *Water Res.* **1972**, *6*, 1259-1274.
8. Edwards, M. *J. AWWA.* **1994**, *86*, 64-78.
9. Vink, B. W. *Chemical Geology*, **1996**, *130*, 21-30.
10. McArthur, J. M.; Ravenscroft, P.; Safiulla, S.; Thirlwall, M. F. *Water Resources Research*, **2001**, *37*, 109-117.

11. Roussel, C.; Néel, C.; Bril, H. *Sci. Total Environ.* **2000**, *263*, 209.
12. McNeill, L. S.; Edwards, M. *J. Environ. Engineering* **1997**, *123*, 453-460.
13. Both, J. V. Jr.; Brown, P. W. *Environ. Sci. Technol.* **1999**, *33*, 3806-3811.
14. Dutré, V.; Vandecasteele, C. *Environ. Sci. Technol.* **1998**, *32*, 2782-2787.
15. Stronach, S. A.; Walker, D. E.; Macphee, D. E.; Glasser, F. P. *Waste Management.* **1997**, 17, 9-13.
16. Meng, X. G.; Bang, S. B.; Korfiatis, G. P. *Water Res.* **2000**, *34*, 1255-1261
17. Harrington, J. M.; Fendorf, S. E.; Rosenzweig, R. F. *Environ. Sci. Technol.* **1998**, *32*, 2425.
18. Langner, H. W.; Jackson, C. R.; McDermott, T. R; Inskeep, W. P. *Environ. Sci. Technol.* **2001**, *35*, 3302.
19. *Medical and Biological Effects of Environmental Pollutants – Arsenic.* Pp. 117-172, National Academy of Sciences, 1977, Washington, DC.
20. David, S. B.; Allison, J. D. *MINTEQA2, An Equilibrium Metal Speciation Model: User's Manual 4.01*, Environmental Research Laboratory, U.S. Environmental Protection Agency: 1999, Athens, GA.
21. Scott, M. J. Kinetics of adsorption and redox processes on iron and manganese oxides: reactions of As(III) and Se(IV) at goethite and birnessite surfaces, EQL report No. 33, California Institute of Technology, May, 1991.
22. Nesbitt, H. W.; Muir, I. J.; Pratt, A. R. *Geochimica Cosmochimica Acta*, **1995**, *59*, 1773.
23. Kim, M. J.; Nriagu, J.; Haack, S. *Environ. Sci. Technol.* **2000**, *34*, 3094.
24. Simon, G.; Huang, H.; Penner-Hahn, J.; Kesler, S. E.; Kao, L. S. *American Minerologist.* **1999**, *84*, 1071.
25. Tallman, D. E.; Shaik, A. U. *Anal. Chem.* **1980**, *52*, 196-199.
26. Nagorski, S. A.; Moore, J. N. *Water Resources Res.* **1999**, *35*, 3441-3450.
27. Aggett, J.; Kriegman, M. R. *Water Res.* **1988**, *22*, 407-411.
28. Stolz, J. F.; Oremland, R. S. *FEMS Microbiol. Rev.* **1999**, *23*, 615.
29. Newman, D. K.; Ahmann, D.; Morel, F. M. M. *Geomicrobiol. J.* **1998**, *15*, 255.
30. Silver, S.; Ji, G.; Bröer, S.; Dey, S.; Dou, D.; Rosen, B. *Mol. Microbiol.* **1993**, *8*, 637-642.
31. Diorio, C.; Cai, J.; Marmor, J.; Shinder, R.; DuBow, M. *J. Bacteriol.* **1995**, *177*, 2050-2056.
32. Cullen, W. R.; Reimer, K. *J. Chem. Rev.* **1989**, *89*, 713-764.
33. Langner, H. W.; Inskeep, W. P. *Environ. Sci. Technol.* **2000**, *34*, 3131.
34. Cummings, D. E.; Caccavo, F.; Fendorf, S.; Rosenzweig, R. F. *Environ. Sci. Technol.* **1999**, *33*, 723.
35. Roden, E. E.; Zachara, J. M. *Environ. Sci. Technol.* **1996**, *30*, 1618.
36. Ahmann, D.; Krumholz, L. R.; Hemond, H. F.; Lovley, D. R.; Morel, F. M. M. *Environ. Sci. Technol.* **1997**, *31*, 2923-2930.

37. Ahmann, D.; Robert, A. L.; Krumholz, L. R.; Morel, F. M. M. *Nature* **1994**, *351*, 750.
38. Laverman, A. M.; Blum, J. S.; Schaefer, J. K.; Phillips, E. J. P.; Lovley, D. R.; Oremland, R. S. *Appl. Environ. Microbiol.* **1995**, *61*, 3556-3561.
39. Zobrist, J.; Dowdle, P. R.; Davis, J. A.; Oremland, R. S. *Environ. Sci. Technol.* **2000**, *34*, 4747-4753.
40. Kuhn, A.; Sigg, L. *Limnol. Oceangr.* **1993**, *38*, 1052-1059.
41. Newman, D. K.; Kennedy, E. K.; Coastes, J. D.; Ahmann, D.; Ellis, D. J.; Lovley, D. R.; Morel, F. M. M. *Arch. Microbiol.* **1997**, *168*, 380-388.
42. Rochette, E. A.; Bostick, B. C.; Li, G.; Fendorf, S. *Environ. Sci. Technol.* **2000**, *34*, 4714-4720.
43. Moore, J. N.; Ficklin, W. H.; Jogns, C. *Environ. Sci. Technol.* **1988**, *22*, 432-437.
44. Osborne, F. H.; Ehrlich, H. L. *J. Appl. Bacteriol.* **1976**, *41*, 295-305.
45. Santini, J. M.; Sly, L. I.; Schnagl, R. D.; Macy, J. M. *Appl. Environ. Microbiol.* **2000**, *66*, 92-97.
46. Wilkie, J. A.; Hering, J. G. *Environ. Sci. Technol.* **1998**, *32*, 657-662.
47. Langner, H. W.; Jackson, T. R.; Inskeep, W. P. *Environ. Sci. Technol.* **2001**, *35*, 3302-3309.
48. Gihring, T. M.; Druschel, G. K.; McCleskey, R. B.; Hamers, R. J.; Banfield, J. F. *Environ. Sci. Technol.* **2001**, (electronic version).
49. Scott, M. J.; Morgan, J. J. *Environ. Sci. Technol.* **1995**, *29*, 1989-1905.
50. Driehaus, W.; Seith, R.; Jekel, M. *Water, Res.* **1995**, *29*, 297-305.
51. Oscarson, D.W.; Huang, P. M.; Liaw, W. K. *J. Environ. Qual.* **1980**, *9*, 700-703.
52. Oscarson, D.W.; Huang, P. M.; Liaw, W. K. *Clays Clay Miner.* **1981**, *29*, 210-225.
53. Bissen, M.; Viellard-Baron, M. M.; Schindelin, A. J.; Frimmel, F. H. *Chemosphere.* **2001**, *44*, 751-757.
54. Yang, H.; Lin, W. Y.; Rajeshwar, K. *J. Photochemistry Photobiology.* **1999**, *123*, 137-143.
55. Hug, S. J.; Canonica, L.; Wegelin, M.; Gechter, D.; Gunten, U. V. *Environ. Sci. Technol.* **2001**, *35*, 2114-2121.
56. Emett, M. T.; Khoe, G. H. *Water Res.* **2001**, *35*, 649-656.
57. Manning, B. A.; Goldberg, S. *Environ. Sci. Technol.* **1997**, *31*, 2005-2011.
58. Lin, Z.; Puls, R. W. *Environmental Geology*, **2000**, *39*, 753-759.
59. Foster, A. L.; Brown, G. E. Jr.; Parks, G. A. *Environ. Sci. Technol.* **1998**, *32*, 1444-1452.
60. Scott, M. J. Ph.D. Thesis, California Institute of Technology, Pasadena, CA, **1991**.

61. Meng, X. G.; Korfiatis, G. P.; Christodoulatos, C.; Bang, S. B. *Water Res.* **2001**, *35*, 2805-2810.

62. Nickson, R. T.; McArthur, J. M.; Ravenscroft, P.; Burgess, W. G.; Ahmed, K. M. *Appl. Geoche.* **2000**, *15*, 403-413.

63. McArthur, J. M.; Ravenscroft, P.; Safiulla, S.; Thirlwall, M. F. *Water Resources Res.* **2001**, *37*, 109-117.

64. Guo, T.; DeLaune, R. D.; Patrick, W. H., Jr. *Environ. International.* **1997**, *23*, 305-316.

65. Meng, X. G.; Korfiatis, G. P.; Jing, C.; Christodoulatos, C. *Environ. Sci. Technol.* **2001**, *35*, 3476-3481.

66. McCreadie, H.; Blowes, D. W.; Ptacek, C. J.; Jambor, J. L. *Environ. Sci. Technol.* **2000**, *34*, 3159-3166.

67. Azcue, J. M. *Environ. Rev.* **1995**, *3*, 212.

68. Nesbitt, H. W. Muir, I. J.; Pratt, A. R. *Geochim. Cosmochim. Acta.* **1995**, *59*, 1773.

Chapter 7

Reactions of Ultrasonically Generated Hydroxyl Radicals with Arsenic in Aqueous Environments

Influence of Oxygen, Argon, Solution pH, and Concentration

Sahar Motamedi, Yong Cai, and Kevin E. O'Shea*

Department of Chemistry, Florida International University,
Miami, FL 33199

The fate, transport, and toxicity of arsenic are dependent on its oxidation state. Hydroxyl and superoxide anion radicals are formed naturally and may contribute significantly to redox processes of arsenic in the environment. We have studied ultrasonically generated hydroxyl radical with arsenic species in the presence of oxygen and argon over a range of solution pH and arsenic concentrations. As(III) reacts via a pseudo-first order process under our experimental conditions. The reaction of hydroxyl radical with As(III) likely yields As(IV), which can react with oxygen to form As(V). The transformation rates are different under oxygen versus argon-saturated conditions and at different solution pH values. Although superoxide anion radical can act as a reducing agent and is formed under our experimental conditions, reduction of As(V) was not observed. The kinetic parameters determined in these studies may be useful for the development of models to predict the fate, redox conversion, and remediation of these species in aquatic environments.

Arsenic is a known carcinogen in humans and has led to major ground-water contamination problems worldwide. Although anthropogenic activities have led to arsenic contamination of ground water, the major source of arsenic contamination is the result of leaching from weathering of minerals in rocks and soils (1,2). It is estimated that over 100 million people are at risk since drinking arsenic-contaminated water is linked to a variety of serious health problems (1).

In the environment, arsenic occurs in numerous chemical forms and its prevalent oxidation states are V, III, 0, and –III. In the aquatic environment, arsenic typically exists in inorganic forms as arsenious acid ($H_3As^{III}O_3$) and its salts $H_2As^{III}O_3^-$, $HAs^{III}O_3^{2-}$, and $As^{III}O_3^{3-}$ (arsenites) and arsenic acid ($H_3As^VO_4$) and its salts $H_2As^VO_4^-$, $HAs^VO_4^{2-}$, and $As^VO_4^{3-}$ (arsenates) (3,4). Arsenite forms, more likely to be found more in anaerobic environments, are generally more toxic and mobile than arsenate forms, which are more likely in aerobic surface waters (5-7). While oxidation of As(III) to As(V) is often required for its removal from ground water (8,9), reduction of As(V) to elemental form serves to immobilize it, and thus constitutes a potential alternative approach to pollution control. The precipitation, adsorption and dissolution of arsenic are dependent on its oxidation state and on the pH of the solution (2,6). Hydroxyl and superoxide anion radicals are naturally occurring species, which can lead to redox conversion of arsenic species ultimately influencing the fate, transport and toxicity of arsenic in the environment.

We have used ultrasonic irradiation to generate and study the reactions of hydroxyl radicals and superoxide anion radicals with arsenate and arsenite. Our results demonstrated that As(III) is readily oxidized to As(V) under hydroxyl radical generating ultrasonic irradiation (640 KHz) of oxygen saturated aqueous solutions. The conversion of As (III) to As(V), monitored by high pressure liquid chromatography- atomic fluorescence spectrometry (HPLC-AFS) analyses, follows a pseudo-first order process. We have evaluated a number of kinetic parameters and assessed the role of oxygen in the oxidation process as reported herein.

Experimental

Materials and Instrumentation

Primary standards were prepared by dissolving 1.320 g As_2O_3 in 100 mL 0.1 N of NaOH to yield a 1000 µg mL^{-1} (ppm) solution of As(III) (arsenite). Arsenate, As(V), primary standards were prepared by dissolving 4.161 g $Na_2HAsO_4 \cdot 7H_2O$ in 100 mL H_2O to make a 1000 (ppm) solution of As(V).

Different concentrations of As(III) and As(V) were prepared by appropriate dilution of the primary standard for ultrasound irradiation. Chemicals were obtained from Aldrich chemical company and were used as received. All glassware was acid-washed prior to use and volumetric glassware was employed in the preparation of all the reaction solutions. Water used to prepare the reaction solutions was filtered through a Culligan system consisting of activated charcoal and two mixed bed ion exchange cartridges. The filtered water is distilled using a Barnstead Mega-ohm B pure system prior to use. Oxygen and Argon are zero grade obtained from Air Products.

Ultrasonic irradiation was conducted with high frequency, high intensity ultrasonic equipment UES Model 15-660, equipped with an aluminum-faced transducer and a specially designed 500 mL capacity focusing-reactor glass vessel (10). A FEP Teflon material membrane from the transducer separates the solution. Power setting for our experiment was set at 50 eV. The pulse duration was 0.62 sec, the pulse repetition was 2.5 sec, and the frequency was 665 KHz. The duty cycle was 0.248 so the delivered energy each time (sec) was 124 J/sec x time (sec). The reaction vessel was placed horizontally in a ten-gallon aquarium with ice water as a cooling medium and the temperature of the irradiated solution was maintained $12 \pm 4\,°C$.

As(III) and As(V) analyses were performed using the PS Analytical Millennium Excalibur system (PSA 10.055) (AFS) coupled to a Spectra Physics HPLC system. The HPLC system comprises a pump, Rheodyne injection valve and loop (100 µL) and a strong anion exchange column (PRP X-100, (250 mm) x (4.6 mm), 10 µm particle size, Hamilton). The column flow was 1.0 mL min[-1] and the injection volume was 100 µL. The mobile phase for HPLC was 1:1 of 0.015 M KH_2PO_4 and 0.015 M K_2HPO_4, pH=5.8. The PSA Millennium Excalibur system is an integrated atomic fluorescence system incorporating vapor generation, gas-liquid separation, moisture removal and atomic fluorescence stages.

Results and Discussion

In the environment, hydroxyl and superoxide radicals are formed by photochemical and/or metal catalyzed reactions (11). We used ultrasonic irradiation of aqueous solutions to generate and study the reactions of hydroxyl and superoxide anion radicals with arsenic. Ultrasonic irradiation of aqueous media with an appropriate frequency leads to homolytic cleavage of the hydrogen-oxygen bond to produce hydroxyl radical and hydrogen atom (12), equation 1.

$$H_2O \qquad \text{-}))))) \rightarrow \qquad HO\bullet + H\bullet \qquad (1)$$

In the presence of oxygen the hydrogen atom rapidly reacts to form superoxide anion radical (*13*), equation 2.

$$O_2 + H\bullet \quad \text{------}\rightarrow \qquad O_2^{\bullet-} + H^+ \qquad (2)$$

An aqueous solution saturated with oxygen and containing 50 ppb As (III) was subject to ultrasonic irradiation. Under these conditions, environmentally relevant $HO\bullet$ and superoxide anion radicals are formed. Control experiments established that under oxygen and argon saturation As(III) is not appreciably oxidized without ultrasonic irradiation. Samples were transferred to brown glass bottles and analyzed within six hours. The concentration of As(III) was monitored as a function of irradiation time as illustrated in Figure 1. Under oxygen saturation, As(III) is readily converted to As(V) with an excellent mass balance throughout the course of the reaction. Similar rates for the disappearance of As(III) and the formation of As(V) suggest As(III) is readily converted to As(V) without any stable intermediate species in an oxygen-saturated environment.

The conversion of As(III) likely involves hydroxyl radical which reacts to form As(IV) at a near diffusion controlled rate (*14*), equation 3.

$$As(III) + HO\bullet \qquad \rightarrow \qquad As(IV) + HO^- \qquad (3)$$

As(IV) is short lived under the reaction conditions (*15*) and can be oxidized to As(V) by a number of competing reactions illustrated in equations 4-6.

$$As(IV) + HO\bullet \qquad \rightarrow \qquad As(V) + HO^- \qquad (4)$$

$$As(IV) + As(IV) \quad \rightarrow \qquad As(III) + As(V) \qquad (5)$$

$$As(IV) + O_2 \qquad \rightarrow \qquad As(V) + O_2^{\bullet-} \qquad (6)$$

As(IV) can be oxidized to As(V) upon reaction with hydroxyl radical, equation 4; by disproportionation with another equivalent of As(IV), equation 5; or by reaction with oxygen (*15*), equation 6. The rate constants for each of these processes are near diffusion-controlled such that concentration becomes the most important factor. Given the relatively low concentrations of As(IV) and hydroxyl radicals compared to oxygen, it is likely that oxygen plays the major role in the conversion of As(IV) to As(V), under oxygen-saturated conditions.

The conversion of As(III) is consistent with pseudo-first order kinetics, equation 7, as demonstrated by the linear relationship between the plot of Ln [As (III)] vs. irradiation time, Figure 2. The pseudo-first order rate constant for this process, determined from the slope of the plots, are 0.041 and 0.051 min^{-1} for oxygen- and argon- saturated conditions, respectively.

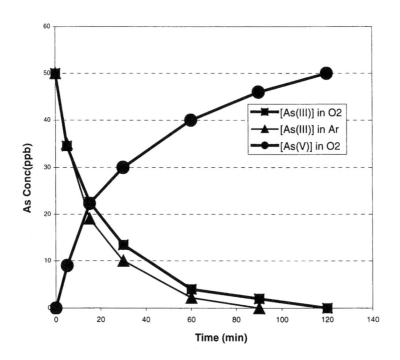

Figure 1. Ultrasonic degradation of As(III) to As(V) in O_2 and Argon gas saturations.

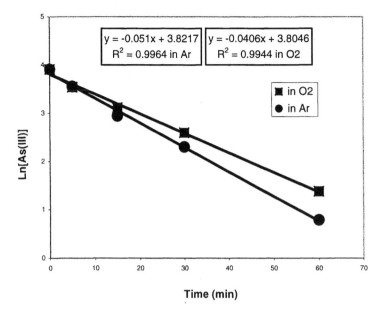

Figure 2. Pseudo-first order kinetics for ultrasonic degradation of As(III) in O_2 and Argon gas saturation.

$$d[As(III)]/dt = -k[As\ (III)] \tag{7}$$

The faster rate of disappearance of As(III) under argon-saturation is consistent with the observation that hydroxyl radical yields are slightly higher under argon relative to oxygen saturation (*16*). We observed that the conversion of As(III) to As(V) reaches a maximum of 50% and slowly decreases upon continued irradiation under argon saturation. Hydroxyl radicals generated by ultrasonic irradiation rapidly convert As(III) to As(IV), but subsequent oxidation to As(V) in the absence of oxygen or at very low oxygen concentration appears to be a minor reaction pathway. The conversion to As(V) would require reactions involving species like hydroxyl radical or As(IV) which are present in very dilute concentrations. Oxidation to As(V) becomes a minor pathway and other reaction processes not including the formation of As(III) and possibly reductive reactions appear to play an important role (*17*).

We also studied the conversion of As(III) at different concentrations under argon- and oxygen-saturated conditions, as shown in Figures 3 and 4. In the concentration range of 10-100 ppb under oxygen-saturation, As(III) is readily converted, but at slightly slower rates than under argon-saturation, as summarized in Table I.

Table I. Pseudo first order rate constants for conversion of As(III) under hydroxyl radical generating ultrasonic irradiation.

Initial [As(III)] (ppb)	O_2 saturation $k(min^{-1})^a$	$(R^2)^b$	Ar saturation $k(min^{-1})^a$	$(R^2)^b$
10	0.054	0.99	0.045	0.99
25	0.052	0.99	0.057	0.96
50	0.041	0.99	0.051	0.99
100	0.020	0.98	0.052	0.99

a. Error in the rate constants is estimated to be ± 10-20 % based on duplicate analyses of individual samples, rate constants were determined for the complete conversion of As(III).

b. The correlation coefficient is based on the best least squares line through the experimental data points.

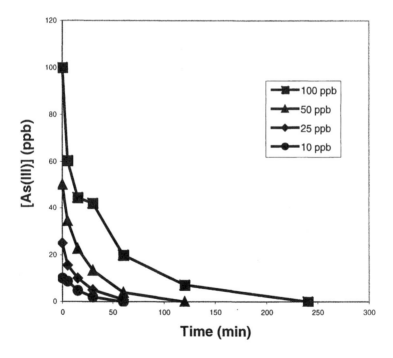

Figure 3. Ultrasonic degradation rates of different initial concentrations of As(III) in O_2 saturation.

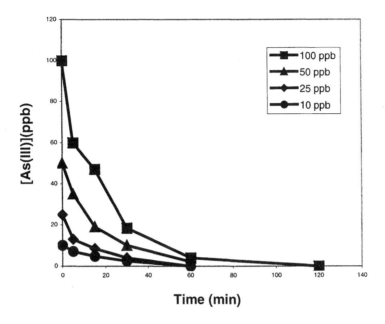

Figure 4. Ultrasonic degradation rates of different initial concentrations of As(III) in Argon saturation.

The rate constant is unchanged within experimental error under argon saturation. At the highest As(III) concentrations the reaction rate constant gradually decreases under oxygen saturated conditions. The observed change in the rate constant may be the result of the lower yields of hydroxyl radicals (versus argon saturation) and the higher concentrations of As(III) such that the relative concentrations are beyond the threshold for observing pseudo-first order kinetics rather than a change in the reaction pathways. In all cases under extended irradiation the level of As(III) is reduced to trace levels 1-2 ppb within ~120 minutes. Control experiments illustrated that adsorption of As(III) and As(V) onto the reaction vessel and analysis vials is not significant in the time frame of the experiments. Although superoxide anion radical may function as an oxidant for arsenic species *(18)*, we do not observe $O_2^{-\bullet}$ functioning as a reducing agent. Therefore As(V) is not converted to As(III), equation 8, under the experimental conditions of ultrasonic irradiation.

$$As(V) \ + \ O_2^{-\bullet} \qquad \rightarrow \qquad As(IV) \ + \ O_2 \qquad\qquad (8)$$

The solution pH can have a pronounced effect on the adsorption and mobility of arsenic (2, 6). We measured the conversion of As (III) at different solution pH in phosphate-buffered solutions. For comparison of rates at different pH we determined the pseudo-first order rate constants, shown in Table II.

Table II. Pseudo-first order rate constants for the conversion of As(III) at different solution pH.

pH^a	initial [As(III)](ppb)	$k(min^{-1})^b$	$(R^2)^c$
3.0	63	0.046	0.93
5.6	60	0.032	0.98
7.6	55	0.015	0.95
10.0	50	0.064	0.98

a. Solution pH was adjusted using phosphate buffer[0.015M-0.025M]. pH did not appreciably change over the course of the reaction.

b. The pseudo-first order rate constants were determined for the entire conversion of As (III) in aerated solution; error in the rate constants is estimated to be ± 10-20 % based on duplicate analyses of individual samples.

c. The correlation coefficient is based on the best least squares fit line through the experimental data points.

Our results illustrate that the rates of conversion are modestly influenced by the solution pH. The pseudo-first order rate constants increase with decreasing pH below pH=7.6. The pK_a values of arsenious acid ($H_3As^{III}O_3$) and arsenic acid ($H_3As^VO_4$) are as follows; $pK_1=$ 9.22, $pK_2=12.13$, and $pK_3=13.4$; $pK_1=$ 2.20, $pK_2=6.97$, and $pK_3=11.53$ respectively (19). The increase in the rate, above pK_1 for As(III), at pH = 10 is attributed to the fact that the $H_3As^{III}O_3$ under such conditions exists primarily in the mono-anionic form, $H_2As^{III}O_3^-$, which is more electron rich and likely more reactive with hydroxyl radical than the neutral $H_3As^{III}O_3$ present at lower pH.

Conclusions

Kinetic analysis of the ultrasonically generated hydroxyl radical oxidation of As(III) was evaluated using pseudo-first order kinetics. The kinetics are influenced by the nature of the saturating gas. Hydroxyl radical is proposed to oxidize As(III) to As(IV), which is rapidly oxidized to As(V) by oxygen. In the absence of oxygen, (under argon saturation) the conversion of As(III) to As(V) reaches a maximum of 50% suggesting other reaction processes, possibly reductive processes, predominate. Within the pH range of 3-9 a modest increase in the oxidation rate of As(III) was observed under acidic conditions. An increased rate above the pKa of $H_3As^{III}O_3$ is rationalized based on the enhanced reaction of the mono-anionic species toward electrophilic hydroxyl radical. Our results also suggests As(V) is not reduced by superoxide anion radical. Ultrasonic irradiation has been used to generate and study the reactions of hydroxyl radical and superoxide anion radical under environmentally relevant conditions. The kinetic parameters from these studies may find applications in predicting the fate and treatment of arsenic in aquatic environments.

Acknowledgements

We thank the National Science Foundation and the Dreyfus Foundation (Henry Teacher Scholar Award to KEO) for their support.

References

1. Christen, K. *Environ. Sci. Technol.* **2001**, *35*, 286A-297A.
2. "Arsenic in the Environment", Part I: Cycling and Characterization" Jerome O. Nriagu, Eds.;John Wiley & Sons, Inc. New York, 1994.
3. Masscheleyn, P. H.; Delaune, R. D.; Patrick, W.H. *J. Environ. Qual.* 1991, *20*, 522-527.
4. Mariner, P. E.; Holzermer, F. J.; Jackson, R. E.; Meinardus, H. W. *Environ. Sci. Technol.* **1996**, *30*, 1645-1651.
5. Ferguson, J. F.; Gavis, J. *Water Res.* 1972, *6*, 1259-1274.
6. Cullen, W. R.; Reimer, K. J. *J. Chem. Rev.* 1989, *89*, 713-764.
7. Meng, X. G.; Bang, S. B.; Korfiatis, G. P. *J. Water Res.* 2000, *34*, 1255-1261.
8. Borho, M.; Wilderer, P. J. *Wat. Sci. Tech.* 1996, *34*, 25-31.
9. Balrama Krishna, M. V.; Chandrasekaran, K.; Karunasagar, D; Arunachalam, J. *J. Hazardous Materials*, **2001**, *B84*, 229-240.
10. Kamat, P. V.; Vinodgopal, K., *Langmuir,* 1996, *12*, 5739-5744.
11. Wallington, C. *ACC. Chem. Res.* 1975, *8*, 125-131.
12. Lorimer, L.P. *J. Sonochemistry*: Theory, Applications and Uses of Ultrasound in Chemistry, Ellis Horwood Limited, Chichester, 1988.
13. Buxton, G. V.; Greenstock, C. L.; Helman, W. P.; Ross, A. P. *J. Phys. Chem.* 1988, *17*, 513-586.
14. Adams, G. E.; Boag, J. W.; Micheal, B. D. *J. Trans. Faraday Soc.* **1965**, *61*, 1674-1677.

15. Klaning, U. K.; Bielski, B. H. J.; Sehested. *J. Inorg. Chem.* **1989**, *28*, 2717-2724.

16. Kim, D. K.; O'Shea, K.E.; Cooper, W. J. *J. Environ. Eng.* 2002, in press.

17. Nagorski, S. A.; Moore, J. N. *J. Water Resources Research*, 1999, *35*, 3441.

18. Hug, S. J.; Canonica, L.; Wegelin, M.; Gechter, D.; Von Gunter, U. *Environ. Sci. Technol.* 2002, *31*, 2114-2121.

19. Wagman, D. D.; Evans, H. H.; Parker, V. B.; Schumm, R.H.; Harlow, I.; Bailey, S. M.; Churney, K. L.; Butall, R. L. *J. Phys Chem.* Ref Data II, *Suppl.* 1982, 2, 392.

Chapter 8

Metal Tolerance, Accumulation, and Detoxification in Plants with Emphasis on Arsenic in Terrestrial Plants

Yong Cai[1] and Lena Q. Ma[2]

[1]Department of Chemistry and Southeast Environmental Research Center, Florida International University, Miami, FL 33199
[2]Soil and Water Science Department, University of Florida, Gainesville, FL 32611–0290

A number of review articles have appeared in the literature recently on metal accumulation by plants. However, reports on the tolerance, accumulation and detoxification of arsenic in plants, especially in terrestrial plants, are limited. In light of the strong needs for study of arsenic biogeochemistry and development of arsenic decontamination techniques, in this paper we review arsenic uptake, accumulation, and detoxification mechanisms in terrestrial plants. In order to discuss the processes involved in arsenic accumulation, we will first provide a brief review on the general understanding about the mechanisms of metal tolerance in plants. We will also discuss the arsenic hyperaccumulators that were recently discovered.

Arsenic is bioactive and potentially toxic. Long-term exposure to low concentrations of arsenic in drinking water can lead to skin, bladder, lung, and prostate cancer (*1,2*). Non-cancer effects of ingesting arsenic at low levels include cardiovascular disease, diabetes, and anemia, as well as reproductive, developmental, immunological and neurological effects. Short-term exposure to high doses of arsenic can cause other adverse health effects (*1-3*). A recent report by the National Academy of Sciences concluded that the current arsenic standard of 50 µg l⁻¹ in drinking water does not achieve U.S.EPA's goal of protecting public health (*2*). As such, EPA has recently lowered its current drinking water standard from 50 to 10 µg l⁻¹. Public awareness of arsenic toxicity and environmental impacts as well as new regulations limiting arsenic in drinking water have resulted in a growing interest in the study of the biogeochemical cycling of arsenic and the development of arsenic decontamination technologies.

Arsenic-contaminated soil is one of the major sources of arsenic in drinking water (*1-6*). The concentration of arsenic in cereals, vegetables and fruits is directly related to the level of arsenic in contaminated soil. Although the remediation of arsenic-contaminated soil is an important and timely issue, cost-effective remediation techniques are not currently available. Phytoremediation, an emerging, plant-based technology for the removal of toxic contaminants from soil and water is a potentially attractive approach (*7-9*). This technique has received much attention lately as a cost-effective alternative to the more established treatment methods used at hazardous waste sites.

A number of plants have been identified as hyperaccumulators (for definition of hyperaccumulator see section Arsenic Hyperaccumulation in Plants) for the phytoextraction of a variety of metals including Cd, Cr, Cu, Hg, Pb, Ni, Se, and Zn, and some of these plants have been used in field applications (*10,11*). However, no arsenic hyperaccumulator has been previously reported (*12*) until the recent finding of Chinese Brake fern, *Pteris vittata*, which efficiently hyperaccumulates arsenic from arsenic contaminated soils (*13*). In addition to its practical application, arsenic uptake, tolerance, and accumulation in plants have been an interesting research topic (*14*). To understand arsenic behaviors in plants, we need to have a better understanding of metal behaviors in plants.

Mechanisms of Metal Tolerance in Plants

One requirement that is of great significance to accumulation of toxic metals is the ability of plants to tolerate the metals that are extracted from the soil. Plants are not unique in having to protect themselves against the toxic effects of metals. Thus, a variety of tolerance and resistance mechanisms have evolved, including avoidance or exclusion, which minimizes the cellular accumulation of metals, and tolerance, which allows plants to survive while accumulating high concentrations of metals (*15*).

In general, in order to survive plants must have developed, on one hand, efficient and specific mechanisms by which metals are taken up and transformed into a physiologically tolerable form, providing the essential elements for the metabolic functions of plants. On the other hand, excesses of these essential elements, or of toxic metals that do not play a role in plant metabolism have to be metabolically inactivated (*16*). The mechanisms proposed for metal detoxification and hyperaccumulation within the plant involve chelation of the metal cation by ligands and/or sequestration of metals away from sites or metabolism in the cytoplasm, notably into the vacuole or cell wall (*9,17,18*). Other possible adaptive responses include activation of alternative metabolic pathways less sensitive to metal ions, modification of enzyme structure, or alteration of membrane permeability by structural reorganization or compositional changes (*18-22*). There is, however, insufficient evidence to support that the latter mechanisms of cellular tolerance are important in hyperaccumulator plants (*18*).

Chelation

It is clear that chelation of metal ions by specific high-affinity ligands reduces the solution concentration of free metal ions, thereby reducing their phytotoxicity (*9*). Much research interest has focused on identifying the intracellular ligands potentially involved in metal accumulation and detoxification. These ligands have been grouped, by Baker and co-workers, into several categories according to the characteristic electron donor centers (18).

Sulfur donor ligands

Organic ligands containing sulfur donor centers form very stable complexes with many metals (*18,23*). Two major sulfur-containing classes of metal chelating ligands have been identified in plants and these may play an important role in plant metal tolerance: phytochelatins (PCs) and metallothioneins (MTs). Phytochelatins are composed of a family of peptides with the general structure (γ-GluCys)$_n$-Gly, where, n = 2 to 11 (*15,24*). The PCs with different n numbers are abbreviated to PC$_n$ and are used in the following discussion. Similar (γ-GluCys)$_n$ peptides with carboxy-terminal amino acids other than Gly have been identified in a number of plants species, but it is likely that they serve the same function as PCs. The γ-carboxamide linkage between glutamate and cysteine indicates that PCs are not synthesized by translation of a mRNA, but are instead the product of an enzymatic reaction (*15,25*). The enzyme PC synthase was first identified by Grill et al (*26*) and has been characterized in a number of subsequent studies (*27-29*). Very recently several research groups simultaneously and independently cloned and characterized genes encoding this enzyme (*111-114*). Isolated from *Arabidopsis*, *S. pombe*, and wheat, these genes, designated as *AtPCS1*, *SpPCS*, and *TaPCS1*, respectively, encode 40-50% sequence-similar 50-55 kDa polypeptides active in the synthesis of PCs from glutathione (GSH). PCs are rapidly induced *in vivo* by a wide range of metal ions, including cations Cd, Cu, Ni, Zn, Ag, Hg and Pb, and oxyanions arsenate-AsO_4^{3-} and arsenite-AsO_3^{3-} (*30*). PCs are able to bind a number of metal ions *in vitro* through thiolate bonds (*15*). It has been suggested (*31,32*) that PCs are directly involved in heavy metal tolerance. However, there is considerable evidence to contradict this view (*18*). The only metal-PC complexes that have been isolated from plants contain ions of Cd, Cu, or Ag (*33*). Both metal-resistant and metal-sensitive plants produce PCs. Several reports have concluded that PCs are not primarily responsible for hyperaccumulation of Zn, Ni, or Pb (*34-36*). Although it is clearly demonstrated that PCs can have an important role in metal detoxification and accumulation in higher plants, formation of metal-complexes provides insufficient explanation for either the metal specificity or species specificity of hyperaccumulation (*18,37*). Therefore, it remains to be determined what exact role PCs play in metal-tolerance mechanisms at the cellular level.

MTs are low molecular weight, cysteine-rich, metal-binding proteins, which are encoded by structural genes (*38,39*). It should be noted that the range of the MT compounds has been extended to include the metal-binding polypeptides, PCs, found in higher plants (*40*). The following discussion on the properties of MTs, however, does not include PCs. It is well known that MTs, present widely in fungi and mammals, are able to function in metal detoxification (*39,41*). Plant MTs, however, have received little attention during the past decade because intracellular metal chelation in plants was thought to be carried out mainly by PCs (*16,18,24,25*) though MTs have been identified and purified by several researchers (*15,23,42,43,115*). The investigation of MT expression levels in *A. thaliana* demonstrated that expression levels of MT2 mRNA strongly correlated with Cu resistance (*43*).

Oxygen donor ligands

In addition to the sulfur-containing proteins and peptides, a large number of low molecular weight organic acids have long been recognized to be involved in plant metal detoxification (*12*). Carboxylic acid anions are abundant in the cells of terrestrial plants and form complexes with divalent and trivalent metal ions of reasonably high stability. Differences in the concentrations of organic acids, such as malate, aconitate, malonate, oxalate, tartrate, and citrate, in leaves of various ecotypes of metal-tolerant plants in their natural habit fostered consideration of these substances as cellular chelators (17). Long before the discovery of PCs in plants, research on several nickel hyperaccumulators from New Caledonia had shown that nickel is predominantly bound to citrate and the amount of citrate produced is strongly correlated with the accumulated nickel (44,45). Since then, there have been a large number of studies in which carboxylic acids have been found to be associated with metals (12,18).

Based on these findings, hypotheses on the mechanisms of metal accumulation and transport within the plants have been proposed. It was proposed that Zn accumulation by several plants is facilitated by the transport of malate-Zn complexes (46). The Zn ions are bound by malate upon uptake into cytoplasm, and the malate then serves as a carrier to transport the Zn ions to the vacuole. The Zn ions are then complexed by a terminal acceptor and the released malate is able to return to the cytoplasm where it is ready to transport more Zn ions. Still and Williams (47) have proposed that Ni accumulation is due to a selective transport ligand in root membrane. This "selector" is restricted to the membrane, so that other organic compounds such as citric and malic acids would be needed to act as "transport" ligands. It is clear that there is a lack of evidence to support these hypotheses. Although the carboxylates are undoubtedly important ligands for metal chelation in the vacuole, their exact roles for the metal detoxification and accumulation are unclear. Carboxylates tend to be present constitutively in the shoots of terrestrial plants and do not seem to account for either the metal specificity or species specificity of plant metal hyperaccumulation (18). It is clear that a great deal of research still remains to be done on how organic acids are involved in metal hyperaccumulation.

Nitrogen donor ligands

Nitrogen donor ligands found in plants generally consist of amino acids and their derivatives. These nitrogen donor centers have relatively high affinity to some metals (23,48). Compared to carboxylic acids, amino acids show even greater stability constants for some metals (e.g. Ni) (48), but reports on the existence and role of amino acids in the accumulation of metals by plants are very limited (49-51). A recent paper published by Krämer et al. (34) made a significant advance in our understanding of nickel binding in plants and provided compelling evidence that nickel is bound to free histidine in *Alyssum* species such as *A. lesbiacum*. These results suggest that histidine is involved in both nickel tolerance and translocation in these hyperaccumulator plants. The lack of studies makes it difficult to evaluate the role of amino acids in the accumulation and tolerance of other metals by plants (18).

Intracellular partitioning

At the cellular level, the main compartments of plants are the cell wall, cytosol, and the vacuoles. It is conceivable that metal tolerance could rely on the ability of a plant to store accumulated excess metals in organs or subcellular compartments where no sensitive metabolic activities take place (52). Plant cell walls are a continuous matrix, which acts as a cation exchanger, holding variable quantities of metal and providing for some metal exclusion (17). The central vacuole seems to be a suitable storage reservoir for excessively accumulated metals. Although some efforts have been made to localize the accumulated metals during the past two decades, the results obtained with classic fractionation techniques must be considered with suspicion, as the probability of redistribution is extremely high (52). Electro probe x-ray microanalysis and scanning electron microscopy are the frequently used techniques to localize metal content at subcellular levels (53-58). In the roots of a Zn and Cd hyperaccumulator (*T. caerulescens*), x-ray microanalysis showed that Cd accumulated mainly in the apoplast and, to a lesser extent, in vacuoles, whereas Zn was principally found in vacuoles and, to a lesser extent, in cell walls (58). In the shoot of the same plant, the highest zinc concentrations again appeared to be in the vacuoles of epidermal and subepidermal cells. Vacuole isolation or compartmental flux analysis, although less frequently used in the study of metal accumulation, may also provide valuable results. In order to determine the localization of Cd and the potential Cd-binding peptides, protoplasts and vacuoles were isolated from leaves of Cd-exposed tobacco seedlings (59). It was found that purified vacuoles contained virtually all of the Cd-binding peptides and Cd found in protoplasts. Brooks et al. (60), using compartmental flux analysis, found that Zn-resistant clones of *Deschampsia caespitosa* were able to actively pump Zn into the vacuoles of root cells, whereas Zn-sensitive clones seemed to have a much lower capacity to do so.

Alternations of cellular metabolism and membrane structure

Toxic metals entering plant cells cause deactivation of cell enzymes, consequently affect plant growth. For the metal-resistant plant to tolerate high concentrations of metals, changes in its metabolism, other than those involved in chelator production or metal compartmentation processes, could play a role (52). In general, enzymes of metal-tolerant plants are as metal sensitive as those of metal-sensitive plants (46), but it is possible that metal sensitivity of enzyme activity could be counteracted by increased enzyme synthesis when plants are exposed to high levels of toxic metals. It is also possible that activation of an alternative metabolic pathway occurs in order to avoid the metal-sensitive metabolic process. However, there is limited evidence to support these hypotheses (*52*).

Metals may affect membrane function, as indicated by a loss in selective permeability possibly due to inhibition of ATPase activity, and passive leakage of K, P, and organic molecules from cells can occur upon metal exposure (52,61). It has been reported that a primary component of cellular resistance to elevated copper concentration in plants appears to be enhanced plasma membrane resistance to, or repair of, Cu-induced membrane damage (*9,19,20,22*). Although limited information is available for Cu (9,22), evidence of such a resistant mechanism is completely unknown for other metals.

Many of the mechanisms proposed thus far fail to explain the metal specificity of metal resistance in higher plants (e.g. cell wall binding or phytochelatin production). This may be taken as an indication that each specific metal-resistance syndrome involves a specific combination of many distinct characteristics (*52*).

Mechanisms of metal translocation

Metals, once taken up by roots, can either be stored in the roots or exported to the shoot. Efficient transport of metals to shoots is an important aspect of plant metal accumulation for phytoextraction. The transport of metals from root to shoot is carried out primarily through xylem with the transpiration stream (*9,62*). Organic acids and amino acids have frequently been reported to be the potential metal chelators, which most likely facilitate metal translocation through xylem (*51,62*). Without being chelated by ligands, movement of metal caions from roots to shoot is expected to be severely retarded as xylem cell walls have a high cation exchange capability. Isolation of a citrato-complex of nickel from the latex of a Ni hyperaccumulator (*Sebertia acuminata*) supports the role of organic acids in metal transport (*44*). Recent evidence from work with a Ni hyperaccumulator from the genus *Alyssum* suggests that increasing concentrations of free histidine in the xylem sap may enhance translocation of this metal to the shoot (*34*). Nicotianamine, a methionine derivative, could be a Cu shuttle in the xylem (*63*). Salt et al. (*64*) demonstrated, using x-ray absorbance fine structure (EXAFS) analysis, that Cd in the xylem sap of *B. juncea* was chelated by oxygen or nitrogen atoms, suggesting the involvement of organic acids in Cd translocation. Mathematical modeling has also predicted the formation of complexes of metal-organic acid or metal-amino acid in xylem sap (*50*).

Although metals may also be transported in plants through phloem sap, the mechanism is poorly documented (*62*). The sole molecule identified as a potential phloem metal transporter is nicotianamine (65). It has been found in a stoichiometry of 1.25 with four metals (Fe, Cu, Zn, and Mn) in the phloem sap, leading the authors to speculate on a united transport mechanism (*66,67*). Such speculation, however, has been argued by other researchers (*62*). Other unknown metal chelators have also been described in the phloem. The chemical identity and the role of these metals complexes, if any, in the phloem metal-transport process remain to be clarified (*62*).

Arsenic Uptake, Tolerance, Accumulation, and Detoxification in Terrestrial Plants

Arsenic Uptake and Tolerance

A prerequisite for plants to accumulate and detoxify arsenic is that these plants must tolerate arsenic present in the surrounding environment, such as soil, and/or the plant tissue. There is a considerable literature on arsenic uptake by a variety of plants, including both arsenic tolerant and non-tolerant species (*33,68-78*). Arsenic tolerant plants often grow in areas of mining and industrial sites that are enriched in arsenic (*13,69,77,79*). For example, in the southwest of England large tracts of land are contaminated with arsenic, either through naturally elevated arsenic associated with a granitic intrusion or through the mining and processing of Cu and arsenic ores (*80*). Only a limited number of species have adapted to grow in the areas because the spoil soils are highly toxic. One of the survived species is Yorkshire Fog (*Holcus lanatus* L.). Arsenic tolerance of this species was investigated based on the measurement of root length for the populations from both arsenic contaminated (mine) and uncontaminated sites (*80*). It was found that plants from arsenic contaminated soils have tolerant individuals ranging from 92-100% of the overall population, but in uncontaminated sites it drops to on average 50%. It was concluded that the high percentage of tolerant plants in arsenic contaminated soils is due to elevated levels of arsenic and not to the mine environment in general. It is interesting to note that soil not contaminated with arsenic in the southwest of England has over 50% of arsenic tolerant plants. It was proposed that this high percentage of tolerance was due to gene flow from arsenic contaminated sites, and/or due to the generally elevated levels of arsenic in soils of the region.

In an effort to assess the ability of *Plantago lanceolata* L., an extremely common weed in the southeastern United States and throughout the world, to adapt to diverse soil toxicities, Pollard (79) found that the populations from sites contaminated with arsenic were more tolerant than a control population from uncontaminated soil. Koch et al. (77) recently determined arsenic concentrations in a number of plants from Yellowknife, Northwest Territories, Canada, where elevated levels of arsenic from historic and recent gold mine operation are of great concern. The plant samples consisted of vascular plants (including flowering plants, grasses, aquatic submergent plants, and emergent plants) and bryophytes (mosses). It was found that mosses contain the highest levels of arsenic (490-1,229 ppm dry weight). Unfortunately, no data on soil arsenic contents were provided. This limits the assessment of the tolerance and accumulation of arsenic by mosses. Concentrations of arsenic in many other plant species collected from both contaminated and uncontaminated sites, as well as their ability to tolerate arsenic can be found in the literature (14).

The mechanisms of plant arsenic uptake are similar to those of phosphorus. Because of the chemical similarity of arsenate and phosphate, it would seem that there is active and competitive uptake for the P uptake system (71,81). It has been reported that in both arsenic tolerant and non-tolerant *Holcus lanatus* L., arsenic uptake uses P uptake system (71-73). It was further proposed that arsenic uptake in the arsenic-tolerant *H. lanatus* is restricted by the altered P uptake system, yet the tolerant plants were capable of accumulating arsenic to high concentration over longer time periods (73). Paliouris and Hutchinson (82) compared the ability of plant arsenic tolerance of *Silene vulgaris* (Moench) growing in Ontario, Canada. The plants were collected from mine tailings near town of Cobalt (arsenic tolerant), and from a uncontaminated site near Baymouth (arsenic sensitive). It was found that reduced arsenic uptake occurred in the roots of tolerant individuals compared with non-tolerant individuals at a low arsenic treatment. At a higher arsenic exposure, however, arsenic entered the roots of tolerant individuals. A partial exclusion or arsenic translocation to the shoots then appeared to operate in the shoots of tolerant individuals, whereas for non-tolerant plants arsenic levels were higher in the shoots.

It is apparent that although some tolerant plants have a restricted arsenic uptake, this is insufficient in itself to explain arsenic tolerance in these species. In addition, it seems unlikely that the restricted arsenic uptake by plant roots is a proper hypothesis for some plant species, which can accumulate arsenic to high levels (69,77) in a very short time period (13). Once being taken up by plants, arsenic may be toxic to plants in a number of ways including interference in aerobic phosphorylation and reduction of arsenate within plant cells to arsenite, which then attacks protein sulfhydryl groups (73,83). For plants that are tolerant and also accumulate arsenic there must exist mechanisms by which arsenic is detoxified within cells.

Arsenic Accumulation and Detoxification

Significant adverse growth effects have been observed in plants on certain soils containing naturally elevated arsenic concentrations (*84*), or on soils contaminated by industrial or agricultural practices (*85*). The general appearance of the vegetation in some localities has been modified by the presence of arsenic anomalies while in some areas where soils contain high concentrations of arsenic a distinctive flora has developed over geologic time (*69,86*). Plants present on arsenic contaminated sites include *Agrostis tenuis, Agrostis stolonifera, Agrostis canina,* and *Jasione Montana* (*69,86*); *Calluna vulgaris, and Holcus lanatus* (*14,69*); and *Ceratophyllum demersum* (*14*), *Pteris vittata* (*13*) and *Pityrogramma calomelanos* (*87*) etc. These species were found to accumulate arsenic to high concentrations, which, in some cases, approached percent levels in dry weight (see details on Arsenic Hyperaccumulator in the following section).

Several important points should be kept in mind when dealing with arsenic accumulation and detoxification. Arsenic generally exists as oxyanions, mainly arsenate and arsenite in most environmental conditions, although small amounts of methylated arsenic species are also present in certain circumstances. This is in contrast to most heavy metals discussed previously. The biochemistry and toxicology of arsenic is complicated by its ability to convert between oxidation states and organo-metalloidal forms both in the environment and in biological systems. These processes cause differences in the relative tissue-binding affinities of the various arsenic species, and they determine both intoxication and detoxification mechanisms (*88*). Finally, as with those transition metals (e.g. Hg, Cu, Cd,), stable complexes may be formed with organic ligands, such as thiols.

It appears that the arsenic detoxification mechanisms depend on a number of parameters, especially on the types of the organisms (*3*). In mammals (*89,90*), fungi, algae, and marine animals (fish, mollusks etc.) (*3,91*), arsenic detoxification usually involves methylation and other biotransformation processes such as incorporation of arsenic into organic molecules by forming arsenocholine, arsenobetaine, or arsenosugars. In bacteria, arsenate is enzymatically reduced to arsenite by ArsC, an enzyme that reduces less toxic arsenate to arsenite. Arsenite is then "pumped" out by the membrane protein ArsB that functions chemiosmotically alone or with the additional ArsA protein as an ATPase (*92*). Methylation of arsenic by bacteria, however, is also reported (*3*). There seems to be a very limited research on the arsenic detoxification mechanisms in plants, especially in terrestrial plants.

Nissen and Bensen (70) reported that freshwater and terrestrial plants differ markedly in their ability to metabolize arsenic. Terrestrial higher plants appear limited in their ability to metabolize arsenic via methylation and biotransformation to arsenic-lipid compounds. It was proposed that arsenic is taken up by plants via the mechanism for P uptake and is readily reduced by the plant to arsenite. Only under conditions of P-deficiency was the arsenite methylated to some extent. No arsenic-lipid formation was observed in plant leaves. Freshwater plants, however, are capable of synthesizing considerable amounts of arsenic-lipid, which are identical to that obtained with marine algae. It has been found that carrots (*93*) and other vegetables (*94*) contain predominately inorganic arsenic. In their recent study with a number of terrestrial plant collected from Yellowknife, Canada, Koch et al. (*77*) observed that for all the terrestrial plant species studied, the major water soluble compounds are inorganic As(V) and As(III). We recently reported that Chinese Brake fern, an arsenic hyperaccumulator, contains almost entirely inorganic arsenic with arsenite as the predominate form in the above ground biomass (*13*, see also the following section). From these studies, it seems unlikely that methylation and biotransformation of arsenic to organo-arsenic compounds are the major routes of arsenic detoxification in terrestrial plants.

Alternative mechanisms are apparently present for arsenic detoxification in terrestrial plants. Unfortunately, information on this aspect is very scarce. Phytochelatins, the heavy metal-binding and thiol-rich polypeptides, have been considered to be a part of arsenic detoxification mechanism in higher plants since some authors have reported PCs accumulation upon exposure to arsenic (*16,24,33,75,78,95,96*). In a paper describing the short- and long-term toxicity of arsenic in *Silene vulgaris*, Sneller et al. (*75*) reported that the short-term PC accumulation (over a 3-day period) was positively correlated with arsenic exposure. Isolation of peptide complexes from prolonged exposed plants showed that PC_2, PC_3 and PC_4 were present, although the later was not present until at least 3 day exposure. Arsenic co-eluted mainly with PC_2 and PC_3. The lack of arsenic in the fractions containing PC_4 was attributed to the probable dissociation of the complexes during extraction or elution in the analysis. It was concluded that PCs provide a detoxification mechanism in *S. vulgaris* through binding of arsenic to their thiol groups. Schmöger et al. (*96*) reported that both arsenate and arsenite efficiently induce the biosynthesis of PCs in vivo and in vitro. The rapid induction of the metal-binding PCs were observed in cell suspension cultures of *Rauvoltia serpentina*, in seeding of *Arabidopis*, and in enzyme preparations of *Silene vulgaris* upon challenge to arsenicals. An approximately 3:1 ratio of the sulfhydryl groups from PCs to arsenic is compatible with reported arsenic-glutathione complexes (96).

The identity of the arsenic-induced PCs and reconstituted arsenic-peptide complexes were demonstrated by different techniques. In an attempt to establish the biochemical fate of arsenic taken up by Indian mustard (*Brassica juncea*), Pickering et al. (*78*) found that after arsenic uptake by the roots, a small fraction is exported to the shoot via the xylem as the oxyanions (arsenate and arsenite). Using x-ray absorption spectroscopy, it was found that the arsenic in both roots and shoots is probably stored as an As(III)-tris-thiolate complex, which is indistinguishable from As(III)-tris-glutathione. It was proposed that the thiolate donors are probably either glutathione or phytochelatins. However, other research failed to demonstrate the formation of arsenic-PC complexes (*33*) although PCs were indeed induced in plants upon exposure to arsenic. This has been attributed to the possible dissociation of the complexes under the experimental conditions used (*96*).

From those very limited results, it appears that as with heavy metals (e.g. Cd, Cu, Zn), arsenic can trigger the formation of PCs in many plants, and the induced PCs may bind with As (III) through their sulfhydryl groups in the cell environment. However, the exact role played by PCs in the accumulation and detoxification of arsenic or other metals in general in plants is still in debate (*36,97*). Toward that end, some of the key points need to be clarified. First, it seems that PCs are the general products induced by exposure to metals and metalloids for many tolerant as well as non-tolerant plant species (*36*). Such a characteristics is unlikely a specific property owned by certain plant species. Therefore, it is difficult to explain the behaviors of hyperaccumulators, which often show a very specific ability of accumulating a special element. Second, although PCs are formed in some plants upon exposure to arsenic and bind with arsenic to some extent, the ratio of the complexes to total arsenic present in the plants is small (*96*), indicating that majority of arsenic is in non-PC bounding form. In fact, it has been proposed that the binding of heavy metal ions to PCs seems to play only a transient role in metal detoxification by plants (*36*). Except for formation of As-PC complexes, other mechanisms may exist and be responsible for arsenic detoxification. Finally, it is well known that arsenite can form very stable complexes with five or six rings with dithiols with vicinal sufhydryl groups (*98-101*), whereas monothiols do not form stable complexes with these arsenite at neutral pH but only in acid environment (*101*). To be able to form reasonably stable complexes between arsenite and PCs, proper special arrangement of PCs may have to be carried out because the sulfhydryl groups in PCs are not vicinal. Information on the structure of the As-PCs complexes and the thermodynamic data for the formation of those complexes does not exist yet.

Arsenic Hyperaccumulation in Terrestrial Plants

Definition of Arsenic Hyperaccumulator

Metal hyperaccumulator was first used by Brooks et al. (*102*) to describe some "strong accumulators" of nickel in 1977. It was defined as those plants containing >1,000 µg g^{-1} (0.1%) metal in dry materials. This term generally represents a concentration much greater (e.g. about 100 times for nickel) than that the highest values to be expected in non-accumulating plants. However, this definition counts only the concentrations of metals in the plant shoots without taking into consideration of the metal contents in the root and the surrounding soils. While some plants can survive in an environment containing extremely high concentrations of metals, they do not show a high ability of accumulating metals. For example, it was reported that *Agrostis tenuis* growing on arsenic mine wastes in southwest of England contained 3,470 µg g^{-1} arsenic. However, arsenic concentration was as high as 26,500 µg g^{-1} in the corresponding soils (*86*). It seems that this species is a "hypertolerant" species, rather than a true hyperaccumulator.

Both bioaccumulation factor (BF) and translocation factor (TF) have to be considered while evaluating whether a particular plant is a metal hyperaccumulator (*103*). The term BF, defined as the ratio of metal concentrations in plant biomass to those in soils, has been used to determine the effectiveness of plants in removing metals from soils. The term TF, defined as the metal concentrations in plant shoot to those in the roots, has been used to determine the effectiveness of plants in translocating metal from the root to the shoot (*104*). Baker (*116*) used leaf:root metal concentration ratios to characterize accumulators and excluders. A hyperaccumulator should be the one that not only tolerates high concentrations of metals, but also accumulates metals to high levels in plant aboveground biomass compared to those in the soils. Therefore, in the case of arsenic hyperaccumulator, it seem more appropriate to define it as the one that has BF>1 and TF>1 as well as accumulates >1,000 µg g^{-1} arsenic in plant biomass. This definition should be applied to other metal hyperaccumulators as well, i.e. metal hyperaccumulators should be evaluated not just on the basis of absolute metal concentrations in the plant biomass.

Discovery of an Arsenic Hyperaccumulator

Recently, Ma and co-workers (*13*) identified the first known arsenic hyperaccumulating plant from an abandoned wood preservation site. This discovery is a significant breakthrough for cleaning up arsenic contaminated sites. The site where the plant was collected was in operation from 1952 to 1962, pressure treating lumber using an aqueous solution of chromate-copper-arsenate (CCA) (*105*). As a result, the site was contaminated with arsenic, with an average soil concentration of 361 µg g^{-1}, compared to background concentrations of 0.42 µg g^{-1} in Florida soils (*106*). Among 14 different plant species collected from the site, only one plant species accumulated a significant amount of arsenic, Chinese Brake fern (*Pteris vitatta; 105*). Chinese Brake fern plants collected from the CCA site during 1999 contained as much as 14,500 μg g^{-1} (~1.5%) arsenic in their aboveground dry-biomass, which is exceptionally high since the corresponding soil arsenic concentration was only 258 μg g^{-1} (*105*). This indicates that Chinese Brake was capable of concentrating as much as 56 times more arsenic in its dry-biomass than the soil where they were growing. Another interesting fact is that the plant appears to suffer no ill effect from the high arsenic concentration in its shoots. Our data are consistent with a recent publication (*107*), where the authors determined arsenic concentrations in Chinese Brake fern collected from arsenic polluted mine tailings. Arsenic concentrations in their fern samples ranged from 103 to 6,030 μg g^{-1}.

Recently, additional Chinese Brake fern samples were collected from the University of Florida (UF) campus. In addition to accumulating arsenic from the CCA contaminated site, Chinese Brake also accumulated significant amounts of arsenic from uncontaminated sites from UF campus, with an arsenic BF as high as 136. It is remarkable that Chinese Brake is capable of bioconcentrating arsenic in its aboveground biomass from soils with low levels (0.47 μg g^{-1}) as well as high levels (1,603 μg g^{-1}) of arsenic, yet it shows little toxicity symptoms.

Data from a greenhouse study using an artificially contaminated soil and the CCA soil are consistent with the field observation (*13*). Chinese Brake was extremely efficient in accumulating arsenic into its abovground biomass in a short period of time, to a level that was not previuosly reported in the literature (> 2% in dry weight). The fact that Chinese Brake survived in a soil containing as much as 1,500 mg kg^{-1} arsenic, which was spiked to the soil, is by itself remarktable.

Arsenic speciation experiment using 1:1 methanol:water solution with HPLC-ICP-MS indicated that, almost all extractable arsenic in the plant biomass is present as relatively toxic inorganic forms, with little detectable organoarsenic species (*13,110*). Furthermore, As(III) is more predominant in the aboveground biomass (up to 100%) than in the roots (8.3%), indicating As(V) is probably converted to As(III) during arsenic translocation from roots to its aboveground biomass in Chinese Brake.

In addition to its remarkable arsenic accumulation capability, Chinese Brake has numerous desirable characteristics that make it an ideal candidate for use in phytoremediating arsenic-contaminated soils (*12,108*). Chinese Brake ferns are extremely efficient in extracting arsenic from soils (extremely high arsenic BF), are versatile and hardy in their growing environment, have large biomass, are fast growing, are easy to reproduce, and are perennial plants (*108*). They have an arsenic BF up to 193 under natural growing condition. They are capable of surviving on a wide range of soil conditions, ranging from limestone surfaces to rocky woodland. The fact that they prefer to grow in alkaline conditions makes them especially attractive since arsenic availability is higher under alkaline conditions (*109*). Unlike most fern plants, Chinese Brake ferns prefer full sun.

They have relatively large biomass, producing fronds that are 30-90 cm in length, with blades of 25-60 cm long and 13-25 cm wide. Furthermore, they can be easily reproduced. Thousands of new plants can be produced in a single growing season from just one plant using spores or root cuttings. Once planted in an arsenic-contaminated site, they come back every year since they are perennial plants, so the aboveground biomass can be harvested season after season until the site is cleaned up without reseeding or replanting.

Other Arsenic Hyperaccumulators

Very recently, another arsenic hyperaccumulating fern, *Pityrogramma calomelanos*, was found in the Ron Phibun district of southern Thailand (*87*). This plant grows readily on arsenic-contaminated soils. Arsenic contents of the soils, where the plants were collected, ranged from 135 to 510 $\mu g\ g^{-1}$. The fern aboveground biomass contained the highest arsenic concentrations (2,760-8,350 $\mu g\ g^{-1}$), while its below ground biomass contained much lower concentrations (88-380 $\mu g\ g^{-1}$). These results are in good agreement with the finding for Chinese Brake fern . It is interesting to note that the speciation pattern of arsenic found in *Pityrogramma calomelanos* is similar to that in *Pteris vittata*. The plant contains predominately inorganic arsenic species. Arsenic in the fern aboveground biomass was present mainly as arsenite, while in fern rhizoids (roots) most of the arsenic was present as arsenate.

The discovery of plants capable of hyperaccumulating arsenic sheds light for the potential use of phytoremediation for arsenic contaminated sites. Effective implementation of the phytoremediation technique requires an understanding of the plant process that controls uptake and translocation of metals from the soil. Currently, however, there is little information available regarding mechanisms of arsenic tolerance, accumulation, and transport in these arsenic hyperaccumulators. Understanding of these processes will also be a critical step for the study of plant genetic engineering to further improve arsenic uptake characteristics. This clearly points out the future research needs for arsenic accumulation by plants.

Acknowledgements

This work was partially supported by NSF grants BES-0086768 and BES-0132114, and by the Foundation/Provost's Office Research Award at Florida International University (FIU). We would like to thank Drs. David Salt, Ken Reimer, and Kelsey Downum for their critical review of this paper. This is contribution # 171 of the Southeast Environmental Research Center at FIU.

References

1. Agency for Toxic Substances and Disease Registry (ATSDR). *Toxicological Profile for Arsenic*; U.S. Department of Health and Human Services: Atlanta, GA, 1999.
2. National Research Council. *Arsenic in Drinking Water*, National Academy Press: Washington, DC, 1999.
3. Cullen, W. R., Reimer, K .J. *Chem. Rev.* **1989,** *89,* 713-764.
4. Arsenic in the Environment, Part 1: Cycling and Characterization; Nriagu, J. O. Ed.; John Wiley & Sons: New York, 1994; pp 1-430.
5. Welch, A. H; Westjohn, D. B.; Helsel, D. R.; Wanty, R. B. *Ground Water,* **2000,** *38,* 589-604.
6. Kim, M. J.; Nriagu, J. *Sci. Total Environ.* **2000,** *247,* 71-79.
7. U.S. EPA, Office of Research and Development. *Introduction to Phytoremediation,* EPA/600/R-99/107. 2000.
8. Dahmani-Muller, H.; van Oort. F.; Gelie, B.; Balabane, M. *Environ. Pollu.* **2000,** *109,* 231-238.
9. Salt, D. E; Smith, R. D; Raskin, I. *Annu. Rev. Plant Physiol. Plant Mol. Biol.* **1998,** *49,* 643-668.
10. Dobson, A. P.; Bradshaw, A. D.; Baker, A. J. M. *Science* **1997,** *277,* 515-522.
11. *Phytoremediation of Contaminated Soil and Water;* Terry, N.; Banuelos, G. Eds.; Lewis Publishers: Boca Raton, 2000; pp 1-389.
12. *Plants That Hyperaccumulate Heavy Metals;* Brooks, R. R., Ed.; University Press: Cambridge, 1998; pp 1-53.
13. Ma, L. Q.; Komart, K. M.; Tu, C.; Zhang, W.; Cai, Y.; Kennelly, E. D. *Nature,* **2001,** *409,* 579.
14. Eisler, R. In *Arsenic in the Environment, Part II: Human Health and Ecosystem Effcets;* Nriagu, J. O. Ed.; John Wiley & Sons: New York. 1994; pp. 185-259.
15. Goldsbrough, P. In *Phytoremediation of Contaminated Soil and Water;* Terry, N.; Banuelos, G., Eds.; Lewis Publishers: Boca Raton. 2000; pp. 221-233.
16. Zenk, M. H. *Gene,* **1996,** *179,* 21-30.
17. Rauser, W. E. *Cell Biochemistry and Biophysics,* **1999,** *31,* 19-48.
18. Baker, A. J. M.; McGrath, S. P.; Reeves, R. D.; Smith, J. A. C. In *Phytoremediation of Contaminated Soil and Water,* Terry, N.; Banuelos, G., Eds.; Lewis: Boca Raton, 2000, 85-107.

19. De Vos, C. H.; Schat, H.; Vooijs, R.; Ernst, W. H. O. *J. Plant Physiol.* **1989**, *135*, 164-169.
20. Strange, J.; Macnair, M. R. *New Phytol.* **1991**, *119*, 383-388.
21. Ernst, W. H. O.; Verkleij, J. A. C.; Schat, H. *Acta Bot. Neerl.* **1992**, *41*, 229-248.
22. Murphy, A.; Zhou, J. M.; Goldsbrough, P. B.; Taiz, L. *Plant Physiol.* **1997**, *113*, 1293-1301.
23. Fraústo da Silva, J. J. R.; Williams, R. J. P. *The Biochemistry of the Elements: the Inorganic Chemistry of Life.* Clarendon Press: Oxford, 1991.
24. Grill, E.; Winnacker, E. L.; Zenk, M. H. *Proc. Natl. Acad. Sci. USA.* **1987**, *84*, 439-443.
25. Robinson, N. J.; Tommey, A. M.; Kuske, C.; Jackson, P. J. *Biochem. J.* **1993**, *295*, 1-10.
26. Grill, E.; Loffler, S.; Winnacker, E. L.; Zenk, M. H. *Proc. Natl. Acad. Sci. USA.* **1989**, *86*, 6838-6842.
27. Klapheck, S.; Schlunz, S.; Bergmann, L. *Plant Physiol.* **1995**, *107*, 515-521.
28. Chen, J.; Zhou, J.; Goldsbrough, P. B. *Physiol. Plant.* **1997**, *101*, 165-172.
29. Cobbett, C. S. *Current Opinion in Plant Biology*, **2000**, *3*, 211-216.
30. Rauser, W. E. *Plant Physiol.* **1995**, *109*, 1141-1149.
31. Jackson, P. J.; Unkefer, C. J.; Doolen, J. A., Watt, K.; Robinson, N. J. *Proc. Natl. Acad. Sci. USA.* **1987**, *84*, 6619-6623.
32. Salt, D. E.; Thurman, D. A.; Tomsett, A. B.; Sewell, A. K. *Proc. R. Soc. London, Ser. B.* **1989**, *236*, 79-89.
33. Maitani, T.; Kubota, H.; Sato, K.; Yamada, T. *Plant Physiol.* **1996**, *110*, 1145-1150.
34. Krämer, U.; Cotter-Howells, J. D.; Charnock, J. M.; Baker, A. J. M.; Smith, J. A. C. *Nature.* **1996**, *379*, 635-638.
35. Shen, Z. G.; Zhao, F. J.; McGrath, S. P. *Plant Cell Enviorn.* **1997**, *20*, 898-906.
36. Leopold, I.; Gunther, D.; Schmidt, J.; Neumann, D. *Phytochemistry,* **1999**, *50*,1323-1328.
37. Homer, F. A.; Reeves, R. D.; Brooks, R. R.; Baker, A. J. M. *Phytochemtsry,* **1991**, *30*, 2141-2145.
38. *Metallothioneins: Synthesis, Structure and Properties of Metallothioneins, Phytochelatins and Metal-Thiolate Complexes;* Stillman, M. J.; Shaw, C. F.; Suzuki, K. T., Eds.; VCH: New York, 1992, pp.1-443.
39. Stillman, M. J. *Coordination Chemistry Reviews.* **1995**, *144*, 461-511.
40. Robinson, N. In *Heavy Metal Tolerance in Plants: Evolutionary Aspects;* Shaw, A. J., Ed.; CRC Press: Boca Raton, 1990, pp 195-214.
41. Hamer, D. A. *Annu. Rev. Biochem.* **1986**, *55*, 913-951.
42. Lane, B.; Kajioka, R.; Kennedy, T. *Biochem. Cell. Biol.* **1987**, *65*, 1001-1005.

43. Murphy, A. S.; Taiz, L. *Plant Physiol.* **1995**, *109*, 1-10.
44. Lee, J.; Reeves, R. D.; Brooks, R. R.; Jaffré, T. *Phytochemistry*, **1977**, *16*, 1503-1505.
45. Lee, J.; Reeves, R. D.; Brooks, R. R.; Jaffré, T. *Phytochemistry*, **1978**, *17*, 1033-1035.
46. Mathys, W. *Physiol. Plant.* **1977**, *40*, 130-136.
47. Still, E. R.; Williams, R. J. P. *J. Inorganic Biochem.* **1980**, *13*, 35-40.
48. Homer, F. A.; Reeves, R. D.; Brooks, R. R. *Current Topics in Phytochemistry.* **1997**, *14*, 31-33.
49. White, M. C.; Decker, A. M.; Chaney, R. L. *Plant Physiol.* **1981**, *67*, 292-300.
50. White, M. C.; Baker, F. D.; Decker, A. M.; Chaney, R. L. *Plant Physiol.* **1981**, *67*, 301-310.
51. Cataldo, D. A.; McFadden, K. M.; Garland, T. R.; Wildung, R. E. *Plant Physiol.* **1988**, *86*, 734-739.
52. Verklaij, J. A. C.; Schat, H. In *Heavy Metal Tolerance in Plants: Evolutionary Aspects;* Shaw, A. J., Ed.; CRC Press: Boca Raton, 1990, pp 179-193.
53. Khan, D. H; Duckett, J. G.; Frankland, B.; Kirkham, J. B. *J. Plant Physiol.* **1984**, *115*, 19-28.
54. Rauser, W. E.; Ackerley, C.A. *Can. J. Bot.* **1987**, *65*, 643-646.
55. Van Steveninck, R. F. M.; Van Steveninck, M. E.; Fernando, D. R.; Horst, W. J., Marschner, H. *J. Plant Physiol.* **1987**, *131*, 247-257.
56. Van Steveninck, R. F. M.; Van Steveninck, M. E.; Wells, A. J.; Fernando, D. R. *J. Plant Physiol,* **1990**, *137*, 140-146.
57. Vázquez, M. D.; Barcelo, J.; Poschenrieder, Ch.; Madico, J.; Hatton, P.; Baker, A. J. M.; Cope, G. H. *J. Plant Physiol.* **1992**, *140*, 350-355.
58. Vázquez, M.D., Poschenrieder, Ch., Barcelo, J. Ultrastructural effects and localization of low cadmium concentrations in bean roots. New Phytol., 1992, 120, 215-226.
59. Vögeli-Lange, R.; Wagner, G. *J. Plant Physiol.* **1990**, *92*, 1086-1093.
60. Brooks, A.; Collins, J. C.; Thurman, D. A. *J. Plant Nutr.* **1981**, *3*, 695.
61. De Vos, C. H.; Schat, H.; Ernst, W. H. O. *J. Cell Biochem.* **1988**, *12A* (Supp. 49).
62. Briat, J. F.; Lebrun, M. *Life Sciences*, **1999**, *322*, 43-54.
63. Pich, A.; Scholz, G. *J. Exp. Bot.* **1996**, *47*, 41-47.
64. Salt, D. E.; Prince, R. C.; Pickering, I. J.; Raskin, I. *Plant Physiol.* **1995**, *109*, 427-433.
65. Stephan, U. W.; Scholz, G. *Physiol. Plant.* **1993**, *88*, 522-529.
66. Stephan, U. W.; Schmidke, I.; Pich, A. *Plant Soil*, **1994**, *165*, 181-188.
67. Schmidke, I.; Stephan, U. W. *Physiol. Plant.* **1995**, 147-153.
68. Rocovich, S. E.; West, D. A. *Science*, **1975**, *188*, 263-264.

69. Porter, E. K.; Peterson, P. J. *Sci. Tot. Environ.* **1975**, *4*, 365-371.
70. Nissen, P.; Benson, A. A. *Physiol. Plant.* **1982**, *54*, 446-450.
71. Macnair, M. R.; Cumbes, Q. *New Phytol.* **1987**, *107*,387-394.
72. Meharg, A. A.; Macnair, M. R. *New Phytol.* **1990**, *116*, 29-35.
73. Meharg, A. A.; Macnair, M. R. *New Phytol.* **1991**, *117*, 225-231.
74. Meharg, A. A.; Macnair, M. R. *New Phytol.* **1991**, *119*, 291-297.
75. Sneller, F. E. C.; Van Heerwaarden, L. M.; Kraaijeveld-Smit, F. J. L.; Ten Bookum, W. M.; Koevoets, P. L. M.; Schat, H.; Verkleij, J. A. C. *New Phytol.* **1999**, *144*, 223-232.
76. Koch, I.; Feldmann, J.; Wang, L.; Andrewes, P.; Reimer, K. J.; Cullen, W. R. *Sci. Tot. Environ.* **1999**, *236*, 101-117.
77. Koch, I.; Wang, L.; Ollson, C.; Cullen, W. R.; Reimer, K. J. *Environ. Sci. Technol.* **2000**, *34*, 22-26.
78. Pickering, I. J.; Prince, R. C.; George, M. J.; Smith, R. D.; George, G. N.; Salt, D. E. *Plant Physiol.* **2000**, *122*, 1171-1177.
79. Pollard, A. J. *New. Phytol.* **1980**, *86*, 109-117.
80. Meharg, A. A.; Cumbes, Q. J.; Macnair, M. R. *Evolution*, **1993**, *47*, 313-316.
81. Asher, C. J.; Reay, P. F. *Australian Journal of Plant Physiology*, **1979**, *6*, 459-466.
82. Paliouris, G.; Hutchinson, T. C. *New Phytol.* **1991**, *117*, 449-459.
83. Ullrich-Eberius, C. I.; Sanz, A.; Novacky, A. J. *Journal of Experimental Botany*, **1989**, *40*, 119-128.
84. Fergus, I. F. *Queensland J. Agric. Res.* **1955**, *12*, 95-100.
85. Woolson, E. A.; Axley, J. H.; Kearney, P. C. *Soil. Sci. Soc. Amer. Proc.* **1971**, *35*, 938-943.
86. Benson, L. M.; Porter, E. K.; Peterson, P. J. *J. Plant Nutrition*, **1981**, *3*, 1-4.
87. Francesconi, K.; Visoottiviseth, P.; Sridokchan, W.; Goessler, W. *Sci. Total Enviorn.* **2002**, *284*, 27-35.
88. Thompson, D. J. *Chem.-Biol. Interactions*, **1993**, *88*, 89-114.
89. Lakso, J. U.; Peoples, S. A. *J. Agric. Food Chem.* **1975**, *23*, 674-676.
90. Aposhian, H. V. *Annu. Rev. Pharmacol. Toxicol.* **1997**, *37*, 397-419.
91. Edmonds, J. S.; Francesconi, K. A. *Nature*, **1981**, *289*, 602-604.
92. Silver, S. *Gene*, **1996**, *179*, 9-19.
93. Helgesen, H.; Larsen, E. H. *Analyst*, **1998**, *123*, 791-796.
94. Pyles, R. A.; Woolson, E. A. *J. Agric. Food Chem.* **1982**, *30*, 866-870.
95. Sneller, F. E. C.; Van Heerwaarden, L. M.; Kraaijeveld-Smit, F. J. L.; Koevoets, P. L. M.; Schat, H.; Verkleij, J. A. C. *J. Agric. Food Chem.* **2000**, *48*, 4014-4019.
96. Schmöger, M. E. V.; Oven, M.; Grill, E. *Plant Physiol.* **2000**, *122*, 793-801.
97. De Knecht, J. A.; Koevoets, L. M.; Verkleij, J. A. S.; Ernst, W. H. O. *New Phytol.* **1992**, *122*, 681-688.

98. Peters, R. A.; Stocken, L. A.; Thompson, R. H. S. *Nature*, **1945**, *156*, 616-619.

99. Vallee, B. L.; Ulmer, D. D.; Wacker, W. E. C. *Archives of Industrial Health*, **1960**, *21*, 132-151.

100. Cruse, W. B. T.; James, M. N. G. *Acta Cryst.* **1972**, *B28*, 1325-1331.

101. Jocelyn, P. G. *Biochemistry of the SH Groups*, Academic Press: New York, 1972.

102. Brooks, R. R.; Lee, J.; Reeves, R. D.; Jaffre, T. *Journal of Geochemical Exploration*, **1977**, *7*, 49-57.

103. Ma, L. Q.; Komar, K. M.; Tu, C.; Zhang, W.; Cai. Y. *Nature*. **2001**, 411, 438.

104. Tu, C.; Ma. L. Q. *J. Environ. Qual.* **2002**, *31*, 641-647.

105. Komar, K. M. Phytoremediation of arsenic contaminated soils: plant identification and uptake enhancement. *M.S. thesis*, **1999**. University of Florida, Gainesville.

106. Chen, M.; Ma, L. Q.; Harris, W. G. *J. Environ. Qual.* **1999**, 28,1173-81.

107. Visoottiviseth, P.; Francesconi, K.; Sridokchan, W. *Environ. Poll.* **2002**, In Press.

108. Jones, D. L. *Encyclopedia of ferns: an introduction to ferns, their structure, biology, economic importance, cultivation and propagation.* Lothian Publishing Company: Melbourne, 1987.

109. Masscheleyn, P. H.; Delaune, R. D.; Patrick, W. H. *Environ. Sci. Technol.* **1991**, 25,1414-1419.

110. Zhang, W. H.; Cai, Y.; Tu, C.; Ma, L. Q. *The Sci. Total Environ.* **2002**, In press.

111. Ha, S. B.; Smith, A. P.; Howden, R.; Dietrich, W. M.; Bugg, S.; O'Conell, M. J.; Goldsbrough, P. B.; Cobbett, C. S. *The Plant Cell*, **1999**, *11*, 1153-1163.

112. Clemens, S.; Kim, E. J.; Neumann, D.; Schroeder, J. I. *The EMBO Journal*, **1999**, *18*, 3325-3333.

113. Vatamaniuk, O. K.; Mari, S.; Lu, Y. P.; Rea, P. A. *Proc. Natl. Acad. Sci. USA*, **1999**, *96*, 7110-7115.

114. Vatamaniuk, O. K.; Mari, S.; Lu, Y. P.; Rea, P. A. *The Journal of Biological Chemistry*, **2000**, *275*, 31451-31459.

115. van Hoof, N. A. L. M.; Hassinen, V. H.; Hakvoort, H. W. J.; Ballintijn, K. F.; Schat, H.; Verkleij, J. A. C.; Ernst, W. H. O.; Karenlampi, S. O.; Tervahauta, A. I. *Plant Physiol.* **2001**, *126*, 1519-1526.

116. Baker, A. J. M. *J. Plant Nutrition*, **1981**, *3*, 643-654.

Chapter 9

Uptake of Arsenic by Plants in Southwest England

Margaret E. Farago, Peter J. Kavanagh, Maria J. Leite,
Julie Mossom, Gill Sawbridge, and Iain Thornton

Environmental Geochemistry Research Group, Imperial College of Science,
Technology and Medicine, Royal School of Mines, London SW7 2BP,
United Kingdom

The literature concerning arsenic in soils and plants is
reviewed briefly. The uptake of arsenic by a number of plants
that colonize sites, contaminated by arsenic from past mining
and smelting activities, is described. Plants growing on soil
with high As were stunted and displayed a red coloration.
Armeria martima operates an exclusion mechanism, with low
root/soil concentration ratios. Uptake of arsenic from soil in
Calluna vulgaris and *Agrostis tenuis* show weak correlations
with Fe in soil and soil pH. The relative accumulation
(leaf/soil ratio) of As in *Calluna* at one site plotted against soil
concentration approximated to a hyberbolic curve.

Arsenic in the Environment

Arsenic in Soil

The sources of naturally occurring arsenic in the environment are primary minerals including arsenopyrite (FeAsS), orpiment (As_2S_3), realgar (As_4S_4) and secondary minerals of arsenic (*1,2*). The geochemistry of arsenic has recently been reviewed (*3-5*), including estimates of the arsenic global cycle (*5*).

Arsenic concentrations in uncontaminated soils have been reported in the range 0.1-40 µg/g worldwide (*6,7*) with an average of 5-6 µg/g (*8*) and 5-100 µg/g for the UK (*9*). High concentrations in soils, relative to these values, can occur in many areas; these arise from (*4*): naturally occurring arsenic-rich minerals and rocks, such as pyrite; waste materials resulting from mining and smelting activities; the burning of arsenic-rich coal; arsenic-containing agrochemicals (*10*), and organoarsenic-based warfare agents (*11*).

Depending on pH and Eh, arsenic is present in soils as As(III) or As(V) (*2*). Microbial activity within the soil gives rise to a number of reactions, including methylation, demethylation, oxidation and reduction (*2*). Under aerobic conditions arsenate(V) is precipitated as ferric arsenate in horizons rich in iron. Both As(III) and As(V) complex with hydrous oxides of Fe, Mn and Al (*12, 13*). The sorption of As by hydrous iron oxides occurs both in acid and in alkaline soils, whereas sorption by hydrous oxides of Al is reported to be important only in acidic soils (*14-16*). Under reducing sulfidic conditions, As(III) associates with metal sulfides, sulfidic minerals being formed under extreme reducing conditions. Clay minerals and organic ligands can also influence the species of arsenic present in soil.

Many authors have sought to discover the species of arsenic present in soil, the leachability and the potential bioavailability of arsenic to biota, including post-ingestion bioavailability to humans. Arsenic-bearing phases in soils and waste materials have been characterized using electron microprobe analysis (EMPA) (*17*) and extended X-ray absorption fine-structure spectroscopy (EXAFS) (*18*). Single (*19*) and sequential (*20-25*), extraction schemes have been developed leading to operationally determined associations between arsenic oxyanions and other soil constituents, many based on the similar chemistry of As and P (*26,27*). The introduction of chromatographic separation coupled with inductively coupled plasma – mass spectrometry (ICP-MS) has lead to the identification of arsenic species in soil extracts and soil solutions, in particular the determination of arsenates III and V (*28-30*) and organoarsenic species (*31*); arsenate(V) is usually found to be the major component. Bioavailability of arsenic in contaminated soil has been tested *in vivo* (*32-34*). An *in vitro* physiologically-based extraction test (PBET) (*35, 36*) is suggested to be a better approach to post-ingestion bioavailability.

Uptake of Arsenic by Plants

Arsenic is a constituent of most plants, although it has not been shown to be essential. "Normal" concentrations of arsenic in terrestrial plants and food crops, are reported (*6,7,37*) rarely to exceed 1 μg/g. Usually terrestrial plants preferentially take up arsenate(V) over arsenate(III) (*38*). Arsenic is translocated to most plant organs, although highest concentrations are often found in roots and old leaves (*38*). Arsenic can be phytotoxic, normally plants tolerate arsenic soil concentration in the range 1-50 μg/g, depending on the plant species, which differ significantly in their response (*39*). Toxicity depends on the chemical form, the bioavailabilty and the concentration of As in the soil (*40*). Other soil factors also influence uptake, which is decreased by high levels of clay, Fe, Ca, P, and N (*39*). Arsenate(III) is more toxic than arsenate(V) (*41*) and dimethylarsinic acid, DMAA, is reported to be more toxic than inorganic arsenic or monomethylarsonic acid, MMAA, to the wetland species *Spartina patens* (*42*). Arsenic toxicity has been shown to result from interference with mineral nutrition balance in *Pisum sativum* (*43*); after a 90 day exposure to sodium arsenate Mn concentration in cotyledons was significantly reduced while that of Zn was increased.

Some plants have been found to accumulate arsenic. Specialised communities are associated with arsenic-contaminated mining areas in southwest England and in an early study of the biogeochemistry of this area, (*44,45*) it was reported that these communities are dominated by *Agrostis* species: *A. capillaris, A. stolonifera and A,canina* which can accumulate As up to 3000 μg/g. The aquatic plant *Potomogeton illinoiensis* is reported to have 4.6 μg/g in shoots and leaves, representing a concentration factor of eight over As levels in the water (*46*). An arsenic hyperaccumulating fern has recently been reported to contain (*47,48*) up to 7,500 μg/g of As in the above ground biomass, a concentration factor of 200. Under laboratory conditions the concentration reached 2%. The arsenic chemical species in 1:1 methanol water extracts of the fern have been investigated by HPLC coupled with ICP-MS (*48*), only As(V) and As(III) and no methylated species were reported. The lichen, *Usnea barbata*, found at forest canopy level in Sri Lanka, is reported to accumulate arsenic from the atmosphere (*49*) with a concentration of 134 ng/g, six times greater than other species in the forest. High concentrations of arsenic have also been found in a number of mushrooms (*50-53*). Dandelion, *Taraxacum officinale*, growing around smelters in Bulgaria has concentrations of As up to 16 μg/g (*54*). The same plant has been used as a biomonitor of urban arsenic pollution in three European cities (*54*).

Uptake of arsenic by crop plants provides an entry into the food chain. A number of studies has been carried out in the arsenic-contaminated areas in southwest England (*55-58*). Soils and vegetables were surveyed in gardens, where As in soils was in the range 155-892 µg/g. Arsenic concentrations in edible tissue of lettuce, onion, beetroot, carrot, pea and bean (*55*), strawberries (*56*) increased significantly with both total and extractable As in soil, although only lettuce exceeded 1 µg/g dry weight. Herbage samples have also been investigated (*57-59*). Mixed herbage samples contained the grass species *Holcus lanatus, Lolium perenne* and *Festuca rubra*. On topsoil with arsenic in the range 334-461 µg/g, As in the herbage was in the range 0.62-0.97 µg/g. On more contaminated soils these values were 210-925 and 2.64-4.85 µg/g respectively (*57*). Another study (*58,59*) reported As in grass samples in the range 0.03-1.1 µg/g where soil content is in the range 19-320 µg/g.

The aim of the present work is to investigate the factors involved in the uptake of arsenic by plants growing on arsenic-contaminated sites with a long term aim of finding suitable species for phytostabilisation of bare soil and unstable waste tips.

Site Description and Methods

The southwestern peninsula of England consists of the counties of Cornwall to the west, and Devon with the River Tamar forming the boundary between the two counties. Centuries of mining and smelting activities have left a legacy of contaminated land, with As- and Cu-rich mine tailings and other wastes (59-65), Metal enrichment is dominated by Sn-Cu-As mineralization and from about 1860 to 1900, the region was the world's major producer of arsenic. In addition, extensive areas of land were contaminated with fallout from the smelting process (59).

Soil and plant samples were collected from the following sites (see Map):
Botallack Workings and *Levant Mine* - old tin and copper mining sites, mineralisation of tin and copper arises from mineral lodes associated with metamorphosed basalts and shales;
Poldice Mine - old tin, copper and arsenic mining/smelting site, mineralisation of copper, tin and arsenic associated with mineral lodes in the metamorphosed granites and slates:
Devon Great Consols (DGC) - mining/smelting site worked for copper, then arsenic. Upper Devonian and Lower Carboniferous rocks underlie most of this area and have been thermally metamorphosed by granitic intrusions resulting in a series of metalliferous lodes.

In addition soil arsenic levels at four uncontaminated control sites from SW England were investigated; Portreath (27-42 µg/g): Combe Martin (18-27 µg/g) (66); Wadebridge (26-67 µg/g) (66) and Cargreen (16-198 µg/g).

On the contaminated sites, there were clusters of vegetation together with bare patches, sampling was random within these clusters: representatives of dominant plant species together with soil from around the roots, were collected and analysed as described previously (55). Elemental concentrations in soil and plant samples were measured by inductively coupled plasma atomic emission spectrometry, ICP-AES. Arsenic was determined by ICP-AES with hydride emission. Data were assessed for accuracy and precision using a quality control program, including reagent blanks, field and analytical replicates, and certified reference materials (61).

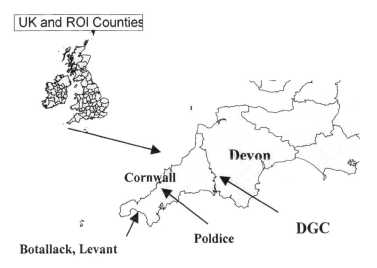

Map. Southwest England, showing contaminated sites.

Results

Armeria maritima was collected from Portreath, a coastal, uncontaminated site, the three contaminated sites: Botallack,, Levant and Poldice. On the contaminated sites, the plants were stunted had displayed a distinct red color.

Although a few *A. maritima* plants were found at the Devon Great Consols (DGC) site, these were very small and were not collected. The Poldice and DGC sites, where both mining and smelting had taken place were re-investigated. Although there are large patches of bare ground at both sites, some plants have colonized the sites. Similar colonization has been reported at an As-contaminated mine site in Australia (*67*). *Calluna vulgaris*, heather, was collected from both sites, and in addition, those collected from Poldice were: *Ulex europeus*, gorse, and *Rumex ulmifolus*, bramble; those from GGC were: the grasses *Holcus lanatus* and *Agrostis tenuis*. Results of the first set of plant and soil analyses are shown in Table I and the second and third sets in Table II.

Discussion

Armeria maritima has long been recognised as an indicator of mineralization (*68-70*). The plant is copper-tolerant and roots of Cu-tolerant plants contain high concentrations of the amino acid proline, whereas this is not so for non-tolerant ecotypes (*71*). In the areas investigated, *Armeria* grows where the contamination from both As (Table I) and Cu are high. The mean Cu concentrations found were: Botallack, 2872; Levant, 1936, Poldice, 1609; DGC, 1902, μg/g respectively. Copper uptake by the plants will be discussed in a future publication. For *Armeria* on these sites, the mean concentration of arsenic in the plants increases significantly with that in soil. Table I shows that there are higher concentrations of arsenic in *Armeria* roots than in leaves for all sites in agreement with the literature (*11,43*). Concentration ratios, or concentration factors are also included in Table I, these indicate that the plant adopts a strategy of exclusion. Arsenate is reported to change the distribution patterns between elements in the pea plant (*43*). Figure 1 shows the concentration ratios for leaf/soil and leaf/root for *Armeria maritima* for the 4 sites of increasing arsenic concentration, for Cu, Zn and Mo. There is a depression of the relative uptake of Cu from soil to leaves with increasing As in soil, with this partly resulting from more As remaining in the roots, relatively less being transported to the leaves. Zinc and Mo move more freely throughout the plant. These points require further investigation, however, cell membranes are reported to play an important role in the uptake of toxic elements, in particular arsenic (*72*), changes in cell membrane structure may be reflected in their selectivity, resulting in altered mineral element balance (*43*).

Calluna vulgaris, heather, colonizes both the contaminated sites at Poldice and DGC. As with *Armeria maritima*, the plants are stunted and display a red coloration. Arsenic concentrations in soils and plant tops are shown in Table II. The maximum concentrations were 317 and 360 μg/g at the two sites. There appears to be no correlation between the concentrations of arsenic in the plant tissues in those in soil on either site. There were, however weak positive

Table I. Arsenic Concentrations (μg/g dry weight) in Soils and *Armeria maritima* (A.m.) from Sites in Southwest England

		Portreath	Botallack	Levant	Poldice
Soil		27-42 (3)	329 (10)	1701 (8)	6713 (33)
A.m. leaves	mean	BDL (3)	2.2 (8)	9.2 (6)	30.5 (20)
	range		1.3 - 3.4	0.7 – 33.3	4.1 – 6.4
A.m. roots	mean	2.86(3)	12.4 (8)	49 (6)	73.5 (20)
	range	2.3 – 3.5	7.1 – 26.4	6.6 - 171	11-454
Ratios					
Leaf/soil		-	0.16	0.004	0.01
Root/soil		0.06	0.08	0.03	0.02
Leaf/root		-	0.19	0.17	0.28

Numbers of samples in parentheses

BDL = Below detection limit

Leaves are green leaves, senescent leaves contain more arsenic

Table II. Arsenic Concentrations (μg/g dry weight) in Soils and *Calluna vulgaris*, *Ulex europeus*, *Rumex ulmifolus*, *Holcus lanatus* and *Agrostis tenuis* from the contaminated sites Poldice and Devon Great Consols

Poldice	Sample	Soil	Calluna	Rubus	Ulex
	n	42	25	10	10
	Range	432-3700	5.95-317	1.46=16	1.5-15.5
	Mean	6181	49.5	5.7	5.9
DGC	Sample	Soil	Calluna	Agrostis	Holcus
	n	25	21	20	3
	Range	186-47900	10.6-360	3.35-203	2.95-85
	Mean	19900	73.5	39	32

correlations with Fe is soil ($R^2 = 0.22$) at DGC, and with pH ($R^2 = 0.27$) at Poldice. When the relative uptake, or plant/soil concentration ratio, is plotted versus arsenic concentrations in the soil, a curve approximating to hyberbolic is obtained (Figure 2). The same ratio platted against Fe in soil gives a similar curve. This type of curve has been found before for grass species growing on As-contaminated land (56). For the grass species, *Agrostis tenuis*, growing at the Devon Great Consols site there is a weak positive correlation between As concentration in the plant tops with pH of the soil and a stronger correlation with Fe in soil ($R^2 = 0.50$).

We have shown (*22,57*) that at the Devon Great Consols site As is strongly correlated with Cu, Fe, Sb and Bi, reflecting the mineralization. From sequential extractions (Table III) the As is predominately bound to the Fe and organic fractions (Table IV). It can also ben seen that the percentage of the total that is soluble in water is very small, however because the total concentrations are so high, the concentrations in water can be appreciable. Arsenic in the water extracts was shown to be in the form of arsenate(V). Extraction with 0.1 M HCl for the four soils in Table IV, showed percentages of arsenic extracted as 13.2, 3.7, 3.1 and 8.3 respectively. The implications for inhaled or ingested contaminated dusts are that although the percentage bioavailabilty may be low, the absolute amount taken in by fauna and humans may be appreciable. It has been shown that juvenile wood mice and bank voles on the Devon Great Consols site accumulated arsenic in liver and kidneys (*74*). Arsenic levels in urine from residents from DGC are elevated (*62*) most likely from the inhalation and ingestion of soils and dusts (*60*).

Arsenic mobility and hence its uptake by, and toxicity to plants is strongly influenced by Fe and Al oxides (*14-16,23,73*). Our limited results (*22*) indicate that generally As in DGC soil is associated less wth Al, thus Fe is perhaps the most important factor in plant uptake. The pH of the soils in Poldice and GGC is acidic, in the range 4-7. Increases in pH by application of lime have been shown to increase arsenic solubility (*14,16*) although this was not reflected in increased uptake by plants, the uptake actually decreasing, suggesting that the solubilized arsenic was not plant available (*16*) Soil texture may also influence arsenic availability in soil (*23*). The soils on the contaminated sites are often very coarse and thus might be expected to display high As mobility. Further work is being carried out on the biogeochemistry of these interesting contaminated sites.

Acknowledgements

PK and JS thank the Social and Economic Research Council and the Science Communication Foundation, respectively, for Studentships. The authors thank the West Devon Borough Council for their logistical support.

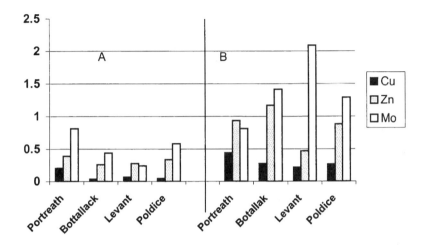

Figure 1. Concentration ratios: A, leaf/soil; B. leaf/root; for Armeria maritima
from Portreath (uncontaminated) and four contaminated sites from SW England

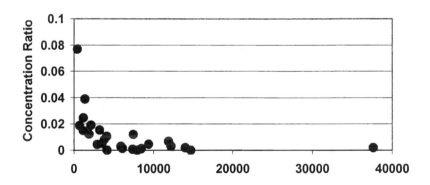

Figure 2. Concentration (plant/soil) ratios for samples of Calluna vulgaris
against arsenic concentrations in soil (x-axis, As concentrations in soil μg/g).

Table III. Summary of Sequential Extraction Scheme (22)

Operation			Operationally determined association
1. Shake with DIW	→	Supernatent	→ Water soluble
2. Residue - 0.5 M NH$_4$F	→	Supernatent	→ Al-associated
3. Residue - 0.1M NaOH	→	Supernatent	→ Fe/organic- associated
4. Residue - 0.25M H$_2$SO$_4$	→	Supernatent	→ Ca-associated
5. Residue - 0.5 M NH$_4$F	→	Supernatent	→ Al-occluded
6. Residue			→ Residual

Table IV. Sequential Analysis for Contaminated Soils from Devon Great Consols: Percentages Extracted for Steps shown in Table III.

Total As	Step 1	Step 2	Step 3	Step 4	Step 5	Step 6
4120	1.2	1.95	68.5	7.2	0.04	21
18655	0.02	0.30	30.5	38	0.06	31
22290	0.02	0.57	66.4	6.1	BDL	27
1200	11.6	17.5	59.9	0.7	0.04	30

Units for total As: μg/g

BDL, below detection limit

References

1. Zemann, J.; Wedepohl, K.H. In *Handbook of Geochemistry*; K.H. Wedepohl Ed.; Springer Verlag, Berlin, 1978, pp 82-90.
2. O'Neill, P. In *Heavy Metals in Soils;* B.J. Alloway Ed.; Blackie, London, 1990, pp 83-99.
3. Thornton, I. In *Environmental Geochemistry and Health*. Fuge, R.; McCall, R. Eds.; Geological Society Publication No 113, London. 1996, pp 153-161.
4. Thornton, I.; Farago, M.E., In *Arsenic; Exposure and Health Effects;* Abernathy, C.O.; Calderon, R.L.; Chappell, W.R. Eds.; Chapman and Hall, London, 1997, pp.1-16.
5. Matschullat, J. *Sci. Tot. Environ.* **2000**, *249*, 297-312.
6. Kabata-Pendias, A.; Pendias, H. *Trace elements in soils and plants*; CRC Press, Boca Raton, FL, 1984, p. 315.
7. Bowen, H.J.M. *Environmental Chemistry of the Elements*; Academic Press, London. 1979.
8. Adriano, D.C. *Trace Elements in the Terrestrial Environment*; Springer Verlag, NY. 1986.
9. Thornton, I.; Plant, J. *J. Geol. Soc. (London)* **1979**, *137*, 575-586.
10. .Murphy, E.A.; Aucott. M. *Sci. Tot. Environ.* **1998**, *218*, 89-101.
11. Pitten, F.-A.; Müller, G.; König, P.; Schmidt, D.; Thurow, K.; Kramer, A. *Sci. Tot Environ.* **1999**, *226*, 237-245.
12. Manceau, A. *Geochim. Cosmochim Acta* **1995**, *59*, 2647-3653.
13. Manning, B; Fendorf, S.E.; Goldberg, S. Environ. Sci. Technol. **1998**, *32*, 2383-2388.
14. Sadiq, M. *Wat. Air Soil. Pollut.* **1997**, *93*, 117-136.
15. Jones, C.A,; Inskeep, W.P.; Neuman, D.R. *J. Environ. Qual.* **1997**, *26*, 433-439.
16. Tyler, G.; Olsson, T. *Plant Soil* **2001**, *230*, 307-321.
17. Davis, A.; Ruby, M.V.; Bloom, M.; Schoof, R.; Freeman, G.; Bergstrom, P.D. *Environ. Sci. Technol.* **1996**, *30*, 392-399.
18. Savage, S.S.; Tingle. T.N.; O'Day, P.A,; Waychunas, A.; Bird, D.K. *App. Geochem.* **2000**, *15*, 1219-1244.
19. Leoppert, R.H. *Pre-prints of Extended Abstracts*, Division of Environmental Chemistry, Am. Chem, Soc. **2001**, *41* No. 1, 472-477.
20. Woolson, E.A.; Axley, J.H.; Kearney, P.C. *Soil Sci. Soc. Am. Proc.* **1973**, *37*, 254-259.
21. Huerta-Diaz, M.A.; Morse, J.W. *Mar. Chem*, **1990**, *29*, 119-114.
22. Kavanagh, P.J. Farago, M.E.; Thornton, I.; Braman, R.S. *Chem. Speciat. Bioavail.* **1997**, *9* 77-81.
23. Lombi.E.; Sletten, R.S.; Wenzell, W.W. *Water Air Soil Pollut*, **2000**, *124*, 319-332.

24. Keon N.E.; Swartz, C.H.; Brabender, D.J.; Harvey,C.; Hemond, H.F. *Environ. Sci. Technol.* **2001**, *35*, 2778-2784.
25. Wenzell, W.W; Kirchbaumer, N.; Prohaska, T.; Stingeder, G.; Lombi.E.; Adriano, D.C., *Anal, Chim. Acta* **2001**, *436*, 309-323.
26. Chang, S.C.; Jackson, M.J., *Soil Sci*, **1957**, *83*, 133.
27. Petersen, G,W,; Corey, R.B. *Soil Sci. Soc. Am. Proc.* **1966**, *30*, 563-565.
28. Onken, B.M.; Hossner, L.R. *Soil Sci. Soc. Am. Proc.* **1996**, *60*, 1385-1392.
29. Manning, B.A.; Martens, D.A. *Environ. Sci. Technol.* **1997**, *31*, 171-177.
30. Manning, B.A.; Goldberg, S. *Soil Sci.* **1997**, *162*, 886-895.
31. Pongratz, R. *Sci. Tot. Environ.* **1998**, *224*, 133-141.
32. Freeman , G.B.; Johnson, J.D.; Killinger, K.M.; Laio, S.C.; Davis, A.O.; Ruby, M.V.; Chaney, R.L.; Lovre, S.C.; Bergstrom, P.D. *Fund. Appl. Toxicol.* **1993**, *21*, 83-88.
33. Freeman , G.B.; Schoof, R.A.; Ruby, M.V,; Davis, A.O.; Dill. J.; Laio, S.C.;Lapin, C.A.; Bergstrom, P.D. *Fund. Appl. Toxicol.* **1996**, *28*, 215-222.
34. Groen, K.; Vaessen, K.;Kliest, J.J.G.; deBoer, J.L.M.; Ooik, T.V.; Timmerman, A.; Vlug, R.F. *Environ. Health Perspect.* **1994**, *102*, 182-184.
35. Ruby, M.V,; Davis, A.O.; Schoof, R.A.; Ebberly, S.: Sellstone, C.M. *Environ. Sci. Technol.* **1966**, *30*, 422-430.
36. Williams, T.M.; Rawlins, B.G.; Smith, B.; Breward, N. *Environ. Geochem. Health*, **1998**, *20*. 169-177.
37. Fergusson, J.E. The Heavy Elements; Chemistry, Environmental Impact and Health Effects; Pergamon Press Oxford 1990, p 614.
38. Tanaki, S.; Frankenberger, W.T. Rev. Environ. Contam. Toxicol. **1992**, 124, 79-110.
39. Jiang, Q.Q.; Singh, B.R. Water Air Soil Pollut. **1994**,74, 321-343.
40. Woolson, E.A. Weed Sci. **1973**, 21, 524-527.
41. Peterson. P.J.; Benson, L.M.; Zeive, R. In Effect of Heavy Metal Pollution on Plants; Lepp, M.W. Ed. 1981, Applied Science Publishers, London, Vol. 1, pp. 299-322.
42. Carbonelli-Barrachina, A.A.; Arabi, M.A.; DeLaune, R.D.; Gambrell, R.P.; Patrick, W.H.Jr. J. Environ. Sci. Health Pt A **1998**, 33, 45-66.
43. Päivöke, A.E.A; Simola,L.K. Ecotoxicol. Environ. Safety Sect, B **2001**, 49, 111-121.
44. Porter, E.K.; Peterson. P.J. Sci.Tot.Environ. **1975**. 4, 365-371.
45. Porter, E.K.; Peterson, P.J. Environ.Pollut. **1977**, 14, 255-265.
46. Compton, A-M. Froust, R.D. Jr.; Salt, D.E.; Ketterer, M.E. Pre-prints of Extended Abstracts, Division of Environmental Chemistry, Am. Chem, Soc. **2001**, 41 No.1, 478-480.
47. Ma, L.; Tu, C.; Kennelly. B.; Komar, K. In *11th International Conference on Heavy Metals in the Environment.* Nriagu, J. Ed. Contribution 1038. University of Michigan School of Public Health. Ann Arbor MI (CD-ROM).

48. Zhang W.; Cai, Y.; Tu, C.; Ma, L.Q. *Pre-prints of Extended Abstracts*, Division of Environmental Chemistry, Am. Chem. Soc. **2001**, *41* No.1, 478-480.
49. Jayasekera, R.; Rossbach, M. *Environ. Geochem. Health* **1996**, *18*, 55-62.
50. Byrne, A.R.; Tušek-Znidarik, M. *Chemosphere* **1983**, *12* 1113-1117.
51. Stijve, T, Bourgui, B. *Deutsch. Lebensm. Rundsch.* **1991**, *10*, 307-310.
52. Vetter, J. *Acta Aliment.* **1990**, *19*. 27-40.
53. Slekovec, M.; Irgolic,KJ. *Chem. Speciat. Bioavail.* **1996.**, *8*, 57-73.
54. Djingova, T.; Kuleff, I. In *Plants as Biomonitors, Indicatirs for Heavy metals in the Terrestrial Environment.* Markert, B. Ed.; 1993, VCH Weinheim pp 435-460.
55. Xu, J.; Thornton, I. *Environ. Geochem. Health* **1985**, *7*, 131-133.
56. Thorsby, P.: Thornton. I. *Trace Subst. Environ. Health* **1979**, *XII*, 93-103.
57. Li, X.; Thornton. I. *Environ. Geochem. Health* **1993**. *15*, 135-144.
58. Abrahams, P.W.; Thornton. I. *Agric. Ecosystems Environ.* **1994**, *48*, 125-137.
59. Abrahams, P.: Thornton, I., *Trans. Inst. Mining and Metall. (B: App. Earth Sci,)* **1987**, *6*, B1-B8.
60. Farago, M.E.; Thornton, I.; Kavanagh, P. J.; Elliott, P.; Leonardi, G. in *Arsenic; Exposure and Health Effects* Abernathy, C.O.; Calderon, R.L.; Chappell, W.R. Eds.; 1997, Chapman and Hall, London, pp. 191-209.
61. Ramsey, M.H.: Thompson, N,: Banerjee, E.K.; *Anal. Proc.* **1987**, 75-100.
62. Kavanagh, P.; Farago, M.E.,;Thornton, I.; Goessler, W.; Kuehnelt, D.; Schlagenhaufen, C.; Irgolic, K.J. *Analyst*, **1998**, *123*. 27-30.
63. Mitchell, P.; Barr D. Environ.l Geochem. Health, **1995**,*17*, 57-82.
64. Thornton, I. In *Arsenic; Exposure and Health Effects*, Chappell, W.R.; Abernathy, C.O. Cothern, C.R. Eds.; 1995, Science and Technology Letters, Northwood, pp. 61-70.
65. Hamilton. E.I. *Sci.Tot Env.* **2000**, *249*, 171-221.
66. Elghali, L. M.Sc. Thesis, Imperial College, University of London, London, UK, 1994.
67. Ashley, P.M.; Lottermoser, B.G. *Aust. J. Earth Sci*, **1999**, *46*, 861-874.
68. Henwood, W.J. *Edinb. New Phil. J.* **1875**, *5*, 61.
69. Farago, M.E.; Mullen, W.A.; Cole, M.M.; Smith, R.F. *Environ. Pollut. Ser.A*, **1980**, *21*,225-244.
70. Farago, M.E.; Mehra, A. In *Metal Compounds in Environment and Life, 4*, Merian E.; Haerdi, W. Eds.; 1992, Science and Technology Letters, Northwood, pp. 161-169.
71. Farago, M.E.; Mullen, W.A. *Inorg. Chim Acta Lett*, **1979**, *32*, 193-195.
72. Meharg, A.A. *Physiol Plant* **1993**, *88*.191-198.
73. Otte, M.L.; Deckers. M.J.; Rozema. J.; Broekman, R.A. *Can. J. Bot.* **1988**, *69*, 2670-2677.
74. Erry, B.V.; Macnair, M.R.; Meharg, A.A.; Shore, R.F. Environ. Pollut. **2000**, *110*, 179-187.

Chapter 10

Volatilization of Metals from a Landfill Site

Generation and Immobilization of Volatile Species of Tin, Antimony, Bismuth, Mercury, Arsenic, and Tellurium on a Municipal Waste Deposit in Delta, British Columbia

Jörg Feldmann

Department of Chemistry, University of Aberdeen, Meston Walk, Old Aberdeen, AB24 3UE Scotland, United Kingdom

In a landfill environment, many mainly main group elements, form volatile organometallic compounds (VMCs). This chapter focuses on the speciation of volatile metal compounds generated in a municipal waste deposit in Delta British Columbia, Canada. Mostly permethylated compounds of Sn, Bi, As, Sb, Se, Hg, and Te have been identified. When landfill gases percolated through the water column of a wetland, water-soluble polar organometallic compounds were identified in the water column as the degradation products of the volatile metal compounds. However, this effect is only marginal since the concentration of the volatile metal compounds above the wetland is not significantly different from the landfill gases in the gas wells in which landfill gas is collected. Stability tests on 12 different volatile species of arsenic, tin and antimony suggests that these compounds have an atmospheric half life, which allows them to volatilize from landfill sites and disperse in their vicinities.

In the biogeochemical cycle of heavy metals solubilization, sequestering and mobilization are key issues. Volatilization of metals is of importance when high temperature processes are considered. For example, volcanic exhalations or industrial processes such as waste incineration are significant sources of heavy elements. The most prominent element in this category is mercury as Hg^o. Information about the volatilization process of heavy elements at ambient temperature i.e. 0-40°C is scarce since only a limited number of elements can form rather stable compounds which have a significant vapor pressure in this temperature range.

A list of compounds with their boiling points is shown in Table I. Some of the listed volatile element species have been identified, and even quantified, in field studies to occur in the environment while others have been identified as volatile metabolites from pure bacteria and fungi cultures in laboratory experiments. The list is by far not complete.

Biotransformation of TBT to DBT, MBT and eventually to inorganic tin is well accepted. However, only a few reports describe the occurrence of methylated butyltin compounds which point to biomethylation of the anthropogenic butylated tin compounds (1). Donard and coworkers (2) have reported the generation of methylated butyltins in polluted sediments in South-West France. Although sediments are very good sinks for TBT, MeBu$_3$Sn has a significantly lower affinity to this phase (1). It could easily be released to the water column and consequently due to the high lipophilicity released to the atmosphere. Other studies have determined the generation of Me$_2$Hg in the mudflats of the heavy polluted River Elbe in northern Germany (3). Many volatile iodine compounds are known to be generated from marine macroalgae. Laturnus et al. (4) have shown that not only methyl-iodine but also ethyl-iodine, in addition to CH$_2$Cl-I, CH$_2$I$_2$, could be detected in the environment. Hansen et al. (5) reported on a constructed wetland which was purpose-built to clean-up refinery process water containing 20-30 µg L^{-1} selenium. They have found that 10-30 % of the selenium was volatilized presumably as Me$_2$Se. The maximum flux was measured to be 190 µg of Se m^{-2} day^{-1}. We have found that arsenic and antimony have been generated by algal mats of geothermal waters in southern British Columbia (6). The water contained up to 300 µg L^{-1} arsenic and only 2 µgL^{-1} antimony, but a direct sampling of the algal mats using a flux chamber demonstrated that AsH$_3$, MeAsH$_2$, Me$_3$As and Me$_3$Sb were also generated from the algal mats. The number of field studies is however very limited and often restricted to special environments which are mostly anaerobic. One reason for this is that the volatile compounds are not persistent so they do not show significant concentrations in the atmosphere. In particular landfill gas contain many different VMCs.

The aim of this paper is to show how volatile metal compounds, generated in a municipal waste deposit, react when they reach the landfill surface. Is a landfill site a significant source of volatile metal compounds or are these compounds too reactive to be volatilized into the atmosphere?

Table I: List of volatile compounds with their boiling points identified in the environment or in microorganism cultures in the laboratory

Element	Species	bp (°C)	Source (ind/bio)	Field/Laboratory	Reference
As	AsH_3	-55	Industry	Reduced (electrolysis)	7
			LG, SG, HS	Anaerobic (F/L)	6, 8
	$MeAsH_2$	2	SG, HS	Anaerobic (F/L)	6
	Me_2AsH	36	LG, SG, HS	Anaerobic (F/L)	6, 8
	Me_3As	50	LG, SG, HS	Anaerobic, (F/L)	8
				Aerobic (L)	9
Sb	SbH_3	-17	LG, SG	Anaerobic (F)	8
	Me_3Sb	81	LG, SG, HS	Anaerobic (F,L)	10,11
				Aerobic (L)	12
Bi	Me_3Bi	109	LG, SG	Anaerobic (F,L)	13
Sn	Me_2SnH_2	35		Aerobic (L)	14
	Me_4Sn	78	LG, SG	Anaerobic	10, 11,15
	Bu_3SnMe		Sediment	Anaerobic (F,L)	1, 2
Te	Me_2Te	83	LG, SG	Anerobic (F,L)	8, 11
Se	Me_2Se	56	LG, SG	Anaerobic (F,L)	8, 11
			WL	Aerobic (F,L)	5
Hg	Hg^o			Aerobic (F,L)	3
	Me_2Hg	91	LG, SG	Anaerobic (F)	8, 3
Pb	Me_4Pb	100	LG,SG	Anaerobic (F)	16
	Me_3PbEt		SG	Anaerobic (F)	16
	Me_2PbEt_2		SG	Anaerobic (F)	16
	$MePbEt_3$		SG, LG	Anaerobic (F)	16
	$PbEt_4$	200	SG, LG	Anaerobic (F)	16
Ni	$Ni(CO)_4$		SG	Anaerobic (F)	17
Mo	$Mo(CO)_6$		LG, SG	Anaerobic (F)	18
W	$W(CO)_6$		LG, SG	Anaerobic (F)	18

F field identification, L laboratory culture experiment, HS algal mats in hot springs, SG sewage gas, LG landfill gas, WL wetlands

Experimental Section

This report will give some data on the process of volatilization on one particular municipal waste deposit in British Columbia (Delta, British Columbia). This site was still in operation in 1997, but older parts were already overgrown and due to the strong solidification through truck driving and the high annual precipitation level of the west coast of Canada, wetlands have been formed. Mainly municipal and similar industrial waste from Greater Vancouver was dumped here. Even when gas is pumped out of the landfill, some gas can easily reach the surface of the landfill after migration through the waste and soil. The gas that migrated through the landfill was visibly bubbling through the water of the wetland. A rich vegetation of typical wetland plants were growing in the water with a depth was about 30-50 cm.

Sampling and location

Gas samples

Gas samples were taken directly from the gas wells in which the landfill gas is collected and transported to the furnace. The gases in the gas wells were sampled directly into Tedlar bags by using a membrane pump (AirPro 6000D, Bios Instrument Corp., NJ). In addition gases which bubbled through the water of an overlying wetland on the landfill site was sampled by using a flux chamber, which was floating on the water surface. Basically the gas was collected under this device and pumped directly into the Tedlar bags. The flow rate was adjusted to 1 L min^{-1}. Five samples each (gas well and from different gas spurts on the wetland) were taken. The Tedlar bags were transported in black bags to the laboratory in order to prevent UV radiation destabilizing the VMCs.

Water samples

The water of the wetland was also sampled into air-tight glass containers and subject to hydride generation methodology in order to identify degradation products of the volatile organometallic compounds, which percolated through the water. The samples (10 mL) were purged with helium (133 mL min^{-1}) and cryotrapped (6 min) prior to the addition of 1 mL 6 % aqueous NaBH$_4$ solution after acidification with 1 mL 1 M HCl. In order to distinguish between trivalent and pentavalent arsenic compound a second run used instead HCl an ammonium acetate buffer. The volatile metal compounds generated were also cryotrapped and analyzed in the same way as the gas samples. No attempt was made to quantify how much of the VMCs are immobilized since it is not known how much gas has been percolated through this water. The emphasis was on the identification of methylated metal species in the water and their ratios. The intensities can directly be compared since the response of the ICP-MS is very similar for the individual species of one element.

Stability tests

In order to test how stable the volatile metal compounds (VMCs) are in an oxic atmosphere, 12 different compounds were chosen to test in air at temperatures 20 and 50°C in the dark (5 replicates were made). The experimental set-up is described more thoroughly elsewhere (19). Briefly, SnH_4, Me_2SnH_2, Me_3SnH, $BuSnH_3$, SbH_3, $MeSbH_2$, Me_2SbH, Me_3Sb, AsH_3, $MeAsH_2$, Me_2AsH, Me_3As were generated using the appropriate involatile metal solutions and employing hydride generation methodology to generate the hydrides. The hydrides were purged by synthetic air into a 4 L Tedlar bag equipped with a septum. In addition the same set-up was done to test the influence of UV light. An ordinary UV lamp was placed so that the gas samples were exposed to approximately 5000 Lux and the temperature of 30°C. 50 mL were sampled from the gas standards and injected directly onto the U-trap, prior to GC-ICP-MS analyses. The gas standards stored in the dark were sampled first on a daily basis for the first week and then on a logarithmic time scale up to 6 months, while the standards exposed to UV light were sampled on an hourly basis. Most of the VMCs follow a first or pseudo second order kinetic, but the study of the kinetics will be published in the near future (20).

Analytical Procedure

The gas samples were analyzed within 24 hours by cryogenic trapping at −80°C using a packed column system filled with non-polar, chromatographic material (SP-2100 on Supelcoport). The trapped gases were deliberated and defocused on a second cryotrap with the same size and material as the previous one (length 22 cm, 6 mm diameter). The packed column was connected online to an ICP-MS (Plasmaquad PQ2+). Helium was used as a carrier gas (133 mL min^{-1}) and mixed with 0.9 L min^{-1} nebulizer gas (Ar). This is thoroughly described elsewhere (18). At this point the author takes the opportunity to thank Prof. W. R. Cullen for his permission to conduct this part of the study in his laboratory at UBC. A semi-quantitative method (21) was used to quantify the volatile metal compounds in the gases. The intensities measured were normalized to the internal standard (10 ng L^{-1} Rh solution nebulized and mixed with the GC gas stream before entering the injector of the plasma torch of the ICP-MS).

The analytical method used for the stability tests was slightly altered; instead of a packed column GC as described before, a capillary GC (CP-Sil 5CB Ultimetal capillary, 25 m x 0.53 mm, i.d., d=1 μm) was used along with a conventional GC (GC95 Ai Cambridge Ltd) and coupled to an ICP-MS (Spectromass 2000), which is also described in more detail elsewhere (19).

Results and Discussion

The landfill gases which percolate through the wetland water contain numerous volatile metal species as shown in Table II. For mercury, Hg^o was the main compound besides minor amounts of Me_2Hg. Me_3As and Me_3Sb dominate the element spectrum for arsenic and antimony, whereas for tin more than five different species have been identified with Me_4Sn being the main tin species. Other studies in which capillary GC was used more than eight different tin species have been recorded (22). The nature of the other tin species are still not very clear. Although we have identified the compound labeled U2 as Me_2SnEt_2 (15), there are some indications for the occurrence of methylated butyltin species in landfill gases (Me_2Bu_2Sn, $MeBu_3Sn$) (23). Apart from tin an ethylated arsenic compound Me_2AsEt was also identified. This is not unexpected since earlier results from a German landfill site have shown the occurrence of a late eluting peak with the an estimated boiling point of 87°C, the very same as the Me_2AsEt (10).

Further, Me_2Te, Me_2Se and Me_3Bi have been recorded in addition to industrial Et_4Pb and small amounts of $MeEt_3Pb$. Relatively pure Et_4Pb was typically used in North America, but it cannot be excluded that small amounts of mixed ethyl-methyl lead have occurred in gasoline, which was eventually dumped on the landfill. In contrast, we have shown previously (8,16), that the main volatile lead compound in sewage gas from a sewage sludge fermentor in Germany was $MeEt_3Pb$. Although this points to the biomethylation of organolead by methanogenic compounds in a fermentor (tetraethyl lead gets de-ethylated and then eventually methylated), it is not unequivocal that a methylation of organolead takes place in this landfill site. All compounds have been identified by standard-addition with appropriate VMC standards. In addition Me_4Sn, Me_3Sb and Me_3Bi have been identified by the 2 dimensional GC-EI-MS (15).

In addition to the gas samples, water samples were analyzed for the water-soluble degradation products of the VMCs. Hydride generation can volatilize most of the ionic organometallic species. Since the metal carbon bond is, for most compounds, robust enough to preserve the nature of the organometallic compound after hydride generation. The soluble metal species, which were

Table II: Relative abundance of volatile metal species in landfill gas percolating through the water and their water-soluble degradation products in the water phase of the wetland.

Element	Gas sample VMCs	% species	Water sample before HG[a]	% species	Reaction
As	AsH_3	4.1	As(III/V)	48.3	Oxidation and
	$MeAsH_2$	15.8	$MeAsO(OH)_2$[b]	1.1	limited
	Me_2AsH	6.4	$Me_2AsO(OH)$[b]	0.5	demethylation
	$Me_3As(III)$	67.8	Me_3AsO[b]	50.1	
	Me_2AsEt	5.9			
Sb	$Me_3Sb(III)$	>98	Sb(III/V)	3.1	Oxidation and
	Sb (U1)	2	$MeSbO(OH)_2$[b]	14.6	demethylation
			$Me_2SbO(OH)$[b]	40.8	
			Me_3SbO[b]	41.5	
Sn	Me_4Sn	87.8	Sn^{4+}	16.4	
	Sn (U1)	4.7	$MeSn^{3+}$	1.6	Demethylation
	Sn (U2)	2.8	Me_2Sn^{2+}	5.9	
	Sn (U3)	3.0	Me_3Sn^+	61.7	Debutylation ?
	Sn (U4)	1.7	Me_4Sn	3.8	
			$BuSn^{3+}$	8.1	
			Bu_2Sn^{2+}	2.5	
Bi	Me_3Bi	100	Bi^{3+}	97.1	Demethylation
			$MeBi^{2+}$	2.9	
Hg	Hg^o	95.7	Hg^{2+}	39.8	Demethylation
	Me_2Hg	4.3	Hg (U1)	12.9	and Oxidation
			$MeHg^+$	39.8	
Se	Me_2Se	100	Me_2Se	100	
Te	Me_2Te	100	Me_2Te	100	
Pb	Et_3PbMe	3.0	N.D.		
	Et_4Pb	97	N.D.		
Mo	$Mo(CO)_6$	100	N.D.		
W	$W(CO)_6$	100	N.D.		

Sn (U1-U4): all R_4Sn, Sb (U1): unknown, N.D. not determined, [a]HG hydride generation at pH 1, [b] species show up as their hydrides so that the acid and oxide forms are only assumed.

volatilized using the above-mentioned method and their relative amounts, are compared to the species identified in the gas samples (Table II) [1]. Figure 1 shows two examples for the element-specific detection (As and Sb) of volatile species in landfill gas and species, which were generated by hydride generation of the water samples.

It can be seen that the volatile species are not the species dissolved in the water; thus transformation processes alter the species. In particular, oxidation and demethylation takes place when the volatile metal compounds percolate through the water column of the wetland. Interestingly, the different elements follow different reaction schemes: Me_3As tends to be oxidized to the pentavalent Me_3AsO which is soluble in water, while Me_4Sn is demethylated to Me_3Sn^+. The large amount of inorganic arsenic can be explained by the occurrence of large amount of arsine, which is not sufficiently trapped with the used method, so that the contribution of AsH_3 is underestimated. Me_3Sb is oxidized from trivalent to pentavalent (Me_3SbO) but also to a much higher degree as arsenic demethylated. Thus, substantially more pentavalent dimethylated and monomethylated antimony compounds (e.g., $Me_2SbO(OH$, $MeSbO(OH)_2$) occur in the water phase. Me_4Sn and other tetraalkyltin compounds are immobilized by demethylation or dealkylation. The Me_3Sn^+ is more polar and therefore more water-soluble. Me_3Bi is completely demethylated to mostly inorganic bismuth. However, it should be noted that the hydride generations have not been sufficiently tested to conclude that only a minority of $MeBi^{2+}$ species had been in the water, since it is not known how ionic $MeBi^{2+}$ species react under the harsh condition for hydride generation.

Mercury is immobilized by the demethylation of Me_2Hg to $MeHg^+$ and by oxidation of Hg^o to Hg^{2+}. For selenium and tellurium compounds, the same compounds were detected in the gas sample as in the water after hydride generation. However, due to lack of knowledge about the behavior of these organometallic compounds using hydride generation, it can only be suggested that tellurium and selenium are maybe oxidized to their tetravalent forms. In general, oxidation can immobilize those metals, which can occur in the higher oxidation state, so that no M-C bonds have to be cleaved in order to form more polar organometallic compounds, while elements such as Sn or Hg have to be demethylated before a polar water-soluble compound can be formed.

It is not known which proportion of VMCs are indeed immobilized by the reaction in the water phase. However a comparison of the gas, which percolated through the water-column, can be compared to the gas transported in the gas

---

[1] The water was also subject to hydride generation at neutral pH. Approximately 40% of the inorganic arsenic occurs as As(III), whereas most of the methylated species occur in their pentavalent forms as $MeAsO(OH)_2$ and $Me_2AsO(OH)$.

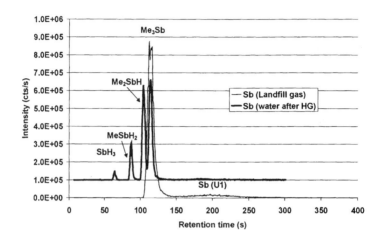

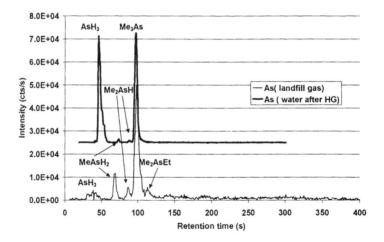

Figure 1: GC-ICP-MS chromatograms of arsenic and antimony compounds in landfill gas and in the water after hydride generation.

wells (see Figure 2). It can be seen that most compounds are in the same order of magnitude so that the immobilzation process by the water is not significant. This is unexpected in particular for those species which are known to be hydrolyzed and oxidized very quickly (e.g., Me$_3$Bi). The reaction time in the water column is probably too short which enables the VMCs to reach the atmosphere in the bubbles without having contact with the water. Since no significant changes in the gas concentration of the gas in the wells, and that which percolated through the water have been determined, it can be concluded that volatile metal compounds generated in municipal waste deposits are released into the atmosphere.

When those VMCs reach the atmosphere, how stable are they? Can they be transported over vast distances? We have determined the atmospheric half-life of twelve different volatile arsenic, antimony and tin compounds in air at a concentration of 10 µg m^{-3}, which can be seen in Figure 3 (19). The atmospheric half-life can be measured in hours and for some even in days rather than in seconds as suggested elsewhere (24), because of the low concentration of the compounds in the air. It should be noted that UV radiation has an enormous impact on the stability of these compounds in the atmosphere while the temperature effect is much smaller.

Conclusion

In can be concluded that municipal waste deposits are significant sources of metal volatilization. Although this paper shows only data from one municipal waste deposit, it has been shown that landfill gas from other sites contain similar concentrations of VMCs. The generated VMCs are stable enough to diffuse through soil, waste and the overlying oxic water column of a wetland and can reach the atmosphere illustrated in Figure 4. Although degradation products of the VMCs have been identified in the water column, this "immobilization" seemed to have no significant effect on the flux out in the atmosphere. According to the estimated atmospheric half-life of VMCs, once in the atmosphere they are stable enough to disperse from the landfill site into the near vicinity. This might have toxicological implications. Since very limited knowledge is available about the toxicity of these compounds in very low concentrations, future studies are necessary to assess the chronic effects of VMCs to mammals.

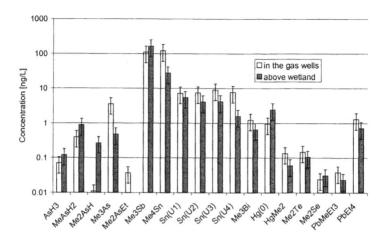

Figure 2: Comparison of volatile metal compounds in landfill gas sampled in the gas wells and above the wetland through which the gas percolated.

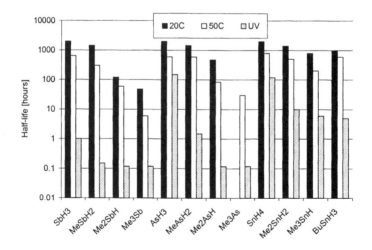

Figure 3:Estimated atmospheric half-life of selected volatile arsenic, antimony and tin compounds (each 10 μ m^{-3}) measured in a synthetic air atmosphere at 20 and 50°C in the dark and at 30°C with UV radiation (5000 Lux) (20).

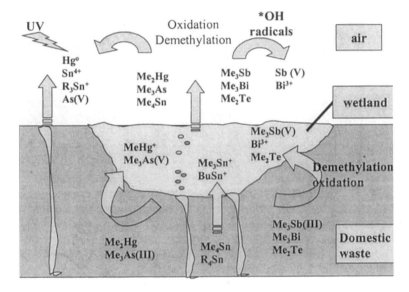

Figure 4. Percolation of landfill gas through the landfill sites and the overlaying wetland with the indication of the predominant species in each compartment.

References

1. Vella, A.J.; Adami, J. P. T. *Appl. Organomet. Chem.* **2001**, *15*, 901-906.
2. Amouroux, D.; Tessier, E.; Donard, O. F. X. *Environ. Sci. Technol.* **2000**, *34*, 988-995.
3. Wallschläger, D.; Hintelmann, D.; Evans, R.D.; et al., *Water Air Soil Poll.* **1995**, *80*, 1325-1329.
4. Laturnus, F. *Environ. Sci. Poll. Res.* **2001**, *8* 103-108.
5. Hansen, D.; Duda, P. J.; Zayed, A.; Terry, N. *Environ. Sci. Technol.* **1998**, *32*, 591-597.
6. Hirner, A. V.; Feldmann, J.; Krupp, E.; Grümping, R.; Goguel, R.; Cullen, W. R. *Org. Geochem.* **1998**, *29*, 1765-1778.
7. Pedersen, B. *Ann. Occup. Hyg.* **1988**, *32*, 385-397.
8. Feldmann, J.; Hirner, A.V. *Intern. J. Environ. Anal. Chem.* **1995**, *60*, 339-359.
9. Challenger, F. *Chem. Rev.* **1945**, *36*, 315-345.
10. Feldmann, J.; Grümping, R.; Hirner, A. V. *Fresenius J. Anal. Chem.* **1994**, *350*, 228-234.
11. Feldmann, J.; Haas, K.; Naëls, L.; Wehmeier, S. In *Plasma Source Mass Spectrometry*; Holland, G., Tanner, S.D., Eds.; Royal Society of Chemistry Proceedings, RSC, London, UK 2001; pp. 361-368.
12. Andrewes, P.; Cullen, W.R.; Feldmann, J.; Koch, I.; Polishchuk, E.; Reimer, K.J. *Appl. Organomet. Chem.* **1998**, *12*, 827-842.
13. Feldmann, et al. *Appl. Organomet. Chem.* **1999**, *13*, 739-748.
14. Donard, O. F. X.; Weber, J.H. *Nature* **1988**, *332*, 339-341.
15. Feldmann, J.; Koch, I.; Cullen, W.R. *Analyst*, **1998**, *123*, 815-820.
16. Feldmann, J.; Kleimann, J. *Korrespondenz Abwasser*, **1997**, *44*, 99-104.
17. Feldmann, J. *J. Environ. Monit.* **1999**, *1*, 33-37.
18. Feldmann, J.; Cullen, W. R. *Environ.. Sci. Technol.* **1997**, *31*, 2125-2129.
19. Haas, K.; Feldmann, J. *Anal. Chem.* **2000**, *72*, 4205-4211.
20. Haas, K.; Feldmann, J. *J. Atmos. Environ.* (in prep.) **2002.**
21. Feldmann, J. *J. Anal. At. Spectrom.* **1997**, *12*, 1069-1076.
22. Haas, K.; Feldmann, J.; Wennrich, R.; Stärk, H. J. *Fresenius J. Anal. Chem.* **2001**, *370*, 587-596.
23. Cullen, W. R. *personal. Comm.* **2001.**
24. Parris, G. E.; Brinckman, F. E. *Environ. Sci. Technol.* **1976**, *10*, 1128-1134.

Chapter 11

Volatile Metal(loid) Species Associated with Waste Materials: Chemical and Toxicological Aspects

Alfred V. Hirner

Institute of Environmental Analytical Chemistry, University of Essen,
45117 Essen, Germany

By using hyphenated analytical techniques, dozens of
organometal(loid) species could be determined in landfill and
sewage gases as well as polluted sediments and soils in
concentrations from the low ng/m^3 (resp. kg) to the mid μg/m^3
(resp. kg) range. Environmental gases are known to contain up
to five species of As, one of Bi, one of Cd, three of Hg, two of
I, five of Pb, two of Sb, three of Se, four of Sn, and two of Te
(altogether 28 compounds of 10 elements) in addition to
volatile methylated silicones; waters and sediments in the
environment show a similar variety of species (additionally
those of Ge). In the light of available toxicological data our
empirical analytical results suggest that certain environmental
scenarios may be of potential toxicological concern (respective
organometal(loid)s given in bracketts): River and harbor
sediments (Sn, As, Hg), soil near ore deposits (Hg), gaseous
emissions from industrial sludge fermenter (Te, As, Bi, Sb),
geothermal water (As, Hg, Sn), sewage gas (As, Bi, Sb, Se,
Te), waste gas (As, Sb, Sn), and leachates (As). Generally it
can be stated that compared to the omnipresent biogenic
background, organometal(loid) emissions from solid waste and
contaminated soils and sediments are at least two magnitudes
higher, and require toxicological evaluations in respect to the
health of people working or living near these sites.

Environmental Biomethylation

Biomethylation is a common process in the biogeochemistry of the earth's surface: The formation of volatile compounds by biological methylation (biomethylation) has been described for natural systems for As, Cd, Ge, Hg, S, Sb, Se, Sn, Te, and possibly Pb as well as for Au, Cr, Pd, Pt, and Tl under laboratory conditions (1,2). Methyl transfer onto metal(loid)s is established not only under anaerobic conditions in recent sediments, but also within macroscopically aerobic milieus (3), and in fossil substances like natural gas or crude oil (4). Another possible route for the formation of volatile metal(loid) species is the generation of hydrides (5). Thus, volatile metal(loid) species may exist under various environmental conditions.

The most important criteria for a high biomethylation potential are

1. an anaerobic atmosphere (at least on a microscale), reducing conditions and slightly acidic pH values in the hydrosphere
2. metal(loid)s in easily accessible forms (e.g. as ions)
3. the presence of microorganisms with biomethylation potential (bacteria, fungi, algae, archeae)
4. the presence of transferable methyl groups.

From 1 to 3 it is evident that biomethylation can happen in geogenic as well as in anthropogenic systems, whenever these conditions are fulfilled. Concerning 4, the source of available methyl groups, biogenic (e.g. methylcobalamine, methyltetrahydrofolate, methyl-coenzyme M or various methionine species) as well as anthropogenic donors (synthetic methylated compounds) exist. In respect to the latter, man-made compounds like $PbMe_4$, $SnMe_4$ or PDMS (polydimethylsiloxanes) may be abundant at polluted sites.

Analytical Methods

In general hyphenated techniques coupling efficient separations by chromatographic methods (gas/liquid/supercritical fluid chromatography and capillary electrophoresis) with highly specific and sensitive multielemental detectors like inductively coupled plasma mass spectrometry (ICP-MS) have been developed, and applied to various environmental matrices like waters, sediments, organisms, urine, and blood (6-8).

In respect to high sensitive multielemental organometal(loid) analysis of environmental samples the instrumental method applied at Essen University is shortly described here (Figure 1): The first step is the trapping of the sample gas at low temperatures (-78^0C). The gas may be directly condensed from wells, pumped through a series of cooled traps by a vacuum pump (typical volume of

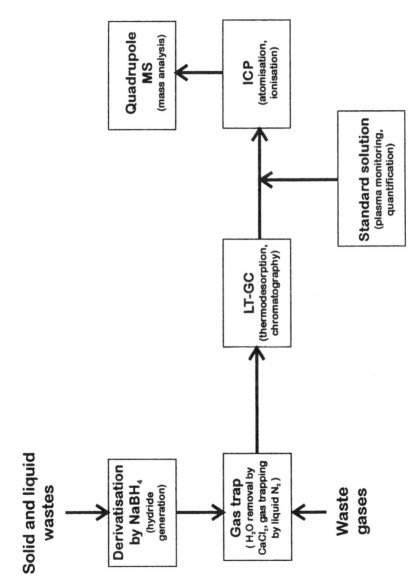

Figure 1. Analytical scheme for metal(oid) speciation in waste gases, leachates and solid wastes

20 L), may simply be collected in gas bags (Tedlar bags) and frozen out in the laboratory later as described above, or may be liberated from liquid and solid samples by derivatization. Until measurement, the trapped gases are stored under liquid nitrogen.

The loaded gas trap is inserted into a low temperature gas chromatograph coupled on-line with a plasma mass spectrometer (LTGC/ICP-MS). By LTGC the condensed species are thermally desorbed within the temperature range from -100 to +165^{0}C, and chromatographically separated in a packed column (Supelcoport 10% SP-2100). The separated gas components are transferred to the plasma torch of the MS, together with a standard aerosol used for monitoring the stability of the plasma as well as enabling a semi-quantitative calibration (reproducibility of results is around ± 30%) applying a special interelement-interphase calibration technique (9). Gas chromatographic retention times and analyte concentrations are calibrated using pure substances.

For species analysis of aqueous samples or sediment suspensions, sample manipulations were generally performed inside a glove box under an argon atmosphere. In the first step, dissolved volatile compounds were purged out with helium without derivatization (e.g. HgMe$_2$, TeMe$_2$). Thereafter, hydride generation methodology was employed by treating 1 mL of sample water with 1 mL of 0.1 N HCl and 3 mL of H$_2$O in a 50 mL flask (1<pH<2), followed by adding 1 mL of freshly prepared (He stripped) 5% solution of NaBH$_4$ (Aldrich, 99% pure). The liberated hydrides together with reduced organometal(loid) species (e.g. trimethylarsine) were transported by a flow of helium of 300 mL/min to a gas trap cooled by liquid nitrogen; the parameters were optimized for multielement speciation analysis.

Instrumental parameters together with further technical details of the instrumental method and the hydride generation methodology are given elsewhere (10-13).

Analytical Results

Gaseous, liquid and solid waste

By using hyphenated analytical techniques, dozens of organometal(loid) species could be determined in landfill and sewage gases (10,11,14) as well as polluted sediments and soils (12,15) in concentrations from the low ng/m^3 (resp. kg) to the mid μg/m^3 (resp. kg) range. The highest concentrations reported so far were >30 mg/kg for tin in tributyltin and 15 mg/kg in tetrabutyltin found at a dockyard site at Bremerhaven (16).

Geogenic environments

Volatile species of As, I, Sb, and Se (e.g. AsH_3, $AsMeH_2$, $AsMe_2H$, $AsMe_3$, $SeMe_2$, $SbMe_3$, and IMe) could be detected in gases over hot springs in British Columbia (Canada) (*17*). Methylated species of As, Ge, Hg, Sb, and Te were also determined in concentrations in the ng/kg- to low µg/kg-range in geothermal waters from Ruapehu, Waimangu, Waiotapu, and Tokaanu (North Island, New Zealand); up to 1% of the total dissolved metal(loid)s are found in methylated forms.

In field screening tests at two ore deposits in Germany organic compounds of several elements (e.g. As, Hg) could clearly be identified in gas and soil samples (*18*). In soil air several metal(loid)organic species could be determined in the ng/m^3-range up to the following concentrations: $AsMe_3$ and $HgMe_2$ approx. 4, $SbMe_3$, $SeMe_2$ and IMe approx. 1, $TeMe_2$ approx. 3. No Pb species could be detected (detection limit approx. 0.1 ng/m^3). Whereas after application of $NaBH_4$ the concentrations of 17 different species of metal(loid) hydrides and organometal(loid)s in the soil samples were found in the low µg/kg range, organic mercury was present at much higher concentrations (0.6 to 4.5 mg/kg); up to 350 mg/m^3 of elemental mercury (after hydride generation) was detected within the soil atmosphere.

Waste treatment

Besides the waste and natural background samples described above, potential emissions from mechanical/biological waste treatment plants were investigated. Essentially the latter are accompanied by similar biomethylation processes as observed in waste deposits or sewage plants (*15*). The only significant difference is the received concentration resembling long-term gas collection in the latter compared to relatively short-time sampling of a steady state degassing during waste processing. Thus concentrations of organometal(loid) compounds in the continuous effluent of treatment plants are only slightly exceeding 1 $µg/m^3$ in case of the elements As, Sb, Se, and Sn only. In particular, in the headspace of a container filled with mechanically/biologically treated residual waste for several years the fully methylated tin (Me_4Sn) was present at concentrations > 1 $µg/m^3$, whereas other partly methylated and higher alkylated (ethylated and butylated) tin species were found in much lower concentrations (unpubl. data). This indicates that - given enough time - most of the available (dissolved) tin was transformed into the species with the highest methylation grade (i.e. peralkylated species). This is an impressive sample where an ordinary metal abundantly present in waste material is eventually transformed into toxic species by natural processes.

Waste fingerprinting

Based on the data reviewed in this paper, the presence of specific organometal(loid) compounds supplement ordinary geochemical fingerprinting parameters (Table I) for tracing the origin of spills/contaminations: As can be seen from Table II, discrete organometal(loid) species are associated with certain environmental scenarios, i.e. the kind of organometal(loid) species present in significant concentrations (i.e. above background concentrations in the low ng per m^3/L/kg range) allows a differentiation between various geogenic and anthropogenic environments (Table II). Even further more detailed differentiations are possible: E.g. waste gases show significant higher concentrations of volatile Sn and Sb species than do sewage gases (*17*).

While the simultaneous occurrence of methylated Sn, Hg, and As compounds may not allow the differentiation between geogenic and anthropogenic environments, Sb and Te will only be biomethylated whenever higher than natural concentrations of these elements are present (mining, industrial products and wastes). Another interesting fingerprinting compound trimethyl bismuth is derived from Bi compounds used in the human environment (cosmetics/medicine), and is consequently found in sewage sludge in relatively high concentrations (*19*).

Differences in the properties of organometal(loid)s have implications for their distribution and accumulation in different compartments of the environment: Partly alkylated compounds are soluble in the waterphase because of their ionic nature, neutral volatile peralkylated compounds will be preferentially found in the atmosphere, at aqueous/solid interfaces or floating on aqueous surfaces (Table III). Because of their lipophilic nature, both partly and fully methylated compounds accumulate in the food chain.

An age determination of organometal(loid) compounds in the environment is only possible, when the formation process or the release of the industrial produced compounds to the environment is stopped.

Degradation processes are of special interest for anthropogenic organometal(loid)s like tributyltin (TBT) or tetraalkyllead. Because these compounds undergo a multistage dealkylation process (tetraalkyl 6 trialkyl 6 dialkyl 6 monoalkyl 6 inorganic species), it is possible to distinguish degradation levels by the relative concentration of the educt and its degradation products. If the rate of degradation of a organometal(loid) compound is known it is possible to estimate the time of introduction of the compound into the environment, e.g. the half-life time of TBT is in the range of a few days to several month depending on the environment (freshwater, seawater or sediment), temperature and TBT concentration (*20*).

Table I. Common chemical fingerprinting techniques in environmental chemistry (examples)

Marker categories	Marker	Sources/processes
Inorganic	As, Se, S	coal burning
	Sb, Br, Zn	traffic emissions (automobiles)
	V, Ni	crude oil leaks
Organic	n-alkanes, pristane/phytane, steranes, hopanes, porhyrins	crude oil, coal and kerogen origin
	petrol additives (MTBE)	refinery, dealer
	DNA	genetic fingerprint
Isotopic	$^{13}C/^{12}C$	biogenic/abiogenic origin terrestrial/marine origin
	$^{34}S/^{32}S$	primary/secondary sulfur
	$^{206}Pb/^{207}Pb$	geogenic/petrol add. lead

Table II. Fingerprinting by environmental scenario with methylated species

Environmental scenario	Hg	Sn	As	Sb	Bi	Se	Te
Wetlands, soil	X						
Geothermal water	X	X	X				
River and harbor sediments	X	X	X				
Waste gas		X	X	X			
Sewage gas			X	X	X	X	X
Gaseous emissions (waste treatment)		X	X	X		X	X
Gaseous emissions (industrial sludge)			X	X	X		X

Table III. Chemical fingerprinting by nature of species

Kind of species	Occurrence/processes	Examples
Peralkylated species	Preferentially in gas phase/aqueous surfaces	$AsMe_3$, $HgMe_2$, $PbEt_4$
Partly alkylated species	Preferentially in aqueous phase	$AsMe^{2+}$, $CdMe^+$, $SnBu^{3+}$
Methylated species	Generated by environmental biol./chem. methylation	$SbMe_3$, $BiMe_3$
Higher alkylated species	Generated by industrial synthesis	$SnBu^{3+}$, $PbEt_4$, $PhHg^+$

Toxicological Evaluation

When the data received in a recent experimental study (21) are applied to environmental systems, the main consequence will be that organisms in waters with methylmercury loads in the lower to middle µg/L range will be expected to receive significant genetic defects: e.g. the genotoxicity of the organometallic compound methylmercury chloride (MeHgCl, up to 1×10^{-5} M) was evaluated by chromosome metaphase analysis in Chinese hamster ovary (CHO) cells treated in vitro for two hours. Structural chromosomal aberrations (CA) and sister chromatid exchanges (SCE) were scored for the assessment of induced genotoxic effects. MeHgCl induced CA and SCE in a dose-related manner at doses exceeding $7,5 \times 10^{-7}$ M. Concentrations above 1×10^{-5} M showed a very high cytotoxic effect without cell surviving. These compound also induced mitotic spindle disturbances.

These results are also in agreement with the observation of specific teratogenic defects when exposing zebrafish to 20 or 30 µg/L methylmercury chloride (22). In respect to these findings it should be noted that organometal(loid) compounds have sometimes been found up to the high µg/L range compared to the just few known 24 h LD_{50} data around 1 µg/L (23).

In the light of this discussion according to our empirical analytical results a "hit list" of environmental scenarios of potential toxicological concern is given in Table IV. The data represent the results received from 85 samples from 20 locations (10-15, 17, 18). Usually the concentrations found were in the ng/m³, ng/L, and ng/kg range in the case of gases, water, and solids, respectively. 40% of the measurements yielded results below the detection limit.

Table IV. Environmental scenarios with significant metal(loid)organic
emission potential

Environment	Organoelement	Max. concentration
River and harbour sediments	Sn	up to> 45 mg/kg
	As,Hg	up to >10 μg/kg
Soil (Hg ore deposit)	Hg	up to 5 mg/kg
Gaseous emissions from	Te	up to 100 μg/m^3
Industrial sludge fermenter	As, Bi, Sb	up to >1 μg/m^3
Geothermal water	As, Hg, Sn	up to >10 μg/L
Sewage gas	As, Bi, Sb	up to >10 μg/m^3
	Se, Te	up to 1 μg/m^3
Waste gas*)	As, Sb, Sn	up to >10 μg/m^3
Leachates*)	As	up to >1 μg/L

*) domestic waste deposits

Conclusions

Many studies have demonstrated that highly mobile and toxic organometal(loid) compounds are present in environmental systems (air, water, soil and sediment), and that the production of such species is possible and likely whenever anaerobic conditions (at least on a microscale) are combined with available metal(loid)s together with methyl donors in the presence of suitable organisms. These necessary conditions exist within geogenic environmental systems (e.g. geothermal waters) as well as within anthropogenic environmental systems (e.g. waste and sewage materials and processing). Compared to the omnipresent biogenic background, organometal(loid) emissions from solid waste, contaminated soils and sediments are at least two magnitudes higher, and require toxicological evaluations in respect to the health of people working or living near these sites.

Last not least it should be mentioned, that contents and emission potentials of special wastes like industrial or electronic waste have never been analyzed for organometal(loid) compounds yet. Also it should be mentioned that in the samples described here nearly as many organometal(loid) species already identified and quantified are still waiting for their evaluation in further studies (*13,15*).

References

1. Craig, P. J.; Glockling, F. *The Biological Alkylation of Heavy Elements;* The Royal Society of Chemistry, London, 1988.

2. Rapsomanikis, S.; Weber, J. H. In *Organometallic compounds in the environment;* Craig, P. J., Ed.; Longman Group Ltd., Harlow, 1986; pp 279-307.

3. Brinckman, F. E.; Bellama, J. M. *Amer Chem Soc Sympos Ser* **1978**, *82*.

4. Irgolic, K. J.; Spall, D.; Puri, B. K.; Ilger, D.; Zingaro, R. A. *Appl. Organomet. Chem.* **1991**, *5*, 117-124.

5. Donard, F. X.; Weber, J. H. *Nature* **1988**, *332*, 339-343.

6. Zoorob, G. K.; McKiernan, J. W.; Caruso, J. A. *Mikrochim. Acta* **1998**, *128*, 145-168.

7. Sutton, K.; Sutton, R. M. C.; Caruso, J. A. *J. Chromatogr.* **1997**, *A 789*, 85-126.

8. Spunar-Lobinska, J.; Witte, C.; Lobinski, R.; Adams, F. C. *Fres. J. Anal. Chem.* **1995**, *351*, 351-377.

9. Feldmann, J. *J. Anal. At. Spectrom.* **1997**, *12*, 1069-1076.

10. Feldmann, J.; Grümping, R.; Hirner, A. V. *Fres. J. Anal. Chem.* **1994**, *350*, 228-234.

11. Feldmann, J.; Hirner, A. V. *Int. J. Environ. Anal. Chem.* **1995**, *60*, 339-349.

12. Krupp, E. M.; Grümping, R.; Furchtbar, U. R. R.; Hirner, A.V. *Fres. J. Anal. Chem.* **1996**, *354*, 546-549.

13. Grüter, U. M.; Kresimon, J.; Hirner, A. V. *Fres. J. Anal. Chem.* **2000**, *368*, 67-72.

14. Hirner, A. V.; Feldmann, J.; Goguel, R.; Rapsomanikis, S.; Fischer, R.; Andreae, M. O. *Appl. Organomet. Chem.* **1994**, *8*, 65-69.

15. Hirner, A. V.; Grüter, U. M.; Kresimon, J. *Fres. J. Anal. Chem.* **2000**, *368*, 263-267.

16. Stichnothe, H.; Thöming, J.; Calmano, W. *8th Int Sympos Interact Sed Water*, Beijing, 1999.

17. Hirner, A. V.; Feldmann, J.; Krupp, E.; Grümping,, R.; Goguel, R..; Cullen, W. R. *Org. Geochem.* **1998**, *29*, 1765-1778.

18. Hirner, A. V.; Krupp, E.; Schulz, F.; Koziol, M.; Hofmeister, W. *J. Geochem. Explor.* **1998**, *64*, 133-139.

19. Feldmann, J.; Krupp, E. M.; Glindemann, D.; Hirner, A. V.; Cullen, W. R. *Appl. Organomet. Chem.* **1999**, *13*, 739-748.

20. Seligmann, P. F.; Maguire, R. J.; Lee, R. F.; Hinga, K. R.; Valkirs, A. O.; Stang, P. M. In *Organotin - Environmental fate and effects*; Champ, M. A.; Seligman, P. F., Ed.; Chapman and Hall, London, 1998.

21. Ehrenstein, C.; Shu, P.; Wickenheiser, E. B.; Hirner, A. V.; Dolfen, M.; Emons, H.; Obe, G. *Chem. Biol. Interact.* (submitted)

22. Samson, J. C.; Shenker, *J. Aquat. Toxicol.* **2000**, *48*, 343-354.

23 Crompton, T.R. Occurence and analysis of organometallic compounds in the environment; J. Wiley & Sons, New York, 1998.

Chapter 12

Biogenic Volatilization of Trace Elements from European Estuaries

Emmanuel Tessier, David Amouroux, and Olivier F. X. Donard

Laboratoire de Chimie Analytique Bio-Inorganique et Environnement, CNRS UMR 5034, Université de Pau et des Pays de l'Adour, Pau, France

This paper describes the importance of volatile chemical forms of selected trace elements in the environment with emphasis on the estuarine aquatic environment. The occurrence, distribution and fluxes of gaseous species of trace elements such as Mercury (Hg), Selenium (Se), Tin (Sn) and Iodine (I), was investigated in three European macro-tidal estuaries in relation with the main biogeochemical parameters. Water and air samples were collected according to ultra-clean methods, in the Gironde, the Rhine and the Scheldt estuaries. On a seasonal basis, this approach allowed us to evaluate the potential pathways of these gaseous species in estuarine waters and to derive their seasonal fluxes to the atmosphere. We were then able to provide an estimation of the global contribution of European estuarine emission of these trace elements and to compare it with anthropogenic sources.

151

Introduction

Many elements are involved in the transfer of gaseous species between earth surface and the atmosphere. Most of the lighter elements (C, N, S etc) are involved in the basic life processes in the environment, including photosynthesis, respiration, and the production of radiatively important ("green house") gases (1). While there have been abundant studies and evidence involving these lighter elements, the atmospheric transfer of the heavier (trace) elements is less recognised.

Heavier elements, as shown in Figure 1, can form volatile species mainly after reduction and/or alkylation processes (2). These pathways are a consequence of biological activity, or indirectly via abiotic reactions. This is important as it provides a transfer mechanism via the atmosphere into ecosystems with potentially harmful or nutritious effects.

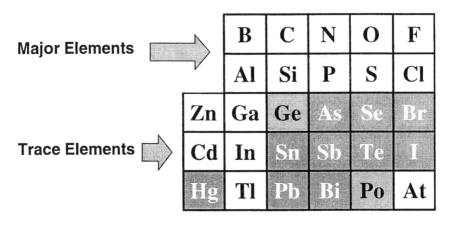

Figure 1 : Trace elements forming volatile gaseous species under natural environmental conditions (elements in grey cells : clear = verified pathways, dark = suggested pathways).

Several investigations made on iodine (I), selenium (Se), mercury (Hg) and tin (Sn) have demonstrated the significance of transfer processes of these heavy elements between earth surface and the atmosphere. As shown in Figure 2, investigation performed in marine and estuarine environments, but also in pristine or polluted continental ecosystems suggest that as for lighter elements, volatilisation of heavy elements through natural processes is a major pathway in their biogeochemical cycles (2).

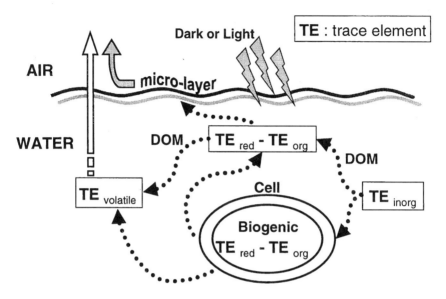

Figure 2 : Main biogeochemical routes for trace elements volatilisation in aquatic environments.

For example, studies in the productive estuarine and coastal waters suggest that trace element fluxes from these areas are equivalent to the open ocean. Like dimethylsulfide (Me_2S), all heavier elements (Se, Te, Po) in the same group of the periodic table have volatile alkyl species (2-5). In the case of ecosystems contaminated by inorganic Se and Hg, or tributylSn (TBT), formation and volatilisation of dimethyl selenide (Me_2Se), elemental mercury ($Hg°$) or methylated butylSn (Bu_3SnMe), respectively, can be an efficient process to remove and remobilize these potentially harmful contaminants via their transfer to the atmosphere (6-8).

In this paper, we present an example of investigation that was achieved during the last four years in the framework of an EU Environment and Climate project focused on the emission of biogenic gases from West-European estuarine waters (BIOGEST). This work presents a good illustration of the scientific strategy that we have developed in the recent years owing to investigation of the formation of transient labile and/or volatile trace element species in the aquatic environments and their impacts on the transfer of such trace elements between the various environmental compartments, including biological cells (bacteria, plankton, fishes...).

Material and Methods

A multi-element analytical method, involving cryogenic trapping, gas chromatography and inductively coupled plasma-mass spectrometry (CT-GC-ICP/MS) for the speciation of dissolved and atmospheric volatile metal compounds allowed us to perform simultaneous investigations on several trace elements such as mercury (Hg), selenium (Se), tin (Sn) and iodine (I) (9-10). Water and atmospheric samples were collected during several cruises on the Gironde (Nov 1996, June 1997, Sept 1997, Feb 1998), the Scheldt (July 1996, Dec 1997, May 1998, October 1998) and the Rhine (Oct 1996, July 1997, Nov 1997, Apr 1998) estuaries along with hydrological parameters.

On each estuary, sampling stations were chosen to collect surface waters every 2.5 salinity unit increments with salinity ranging between 0 and 34 (practical salinity scale, PSS). Surface waters were sampled at a 3m depth using a PTFE coated Go-Flo non-metallic sampler (General Oceanic) to avoid ship contamination and microlayer surface water contamination. After sampling, collected water was immediately transferred through a silicone tubing into a gas-tight, PTFE-lined 1L Pyrex bottle until shipboard treatment. Water sample treatment including a cryofocusing step was detailed elsewhere by Amouroux *et al.* (*10*) and Tseng *et al.* (*11*). Within 30 minutes after collection, 1 Litre of sample is purged for 1 hour with 600 to 700 ml min^{-1} He flow. Water vapour is removed from the gas stream through a moisture trap maintained at -20°C. Volatile compounds are subsequently cryo-trapped into glass tube filled with glass wool and immersed in liquid nitrogen (-196°C). Most of the volatile compounds extracted from the aqueous sample can, therefore, be concentrated and stabilised into the cryogenic trap. Those traps are then sealed, stored and transported into a dry atmosphere cryogenic container (-190°C) until analyzed. The general operating parameters for the whole analytical procedure are described in detail elsewhere by Amouroux *et al.* (*10*). The analytical set-up developed for simultaneous determination of volatile species using ICP/MS detection is described in Figure 3. In the laboratory clean room, the cryogenic traps are flash desorbed by a fast-heating furnace. The volatile species are then cryofocused onto the head of the chromatographic column, immersed in liquid nitrogen (-196°C), before gas chromatographic (GC) separation. The chromatographic column is packed with Chromosorb W-HP (60/80 mesh size) coated with 10% Supelco SP-2100 and is then silanized with about 200µl of Hexamethyldisilazane. The desorption and separation of the analytes (non-polar organometals) can then be operated by applying a temperature gradient (from –

100°C to 250°C) allowing the volatile compounds to elute according to their boiling point. In order to avoid plasma disturbances due to the presence of desorbed gas or carbon-containing compound, such as carbon dioxide, a PTFE 3-way valve is positioned at the output of the chromatographic column to allow to vent the gas excess. Furthermore the transfer lines between the desorption cell and the chromatographic column on the one hand and between the column and the detector on the other hand are heated during the desorption steps allowing us to avoid any water condensation in the analytical chain. Finally, the detection is performed by inductively coupled plasma mass spectrometry (ICP/MS). This detector allows one to identify and quantify the volatile compounds of up to 30 isotopes of different trace metals and metalloids with very low detection limits and high sensitivity and selectivity (9). Absolute detection limit, defined as $3\sigma_{noise}$, ranges from 0.4 to 10 fmol for the different trace elements analysed.

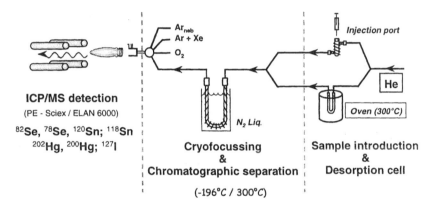

Figure 3: Experimental set-up for the simultaneous determination of volatile species of several elements in the environment using ICP/MS detection.

Air samples were collected using a similar method described by Pécheyran *et al.* (9). Air is pumped during ½ hour at c.a. 800 ml min⁻¹ from the top or the bow of the ship to avoid any contamination. The aerosols and water vapour are removed from the gas stream using on-line 0.1 µm quartz filter (Millipore) and moisture trap (-20°C), respectively. The volatile species are then trapped onto a cryogenic trap (-170°C) and stored in cryogenic container as described below. In the laboratory, the cryogenic traps are desorbed and analysed similarly to the purge and trap samples.

Results and Discussion

Identification of Volatile Trace Element Species

The analytical method described below has been applied for the determination of volatile metal and metalloid species in estuarine waters and the atmosphere. During the Biogest cruises, we focussed our attention on four trace elements, mercury (Hg), selenium (Se), tin (Sn) and iodine (I). Examples of chromatogram obtained from estuarine water samples are presented in Figures 4 and 5. We observed and identified several ubiquitous chromatographic peaks from the various samples collected in the Gironde, the Rhine and the Scheldt estuaries. The major volatile compounds encountered are outlined in Table I.

Table I. Volatile species of the investigated trace elements identified in estuarine water and atmosphere

Elements	Hg	Se	Sn	I
Isotopes	200/202	78/82	118/120	127
Water	Hg°	Me_2Se	Me_nSnH_{4-n}	MeI
	Me_2Hg	MeSSeMe	Me_nSnBu_{4-n}	(R-I)
	(MeHgCl)	Me_2Se_2	(n=1-4)	(R= Et, Pr, Bu)
Atmosphere	Hg°	Me_2Se	Me_nSnH_{4-n}	MeI
	(MeHgCl)		(n=0-4)	(R-I)
				(R= Et, Pr, Bu)

Compounds in parenthesis are subject to further investigations; Me=Methyl group, Et=Ethyl group, Pr=Propyl group, Bu=Butyl group.

We identified the potential molecular species by comparison with the retention time obtained with standards solutions (10). The sample quantification for the volatile species was then based on aqueous standards, commercially available, and screening the mass spectrum of the observed compounds (i.e. Me_4Sn, Et_4Sn and Bu_4Sn; Hg°, Me_2Hg and Et_2Hg, Me_2Se and Me_2Se_2; MeI for volatile tin, mercury, selenium and iodine species, respectively).

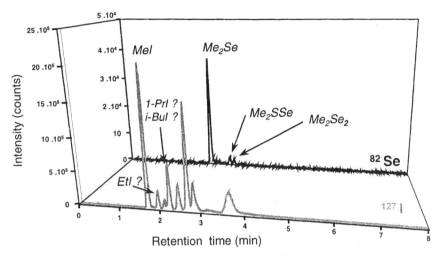

Figure 4: Example of chromatograms for ^{127}I and ^{82}Se, obtained from a surface water sample collected in estuarine environment

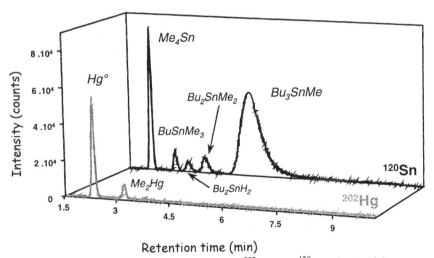

Figure 5: Example of chromatograms for ^{202}Hg and ^{120}Sn, obtained from a surface water sample collected in estuarine environment.

Saturation Ratio and Flux Calculation

In a second step, air and water measurements of these volatile species allowed us to calculate fluxes between estuarine water and the atmosphere. First, the saturation ratio (SR) estimation is necessary to validate the transfer of volatile species at natural interfaces. For the water to air exchange, the saturation ratio (SR) is expressed as: $SR = (C_w \times H) / C_a$. Where C_w and C_a are the concentrations measured in the water column and in overlying air, respectively. H represents the dimensionless Henry's law constant of the considered volatile compound. H values used in our calculations were obtained from literature available constants for volatile Hg and I compounds and from a quantitative structure-activity relationship (QSAR) between Henry's law constants available in the literature for alkylated organometals species (*12*) and molecular total surface area (*13*) for volatile Se and Sn compounds. This simple method was found to provide a very good approximation of Henry's constant for volatile alkylated compounds (*14*). In order to apply these calculations to real environmental processes, we have previously postulated that the volatile species investigated exhibit a low solubility in water and are likely to be unreactive in the aqueous substrate (*15*).

Flux calculations are then based on the Fick's first law of diffusion with the assumption that the concentrations of the studied compounds in the water column are in steady state compared to their transfer velocities at the interface. The flux density (mol m^{-2} d^{-1}) is expressed as:

$$F = K \times \left(C_w - \frac{C_a}{H} \right)$$

where K is the transfer velocity (cm h^{-1}) at the air-water interface. Considering the low water solubility of the studied compounds we can assume that K is equal to k_w, the water gas transfer velocity (*16*). We can then calculate the k_w for the various volatile compounds with the model proposed by Clark *et al.* (*17*), developed for the tidal estuarine system of the Hudson bay. The transfer velocity is given by the following expression:

$$k_w = \left(\frac{Sc}{600} \right)^{\frac{1}{2}} \times \left(2 + 0.24u^2 \right)$$

where Sc is the solute Schmidt number, inversely proportional to the diffusion coefficient (D_0) and u is the wind speed (m/s) recorded at a height of 10 m. The free-solution diffusion coefficient D_0 ($m^2 s^{-1}$) was obtained from the Wilke & Chang equation (*18*) and was then corrected for the temperature and salinity according to Saltzman et al. (*19*). We then implemented the Clark's model with in situ wind speed values measured during each sampling point. The application of the gaz exchange model was performed assuming that the half-life of the volatile species in the aqueous phase is greater than their respective transfer time through the air-water interface.

Overall Distribution and Air-Water Exchange of Volatile Trace Element Species

Seasonal concentration profiles versus salinity, for the major volatile species observed in the three estuaries investigated, are presented in Figures 6 and 7. Volatile Hg was mainly found as elemental Hg (Hg°) in both aquatic and atmospheric compartments (Hg° range 50-1500 fmol L^{-1} and 1-13 ng m^{-3} for water and air respectively). In the Gironde and the Scheldt estuaries, dimethyl mercury (Me_2Hg) and probably monomethyl mercury chloride (MeHgCl) were also evidenced. Dissolved Hg° variations in estuarine waters could be negligible such as in the Gironde estuary, or display significant ranges like in the Scheldt or Rhine estuaries. Concentrations in water are increasing at high salinity (*20-35*) during the summer period, and concentration maxima were observed at lower salinity (0-5). Hg° distribution was found to be well correlated to the potential "active chlorophyll" during the warmer seasons (i.e. ratio [chlorophyll a]/([chlorophyll a]+[Phaeopigments])). Moreover, seasonal variations of the Hg° mean concentrations present the same pattern in the three estuaries with higher values in spring and summer. These results suggest that biological turnover and seasonal pattern control Hg° production in surface water involving the reduction of inorganic mercury under light-induced biological or chemical processes. Hg° was then found to be supersaturated in water compared to the atmosphere during the warmer seasons, and near to the saturation during colder seasons. Estimated fluxes to the atmosphere are then seasonally dependent and range between 5 to 7500 pmol $m^{-2} d^{-1}$.

Volatile Se in estuarine waters was mainly detected as dimethylselenide (Me_2Se) and concentrations were also found significantly higher in the Scheldt and the Rhine estuaries (ca. 200-25000 fmol L^{-1}) than in the Gironde estuary (ca. 300-5000 fmol L^{-1}) (*20*). Amounts of dimethyldiselenide (Me_2Se_2) and dimethylselenide sulphide (Me_2SSe) have also been observed in the three estuaries investigated. In the whole air samples the concentrations measured were near to the method detection limit (ca. < 0.2 ng m^{-3}). Me_2Se in surface waters exhibits concentration maxima in the 0-10 salinity range and a decrease by estuarine mixing processes for higher salinity. Me$_2$Se distribution in all estuaries investigated was found closely anticorrelated with "active chlorophyll", showing that its formation may be related to plankton biomass degradation. Formation of volatile Se compounds is then mainly due to riverine and marine plankton inputs and the associated microbial activity. Concerning Me$_2$SSe, this compound was found to be produced under abiotic conditions (21). In all seasons for all estuaries the water column was found supersaturated with Me$_2$Se and estimated fluxes to the atmosphere are ranging between 0.4 and 160 nmol m^{-2} d^{-1}.

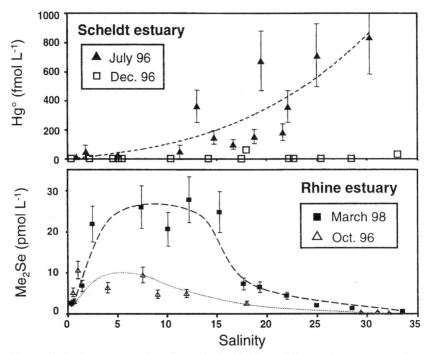

Figure 6: Typical seasonal profiles of volatile Hg and Se species concentrations as function of water salinity in estuarine environment.

Significant concentrations of volatile tin compounds have been discovered by Amouroux *et al.* in the different estuaries investigated (8). These species are probably originating from both methylation of natural or anthropogenic inorganic tin and tributyltin (TBT) released by ship antifouling paintings.

Tributylmethyl tin (MeBu$_3$Sn) is the most ubiquitous compound observed in water and exhibits significantly higher concentrations in the Scheldt (ca. 75-2000 fmol L^{-1}) than in the Rhine and the Gironde (ca. 5-125 fmol L^{-1} and 5-90 fmol L^{-1}, respectively) (22). In air samples, only tetramethyl tin (Me$_4$Sn) was detectable in the overlying atmosphere of the Scheldt and the Rhine estuaries with concentrations ranging between 1 pg m^{-3} and 20 pg m^{-3}. Nevertheless MeBu$_3$Sn overall distribution in all seasons and all estuaries always displays the same pattern with concentration maxima between salinity 10 and 15 which then decrease in lower and higher salinity. Volatile organotin compounds seem then to be produced in the estuary at intermediate salinity, within the estuarine maximum turbidity zone characterised by a strong sedimentation rate and potential remobilization of the particulate material. The occurrence of these compounds seems also to be directly related to the anthropogenic load of the investigated area. The Scheldt estuary appears, then, to be contaminated by these species compared to the Rhine and the Gironde.

Moreover, concentrations encountered in anoxic estuarine sediments of the Scheldt estuary are a thousand times higher than in the water above. Methylated butyl-Sn probably originates from the chemical and/or biological methylation in the sediment of anthropogenic butyl-Sn released in the estuarine environment. These results suggest that microbially mediated methylation mechanisms are likely to produce volatile organotin tin species into the estuary (8). These processes represent a significant remobilization pathway into the water column and then to the atmosphere of toxic volatile tin compounds accumulated in the sediment. Bu$_3$MeSn was found supersaturated in all seasons for all estuaries. Estimated fluxes are then strongly dependent on the contamination levels of the estuaries and range from 5 to 2400 pmol m^{-2} d^{-1} (22).

Volatile Iodine was mainly detected as iodomethane (MeI) and numerous other volatile iodine compounds (iodo-alkanes) were also observed in both air and water samples (20). MeI represents around 50 % of the total volatile iodine species encountered. Concentrations in estuarine waters are ranging from 1 to 15 pmol L^{-1} for the Rhine and the Gironde and from 1 to 100 pmol L^{-1} for the Scheldt estuary. Atmospheric sample concentrations display more homogeneous values (ca. 1-20 ng m^{-3}) in the three investigated areas. Here again the contamination level in MeI of natural waters seems to be dependent on the anthropogenic load of the studied environment. Water concentrations increase along the salinity gradient, with higher values during the warmer seasons and particularly in spring for the three estuaries. We also observed a significant spike of MeI concentrations at the mouth of the Scheldt estuary during spring phytoplanktonic bloom suggesting that plankton or algae primary productivity is directly involved in MeI production. Moreover the MeI distribution within the three estuaries shows a strong correlation with "active chlorophyll" and particularly during spring and summer. Biological light induced methylation mechanisms are probably involved in MeI production in estuarine waters. MeI was found supersaturated in all seasons and all estuaries with estimated fluxes to the atmosphere ranging between 0.5 and 170 nmol m^{-2} d^{-1}.

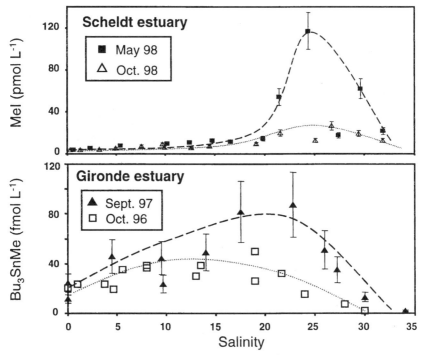

Figure 7: Typical seasonal profiles of volatile I and Sn species concentrations as function of water salinity in estuarine environment.

We then extrapolated our flux estimations to the whole European estuaries in order to evaluate on a more global scale the potential contribution of European estuarine emission of Mercury, Selenium, Tin and Iodine compared to natural and anthropogenic sources (see Table II). European estuarine flux estimations were based on the average flux densities of the three investigated estuaries, and were then extrapolated to the total European estuarine surface area estimated to be 111,200 km^2 (*23*). Our results suggest that estuaries are a significant source of those trace elements to the atmosphere, when compared to anthropogenic sources such as fossil fuel combustion.

Table II. Average flux density (FD, in nmol m^{-2} d^{-1}) and Water to air flux (in t yr^{-1}) of the investigated trace elements

	Hg°		$TVSe^a$		$TVSn^b$		$MeI\ (TVI)^c$	
	FD	Flux (t/yr)	FD	Flux (t/yr)	FD	Flux (t/yr)	FD	Flux (t/yr)
Scheldt	0.5	0.01	32	0.25	0.99	0.0114	16 (46)	0.20 (0.60)
Gironde	0.7	0.03	3	0.06	0.04	0.0011	5 (11)	0.10 (0.30)
Rhine	0.7	0.01	13	0.07	0.04	0.0004	6 (11)	0.05 (0.09)
European Estuaries		5		52		2		45 (116)

Flux calculations based on:

[a] TVSe concentrations (Total sum of the volatile Selenium species observed).

[b] TVSn concentrations (Total sum of the volatile Tin species observed).

[c] extrapolated TVI concentrations (Total sum of volatile Iodine species extrapolated from the relative abundance of iodomethane).

Conclusion

The three European estuaries investigated were found to be a source of gaseous species of Hg, Se, Sn and I. Seasonal results have been integrated and annually derived emission rates indicate that estuaries could contribute to a significant input of Hg and Se into the atmosphere. Since the larger uncertainty on flux estimations comes from the transfer velocity determination, our final flux calculations provides an order of magnitude of the natural evasion process. Available literature data for anthropogenic emission of trace elements are obtained within the same range of accuracy. Thus, the comparison between natural and anthropogenic emissions is suitable for biogeochemical assessments. In our estimation, volatile mercury and selenium emitted from European estuaries to the atmosphere represent 4 and 14% of European emissions from fossil fuel combustion, respectively [(132 tons of Hg in 1992 (24), 373 tons of Se in 1979 (25)]. We did not find available anthropogenic emission data concerning MeI, in order to evaluate the significance of Iodine evasion from European estuaries. For tin, fluxes do not appear to have a consequent impact on atmospheric input (26), but the importance of these volatile tin species has to be reassessed in term of contaminant remobilization and risk assessment in aquatic environments. In order to understand the biogeochemical pathways involved and predict trace elements volatilisation processes, we have now to focus on turbid and low oxygen areas of the estuary (sediments, maximum TZ), which probably represent privileged sites for microbial methylation (e.g. Bu_3SnMe, Me_2Se). Estuarine plumes with large exchange surface with the atmosphere exhibit intense photobiological and photochemical processes which should also be investigated to accurately estimate fluxes from the whole estuarine system (e.g. $Hg°$, MeI).

References

1. *Global Biogeochemical Cycles;* Butcher, S.S.; Charlson, R.J.; Orians, G.H.; Wolfe, G.V., Eds.; Academic Press: London, 1992; 379 pp.
2. Thayer, J.S. Environmental Chemistry of the Heavy Elements: Hydrido and Organo Compounds; VCH Publishers: New York, 1995; 145 pp.
3. Amouroux, D.; Donard, O.F.X. *Mar. Chem.* **1997,** *58,* 173-188.
4. Kim, G.; Hussain, N.;Church, T.M. *Tellus* **2000,** *52B,* 74-80.
5. Amouroux, D.; Liss, P.S.; Tessier, E.; Hamren-Larsson, M.; Donard, O.F.X. *Earth Planet. Sci. Let.* **2001,** *189,* 277-283.
6. Cooke, T.D.; Bruland, K.W. *Environ. Sci. Technol.* **1987,** *21,* 1214-1219.
7. Mason, R.P.; Fitzgerald, W.F.; Morel, F.M.M. *Geochim. Cosmochim. Acta* **1994,** *58,* 3191-3198.
8. Amouroux, D.; Tessier, E.; Donard, O.F.X. *Environ. Sci. Technol.* **2000,** *34,* 988-995.
9. Pécheyran, C.; Quétel, C.R.; Martin, F.M.; Donard, O.F.X. *Anal. Chem.* **1998,** *70,* 2639-2645.
10. Amouroux, D.; Tessier, E.; Pécheyran, C.; Donard, O.F.X. *Anal. Chim. Acta,* **1998,** *377,* 241-254.
11. Tseng, C.M.; De Diego, A.; Pinaly, H.; Amouroux, D.; Donard, O.F.X. *J. Anal. At. Spectrom.* **1998,** *13,* 755-764.
12. De Ligny, C.L.; Van Der Veen N.G. *Rec. Trav. Chim.* **1971,** *90,* 984-1001.
13. Craig, P.J. In *Organometallic compounds in the environment;* Craig, P.J., Ed.; Longman: London, 1986; pp 37-64.
14. Amouroux, D. Ph.D. thesis, University of Bordeaux I, Bordeaux, France, 1995.
15. Duce, R.A; Liss, P.S.; Merrill, J.T.; Atlas, E.L.; Buat-Menard, P.; Hicks, B.B.; Miller, J.M.; Prospero, J.M.; Arimoto, R.; Church, T.M.; Ellis, W.; Galloway, J.N.; Hansen, L.; Jickells, T.D.; Knap, A.H.; Reinhardt, K.H.; Schneider, B.; Soudine, A.; Tokos, J.J.; Tsunogai, S.; Wollast, R.; Zhou, M. *Global Biogeochem. Cycles* **1991,** *5,* 193-259.

16. Liss, P.S.; Merlivat, L. In *The role of air-sea exchange in geochemical cycling;* Buat-Ménard P., Ed.; D. Reidel Publishing Company: Dordrecht, Netherlands, 1986; pp 113-127.

17. Clark, J.F.; Schlosser, P.; Simpson, H.J.; Stute, M.; Wanninkof, R.; Ho, D.T. In *Air-water gas transfer;* Jaehne B. and Monahan E., Eds.; Aeon Verlag: Hanau, Germany, 1995; pp 785-800.

18. Wilke, C.R.; Chang, P. *AlChE J.* **1955,** *1,* 264-270.

19. Saltzman, E.S.; King, D.B. *J. Geophys. Res.* **1993,** *98,* 16481-16486.

20. Tessier, E.; Amouroux, D.; Abril, G.; Etcheber, H.; Donard, O.F.X. *Biogeochem.* **2001,** in press.

21. Amouroux, D.; Pécheyran, C.; Donard, O.F.X. *Appl. Organomet. Chem.* **2000,** *14,* 236-244.

22. Tessier, E.; Amouroux, D.; Donard, O.F.X. *Biogeochem.* **2001,** in press.

23. Frankignoulle, M.; Abril, G.; Borges, A.; Bourge, I.; Canon, C.; Delille, B.; Libert E.; Théate, JM. *Science (Washington D.C.)* **1998,** *282,* 434-436.

24. Pirrone, N.; Keeler, G.J.; Nriagu, J.O. *Atmos. Envir.* **1996,** *30,* 2981-2987.

25. Pacyna, J.M. *Atmos. Envir.* **1984,** *18,* 41-50.

26. Nriagu, J.O.; Pacyna, J.M. *Nature (London)* **1988,** *333,* 134-139.

Chapter 13

An Approach for Characterizing Arsenic Sources and Risk at Contaminated Sites: Application to Gold Mining Sites in Yellowknife, NWT, Canada

Kenneth J. Reimer, Christopher A. Ollson, and Iris Koch

Environmental Sciences Group, Royal Military College of Canada, Kingston, Ontario K7K 7B4, Canada

Yellowknife, Canada has an extensive soil arsenic contaminant problem as a result of 60 years of gold mining activity. Multivariate statistics (PCA) were used to determine that the natural concentration of arsenic in the area is also elevated (up to 150 ppm). As total arsenic concentrations may overestimate the actual risk posed to ecological and human health, sequential selective extraction (SSE) and a simulated gastric fluid extraction (GFE) were used to assess environmentally available and bioavailable fractions in soils. Subjecting various soil types to these techniques confirmed this hypothesis. The high arsenic content in crushed mine rock was neither environmentally available (<10% for SSE) or bioaccessible (<12% for GFE). Conversely, the low arsenic content in organic soils is more environmentally available (10-50% for SSE) and bioaccessible (>30% for GFE). Collectively, these techniques can be used to identify actual risks and develop effective remediation strategies.

Introduction

To the general public, arsenic is virtually a synonym for poison. Indeed, it has been ranked the number one priority hazardous substance by the Agency of Toxic Substances and Disease Registry (ATSDR) and the US-EPA, amongst all hazardous substances found at facilities on the National Priorities List (NPL) (*1*).

In most cases, soil is the contaminated matrix of concern, and its remediation is required to meet regulations. The tools to select site specific soil remediation guidelines, in an effort to minimize risk to human health from arsenic contamination, are currently not well defined. A recent survey of Records of Decision (RODs) concerning arsenic cleanup goals illustrates the lack of unification in standards and methods across the U.S. (*2*). The survey of 69 RODs found that 16% of cleanup goals were based on the attainment of background levels of arsenic, 75% were based on not exceeding a specific human health risk (ranging from 10^{-4} to 10^{-6}) and 9% were based on ecological or other risk considerations. Clearly these differences in the decision-making process indicate that a widely accepted and rigorous approach, applicable to all contaminated sites, does not yet exist.

The typical North American background concentration range of arsenic is between 5 and 14 ppm in soils. In areas associated with arsenic bearing minerals natural arsenic levels have been reported to be much higher, with background concentrations reaching 250 ppm (*3*). In the earth's crust arsenic is often associated with sulfide minerals, such as arsenopyrite (FeAsS) (*4*). This arsenic-bearing mineral is relatively stable under ambient environmental conditions and arsenic is only released through weathering over the geological time scale. However, the mining and smelting of this mineral, which is closely associated with gold and other precious and base metals, anthropogenically releases arsenic. It is not uncommon to find arsenic levels on these contaminated sites in the thousands of parts per million (ppm).

The first step in characterizing arsenic contaminated sites is to establish a natural background level in soil. Natural and anthropogenic sources of arsenic must be differentiated, involving difficult and controversial procedures. The process is especially problematic, yet necessary, when natural geology contributes to environmental arsenic levels that are substantially higher than guidelines. When little or no pre-anthropogenic input information is available, and when an area is geologically isolated, the process is complicated further.

Others have noted that basing a soil cleanup level on background levels may result in over-remediation and unnecessary cleanup costs (*2*). The establishment of the natural background level of arsenic on contaminated sites merely sets a benchmark below which remediation is not warranted. Risk to human or ecological health, as a more relevant basis, is then crucial for establishing the level to which the site should be cleaned up.

Traditional risk assessment methods incorporate only total arsenic into their calculations (5). This may seriously overestimate risk because the following assumptions are made: (1) that all arsenic in a matrix is toxic, and (2) that all the arsenic is available for exposure. When dealing with soils the first assumption is incorrect when one considers the chemical fraction in which the arsenic is bound, since not all forms of arsenic are toxic (e.g., arsenopyrite is non-toxic). The second assumption is also incorrect when bioavailability is considered. For example, bioavailabilities of arsenic from soils, tailings, slag and household dust in test animals ranged from 0 to 98% with a mean of 30% (bioavailability calculated relative to toxicity tests) (6). The use of one or both of these assumptions overestimates exposure, and consequently, risk.

The first section of this chapter describes the use of principal components analysis to distinguish between soils that are naturally elevated in arsenic and those that have been anthropogenically influenced by mining activity, in Yellowknife, Canada. A case study of a townsite associated with one of the local mines demonstrates the evaluation of the chemical nature and potential bioavailability of arsenic in soil. This combined approach to characterizing arsenic mobility and bioavailability is essential when establishing the true risk that arsenic contamination poses to human or ecological health.

Study Site

Yellowknife has been an active gold mining community since 1938. The Miramar Giant mine is five kilometres north of the city and the Miramar Con mine borders the southern city limit (Map 1). For several years it has been known that gold mining has increased the levels of arsenic in several areas of Yellowknife. Since the gold is found with arsenopyrite, the milling of this arsenic-rich ore has generated a considerable amount of arsenic waste in the form of aerial emissions, solid waste (e.g., tailings) and liquid effluent.

Yellowknife was chosen as the study site because environmental contamination results from both naturally and anthropogenically introduced arsenic, which is the sole contaminant of concern (7). More importantly, public concern about arsenic contamination was, and continues to be, the driving force behind the characterization of risk and associated research in Yellowknife.

169

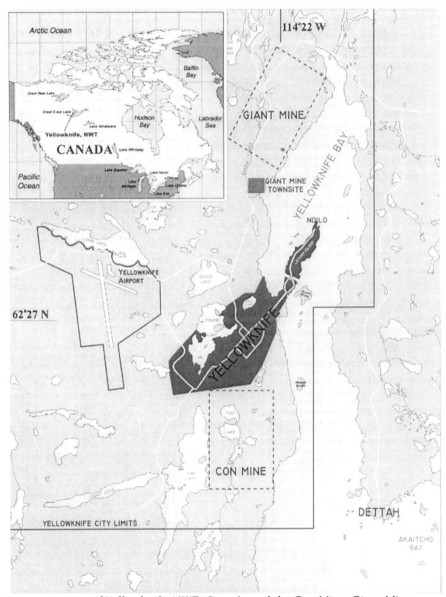

Map 1. Location of Yellowknife, NWT, Canada and the Con Mine, Giant Mine and the Giant Mine Townsite.

Methods

Total Arsenic in Soils: The fieldwork was carried out in Yellowknife in the summers of 1997 to 2000, using a targeted approach to obtain samples representative of the various geographical areas. Methods for soil sampling and analysis by neutron activation analysis (As, Sb, Fe, Na, K and Au) and ICP-OES (Zn, Cu, Ni and Mn) or AAS have been described in detail elsewhere (7,8).

Sequential Selective Extraction: Samples (<80 mesh sieved fraction) were submitted to ALS Chemex for sequential selective extraction using methods detailed by Hall (9,10,11).

Gastric Fluid Extraction: The gastric fluid extraction method was based on methods described in published literature (12,13). The <80 mesh sieved soil was used in this experiment, to compare with the sequential extraction. A 1:100 ratio of soil to gastric fluid (pH 1.8 with HCl, NaCl, pepsin and organic salts) was used in a 37 °C extraction for one hour, and then for four hours (following adjustment to mimic intestinal conditions). The supernatant was analyzed by ICP-OES or hydride generation-AAS (following acid digestion).

Statistical Analysis: The solid-phase elemental compositions of samples (using the ten elements As, Sb, Fe, Au, Ni, Cu, Zn, Mn, K, Na) were compared using the multivariate statistical technique, principal components analysis (PCA), with the statistical program SYSTAT® 8.0. The data for each metal concentration and sample set were normalized (typically using log, log10, and square root functions) to eliminate the effect of a large range of metal concentrations over the data set. An ANOVA was used to compare the means from the three different groups identified within the PCAs. Linear regression and correlation were examined with Excel.

Quality Assurance and Quality Control (QA/QC): Rigorous QA/QC measures, consisting of field duplicates, lab duplicates, soil standard reference materials and blanks, were included in the field and analytical phases of the project. The results were within acceptable controls and limits, in accordance to US EPA guidelines (14), indicating that the data are reliable.

Results

Distinguishing Between Natural and Anthropogenic Arsenic

The existence of little or no pre-mining environmental information and the ensuing absence of a baseline natural level of arsenic in Yellowknife complicates the differentiation of natural and anthropogenic arsenic. Thus the first step was to determine total arsenic concentrations in soils for the Yellowknife area. The highest mean concentrations were not surprisingly found in the tailings ponds on both mine sites, and elevated levels were found in the land surrounding the mine properties. The highest concentration of arsenic in a single sample (89 000 ppm) was found at the base of the Giant Mine roaster stack. The average arsenic concentration in the City of Yellowknife was 32±34 ppm. The levels are summarized in Table 1 (*7*, *15*, *16*, *17*).

Table 1. Arsenic concentrations in soils in the Yellowknife area (SD =standard deviation).

Location	Low concentration (ppm)	High concentration (ppm)	Mean concentration (ppm±SD)
Giant Mine Tailings	1740	4840	3260±950
Giant Mine Property	41	89 000	350±69
Con Mine Tailings	1400	25 000	6311±7100
Con Mine Property	5	1170	118±350
City of Yellowknife	3	148	32±34

Principal Components Analysis

Principal components analysis was used to examine the samples summarized in Table 1, which contain arsenic derived from both natural processes and mine activities. PCA is a powerful technique that can be used to examine relationships of samples based on their elemental composition (*8*). Each sample is characterized by its position on a reduced (usually two- or three-dimensional) plot. The axes of the plot are linear combinations of the original *n* variables. The proximity of samples to each other on this plot indicates similarity in their elemental makeup. The selection of the ten elements was based on those that are environmentally important and whose concentrations are influenced by the mining and milling process. For example, arsenic, antimony and gold are components of the mined ore; iron and zinc are additives in the milling and effluent treatment process. The positions of all samples are shown on a 2-D plot in Figure 1.

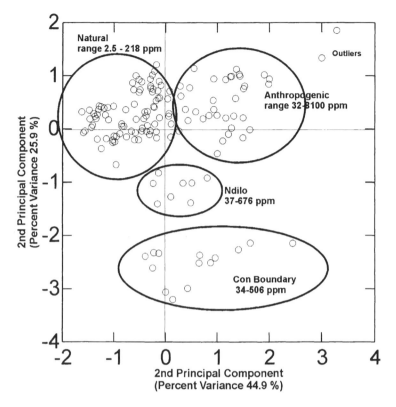

Figure 1. Principal components analysis biplot of soil and tailings samples from the Yellowknife area.

Examining the PCA for samples that cluster together reveals four distinct groupings, from a total of 174 samples, with the two clearest ones on the right side (tailings and mine-impacted locations, or anthropogenic samples) and left side of the plot (mine-unimpacted locations, or background samples). The results of ANOVA statistical analysis proved that there was a significant difference between these groupings ($p<0.01$).

The arsenic soil concentrations in the left ellipse (background) range from 2.5 to 218 ppm (median 33 ppm). From this observation, and since greater than 99% of samples contain <150 ppm arsenic, the typical background concentration of arsenic in Yellowknife was determined to be 3-150 ppm.

The ellipse on the right (anthropogenic) contains samples with arsenic concentrations ranging from 22 to 31 000 ppm. These samples consist of mine tailings, crushed overburden rock fill and soils that have been directly impacted by fugitive arsenic emissions from mining operations.

The two other loose groupings include sample locations from Ndilo and the Con Mine boundary that contain elevated levels of arsenic. These areas are intermediate between the city and those that are mine-impacted. These soil samples may have been influenced by aerial emissions or they may reflect a higher natural surficial geological expression of arsenic.

Giant Mine Townsite: A Case Study for the Characterization of Arsenic Form and Mobility

The Giant Mine Townsite samples were used to assess the risk posed by arsenic in the soils. The Townsite was built in the 1950's to house mine workers and their families. Currently, 22 residences surrounded by lawns as well as a central playground built on sand are situated on the property. The Townsite roadways are all constructed of crushed rock fill (overburden) from the mine. Recently, regulators raised concerns that arsenic contamination in the area may be potentially affecting the health of the residents. Not surprisingly, this concern focused only on the fill introduced from the mine site.

Total Arsenic concentrations in Soil Samples

The total arsenic concentrations in soil samples on the Giant Mine Townsite average 1493±1226 (range 52 to 3705 ppm). The reason for the large variability is that there are distinct soil types. The highest concentrations of total arsenic are present in the crushed rock fill material (samples locations 6 – 15 on Fig. 2, 2343±63 ppm), while the lawn samples have a much lower total concentration of arsenic (sample locations 1-3, 62±7.5 ppm). A student's t-test verified a statistical difference between the means (df=8, t=9.233, p<0.05). Two other samples, different from the aforementioned ones, complete the data set. They are sand from the central playground (sample location 4, 272 ppm arsenic) and organic soil overlying arsenopyrite bearing rock (sample location 5, 784 ppm arsenic).

Two methods assessing the form and mobility of arsenic in these soil samples are described in the following sections. Sequential selective extraction (SSE) is an accepted method commonly used to provide information about the mobility and solubility of arsenic from soils and sediments (9, 10, 11, 18). Gastric fluid extraction (GFE) is a new approach that is used to assess bioaccessibility (12, 13, 19).

Sequential Selective Extraction

Sequential selective extraction (SSE) involves subjecting a soil sample to successive chemical extractions that are increasingly effective in dissolving the sample. SSE results can reveal the solid phase to which arsenic is bound, based on its solubility in the successive extractants, in seven phases as follows. The first three phases are adsorbed/exchangeable, carbonates, and organics, and are considered to be potentially environmentally available under ambient conditions. The next two phases are amorphous iron oxyhydroxides and crystalline iron oxide and are potentially environmentally available under specific Eh-pH conditions. The last two phases, sulfides and residual, are not considered to be environmentally available under most conditions.

Results for soils subjected to SSE are summarized in Figure 2. Lawn soils, with lowest arsenic content, contain the greatest proportion of environmentally available (i.e., adsorbed/exchangeable, carbonates and organics) arsenic. The remainder of arsenic is bound in iron oxide (less available) and sulfide (unavailable) fractions. On the other hand, the higher levels of arsenic in crushed rock are mostly found in sulfide fractions. A smaller proportion of the arsenic is found in iron oxides, and very little is environmentally available. Arsenic in sand (sample location 4) is mostly in the sulfide form, although approximately 40% is environmentally available.

Gastric Fluid Extraction

The second method, GFE, is conducted under environmental conditions and with solvents that mimic the human gastro-intestinal system. GFE is helpful in assessing arsenic bioaccessibility, which has been linked to bioavailability, in soil (*12, 13*) that might be injested.

Arsenic concentrations resulting from GFE of Townsite soils ranged from 21 ppm to 343 ppm with respect to the original sample (corresponding to 5-73% of total arsenic). Greater than 30% , and up to 73%, of the total arsenic in lawn and organic samples was extracted by GFE (Figure 3).

A marked decrease was observed in the percent of total arsenic extracted from crushed rock material compared with extraction from lawn soils. The proportion of arsenic extracted ranged from 5 to 23% with less than 12% extracted from the majority of crushed rock samples.

Comparison of SSE and GFE Results

A comparison of the percent arsenic extracted with GFE and the various phases of SSE reveals interesting trends. For example, although the arsenic content is high in the crushed rock, very little is environmentally available (<10% for SSE) or bioaccessible (<12% for GFE). On the other hand, the low arsenic content in lawn soils is more environmentally available (10-50% for SSE) and bioaccessible (>30% for GFE).

The relationship between percent arsenic in adsorbed/exchangeable, carbonate and organic fractions, and percent arsenic from GFE is illustrated in Figure 4 (R^2=0.77, y=0.58x–0.50). Although a reasonable relationship exists between the environmentally available fraction predicted by SSE and the bioaccessible fraction predicted by GFE, the slope of 0.58 indicates an underestimation of bioaccessibility by these fractions of the SSE. In other words, the bioaccessible fraction includes not only adsorbed/exchangeable, carbonate and organic phases, but perhaps iron oxides as well. Adding the iron oxides to the SSE predicted fraction, however, causes the slope to increase to 3.6-4.5, indicating that this composition far overestimates the bioaccessible fraction.

Conclusion

The establishment of background levels, ranging up to 150 ppm arsenic in Yellowknife, clearly demonstrates that cleaning up below this value would not be practical. Potential arsenic exposure at the Townsite, as measured by GFE, reveals some interesting trends. The lawn soils, despite a high percentage of extractable arsenic, present the lowest exposure risk. Sand from the playground area, with a total arsenic concentration of 272 ppm, would typically alarm local residents, but actual arsenic exposure is the same as for the lawn soils. The highest exposures are equally attributable to the crushed rock fill and the organic soil overlying native bedrock. Further assessment is required to determine if such substrates constitute actual risk, but the application of PCA, SSE and GFE techniques clearly provides a better foundation for risk assessment and the development of effective remediation strategies.

References

1. Agency for Toxic Substances and Disease Registry (ATSDR) and US-Environmental Protection Agency (EPA), 1999 CERCLA List of Priority Hazardous Substances, 1999, http://www.atsdr.cdc.gov/99list.html, accessed 11 October 2001.
2. Davis, A., Sherwin, D., Ditmars, R., Hoenke, K.A. *Environ. Sci. Technol.* **2001**, 35, 2401-2406.
3. Tanaka, T. *Appl. Organomet. Chem.* **1988**, 2, 283-295.

176

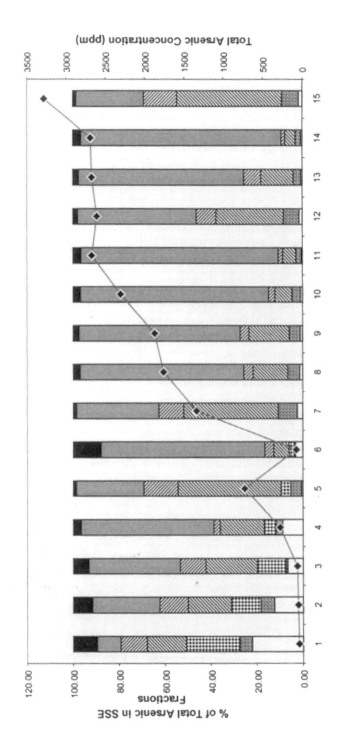

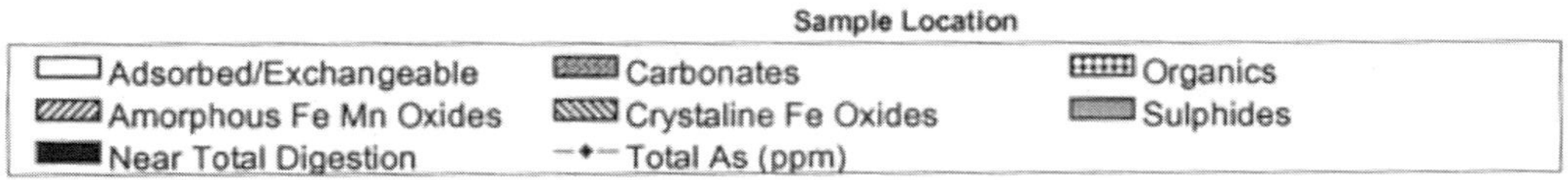

Figure 2. Percent arsenic in sequential selective extractions in soil samples from the Giant Mine Townsite. The lawn soil samples (1 to 3) are shown on the left hand side of the graph with the sandbox (4) and the organic sample overlying bedrock (5). The crushed rock fill materials are on the right side of the graph (6 to 15).

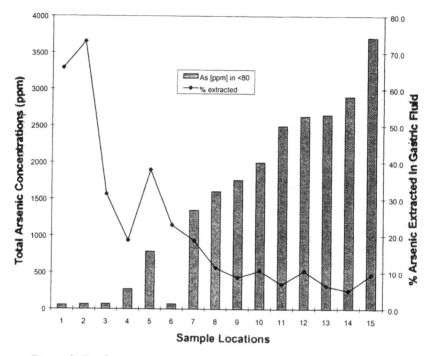

Figure 3. Total arsenic concentrations (ppm), shown by the bars and represented on the left hand side of the plot, and the percent arsenic extracted using gastric fluid analysis, indicated by the line, for the different samples.

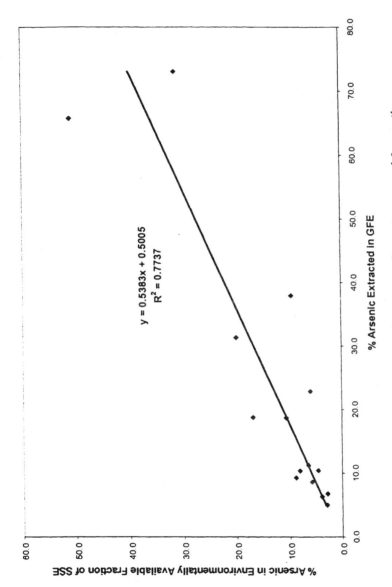

Figure 4. Correlation between the percent of arsenic extracted from soil samples using GFE and the environmentally available fraction of SSE.

4. Cullen, W.R., Reimer, K.J. *Chemical Review* **1989**, 89, 713-764.
5. "A Protocol for the Derivation of Environmental and Human Health Soil Quality Guidelines." 1996. Canadian Council of Ministers of Environment (CCME), CCME EPC-101E.
6. Ruby, M.V., Schoof, R., Brattin, W., Goldade, M., Post, G., Harnois, M., Mosby, D.E., Casteel, S.W., Berti, W., Carpenter, M., Edwards, D., Cragin, D., Chappell, W. *Environ. Sci. Technol.* **1999**, 33, 3697-3705.
7. Ollson, C.A., Koch, I., Reimer, K.J., Walker, S.R., Jamieson, H.E. *Characterization of arsenic in solid phase samples collected on the Giant Mine Townsite, Yellowknife, NWT.* Prepared for Royal Oak Project Team, Indian and Northern Affairs Canada. 2001. ESG, Royal Military College of Canada, Kingston.
8. Bright, D.A., Coedy, B., Dushenko, W.T., Reimer, K.J. *Environ. Sci. Technol.* **1994**, 155, 237-252.
9. Hall, G.E.M., Gauthier, G., Pelchat, J.C., Pelchat, P., Vaive, J.E. *J. Anal. At. Spectrom.* **1996**, 11, 87-96.
10. Hall, G.E.M., Vaive, J.E., Beer, R., Hoashi, M. *J. Geochem. Explor.* **1996**, 56, 59-78.
11. Hall, G.E.M., Vaive, J.E., MacLaurin, A.I. *J. Geochem. Explor.* **1996**, 56, 23-36.
12. Ruby, M.V., Davis, A., Schoof, R., Eberle, S., Sellstone, C.M. *Environ. Sci. Technol.* **1996**, 30, 422-430.
13. Rodriguez, R.R., Basta, N.T., Casteel, S.W., Pace, L.W. *Environ. Sci. Technol.* **1999**, 33, 642-649.
14. "Guidance on Data Quality Indicators." 2001. U.S. Environmental Protection Agency. Quality Staff. Office of Environmental Information. Washington, DC, EPA QA/G-5i.
15. Mace, I. M.Eng. Thesis, Royal Military College of Canada, 1998.
16. Ollson, C.A. M.Sc. Thesis, Royal Military College of Canada, 2000.
17. Environmental Sciences Group (ESG). *Arsenic Levels in the Yellowknife area: Distinguishing between natural and anthropogenic inputs.* Prepared for the Yellowknife Arsenic Soil Remediation Committee (YASRC). 2001. ESG, Royal Military College of Canada, Kingston, RMC-CCE-ES-01-01.
18. Keon, N.E., Swartz, C.H., Brabander, D.J., Harvey, C., Hemond, H.F. *Environ. Sci. Technol.* **2001**, 35, 2778-2784.
19. Hamel, S.C., Buckley, B., Lioy, P.J. *Environ. Sci. Technol.* **1998**, 32, 358-362.

Chapter 14

Arsenic and Zinc Biogeochemistry in Acidified Pyrite Mine Waste from the Aznalcóllar Environmental Disaster

Ángel A. Carbonell-Barrachina, Asunción Rocamora,
Carmen García-Gomis, Francisco Martínez-Sánchez,
and Francisco Burló

División Tecnología de Alimentos, Departamento de Tecnología
Agroalimentaria, Universidad Miguel Hernández, Carretera de Beniel, km
3.2, 03312 Orihuela, Alicante, España

This laboratory experiment systematically examines arsenic, iron, zinc, and sulfate solubility and fractionation in pyrite mine waste suspensions as affected by redox potential (Eh). Under aerobic conditions, As solubility was low, however, under moderately reducing conditions (0-100 mV), As solubility significantly increased due to dissolution of iron oxy-hydroxides. Upon reduction to -250 mV, As solubility was controlled by the formation of insoluble sulfides, and as a result soluble As contents dramatically decreased. Soluble Fe concentration increased with time under anaerobic conditions, whereas, it decreased under aerobic conditions likely due to formation of insoluble oxy-hydroxides. Under aerobic conditions, soluble Zn significantly increased with incubation time and reached concentrations as high as 800 mg kg^{-1} waste. Zinc was initially present as insoluble zinc sulfides. However, after further oxidation, sulfide was transformed to sulfate and Zn^{2+} was then released into the waste solution. Selective extraction of incubated wastes illustrated that arsenic biogeochemistry was mainly controlled by As bound to: a) amorphous Fe oxy-hydroxides and b) insoluble organics and sulfides. Remediation of a site polluted by both arsenic and zinc is quite complicated because the redox conditions favoring insolubility of arsenic favors maximum solubility of zinc, and *vice versa*. Therefore, the best way for the remediation of an arsenic- and zinc-polluted environment, in our opinion, is: a) phyto-remediation with plants accumulating large amounts of Zn in their tissues, and b) simultaneous addition of: 1) amorphous iron oxy-hydroxides (aerobic Eh) or 2) organic matter rich in S compounds (anaerobic Eh).

The southwest of Spain is rich in sulfide deposits; in particular, at Aznalcóllar (Seville) zinc, lead, copper and manganese-rich pyrite deposits (FeS_2) are mined. In these mines, sulfuric acid is used to extract metals from floated pyrite, and the highly acidic metal-contaminated waste from this process is stored in massive tailings ponds (*1*).

In April 1998, part of the tailings pond dike of Los Frailes zinc mine [Aznalcóllar, Seville (Spain)] collapsed, releasing an estimated 5,000,000 m^3 of acidic waste rich in toxic metals and metalloids, including As (up to concentrations of 5000 mg kg^{-1}), posing a long-term threat to the Guadalquivir ecosystem. This ecosystem encompasses the World Heritage site of the Doñana National and Natural Parks (one of the largest wetland areas in Europe) (*1*).

When environmental accidents occur, fast responses from authorities and the scientific community are required. In order to choose the most appropriate solutions, it is necessary to have a detailed and wide knowledge on the processes controlling the biogeochemistry of potential contaminants, including heavy metals and metalloids. These processes include metal adsorption/desorption reactions onto soils, precipitation of insoluble compounds and metals release into the soil solution, among others.

There is a dearth of information in the literature regarding reasonable ways of decontaminating As polluted sites because As is not easily accumulated in plant tissues. The use of hyper-accumulators is one of the most used methods for the removal of heavy metals from polluted lands (*2*).

Once arsenic is present in a particular environment, in this particular case when the waste has reached agricultural soil solution and/or river sediment solution, it can follow any of the reactions described in Figure 1. Our main goals

183

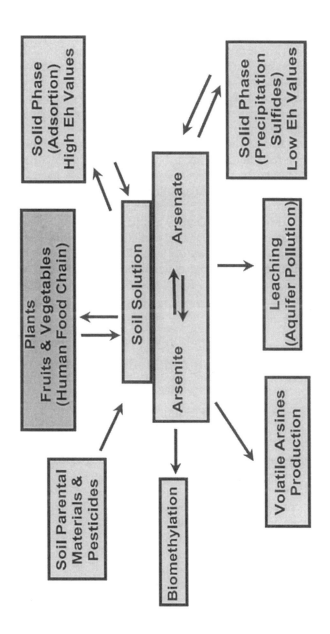

were: a) to keep a reasonable level of soluble As for facilitating As accumulation in plant tissues (bio-remediation: reaction 7), or under any other circumstances b) to minimize As solubility by implementing adsorption onto iron and manganese oxides (reaction 1) or precipitation reactions (reaction 2), forming insoluble compounds or complexes with sulfide and/or organic matter.

The two chemical species controlling As chemistry and toxicity in soils, sediments and wastes are arsenate and arsenite. Arsenate is the thermodynamically stable form under aerobic conditions; it is a chemical analogue of phosphate (3) and it is mainly adsorbed onto iron and manganese oxides. Arsenite is the predominant species under anaerobic conditions; it is a neutral species at natural pH values and is more soluble, mobile and phytotoxic than arsenate (4). Arsenic is subject to chemically and/or microbiologically mediated oxidation-reduction, and methylation reactions in soils, wastes, and natural waters (5). The most fundamental interactions among As chemical species are redox reactions leading to transformation of species.

In this study, a laboratory experiment was conducted that allowed the analysis of As, Fe, Zn and SO_4^{2-} solubility and partitioning in pyrite mine waste suspensions under controlled redox and pH conditions. The study focused on the influence of redox potential (Eh) on the solubility of As in mine waste from Aznalcóllar (Seville) to identify redox conditions that can limit or enhance arsenic solubility as desired in order to: a) immobilize it (insoluble compounds) in this specific material or b) facilitate As uptake and accumulation by trace element accumulating plants (bio-remediation).

Materials and Methods

Acidified Pyrite Mine Waste Equilibrium

Acidified pyrite mine waste suspensions were equilibrated (at 25 ± 2 °C) in laboratory microcosms at various redox conditions using a modification (6) of the redox-pH control system developed by Patrick et al. (7). In this system, the Eh is automatically maintained at a pre-selected potential level. Platinum electrodes in the suspension were connected to a potentiometer (Knick Stratos 2401) to provide a continuous measurement of the waste suspension Eh.

In the absence of added O_2, chemical and microbial processes caused the Eh to decrease. Whenever the Eh dropped below the desired level a small amount of O_2 (via air) was added to the system to maintain the desired Eh. Nitrogen gas was effective in purging excess air and in preventing a buildup of gaseous decomposition products such as CO_2 and H_2S. Using this system, the desired Eh was maintained within ± 10 mV of the target level.

Suspensions were prepared using 1500 mL of deionized water plus 200 g of acidic waste from the Aznalcollar spill out, sampling area Q (Figure 2). The main properties of this waste are given in Table I together with the maximum concentrations of trace elements found in the literature.

Table I. Concentration of Main Trace Elements in the Pyrite Mine Waste Used in This Experiment and Maximum Concentrations Found in this Same Spill Out (8).

Trace Element	Concentration (mg kg^{-1})	
	Literature (8)	This Experiment
Zinc	8000	6272
Lead	8000	-
Arsenic	5000	1741
Copper	2000	1364
Cadmium	28	-
Mercury	15	-
Calcium	-	1599
Magnesium	-	4138
Potassium	-	651
Iron	-	140500
Manganese	-	456

In this experiment, waste suspensions were equilibrated under controlled redox and pH conditions. The redox conditions selected were: a) *highly aerobic*: atmospheric air was continuously introduced into the system; b) *moderately anaerobic:* 0 mV; air or oxygen-free nitrogen were introduced as necessary to keep the Eh at the pre-selected level; c) *highly anaerobic*: oxygen-free nitrogen was continuously introduced into the system. Microcosms were set up in duplicate.

The initial pH of the waste was 2.4, however it was manually adjusted twice a day to pH 7.0 by additions of 2 M HCl or NaOH, as required. This adjustment was made in order to have only one parameter under study, the Eh; in further studies combinations of both Eh and pH will be carried out. The addition of HCl will not affect As partitioning because As sorption/desorption processes and oxidation/reduction reactions are not strongly related to variations in concentrations of Cl$^-$ (9).

After an acclimatization period of 7 days and until the end of the experiment (3 months), an aliquot was withdrawn weekly from each microcosms,

186

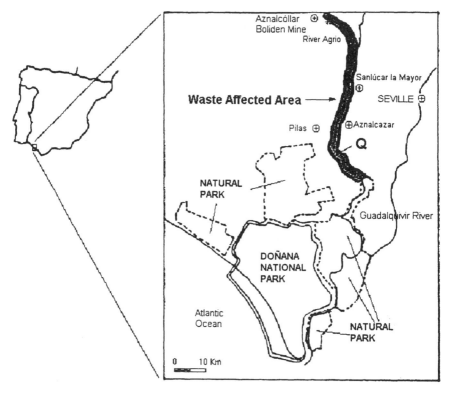

Figure 2. Map of the affected area, including Doñana National and Natural Parks. The sampled area is marked in the map by the letter Q and corresponds to the Quema area.

centrifuged (30 min at 5000 rpm, Suprafuge 22, Heraeus Sepatech, Sorvall SA-600 rotor, Am Kalkber, Germany), and filtered through a 0.45-µm micropore filter (Whatman Inc., Clinton NJ, USA) under an inert N_2 atmosphere using a pressure-vacuum system (*10*). Concentrated HCl (200µL/20 mL) was added to preserve this sample aliquot. Concentrations of selected soluble metals: As, Cu, Fe, Mn, Zn, Ca and Mg were determined by atomic absorption spectrometry (AAS) and K and Na by atomic emission spectrometry (AES).

Soluble sulfate was also determined in this same aliquot. Sulfate analysis was performed in a Dionex Ion Chromatograph Model 2010I equipped with a Dionex AS45 4 mm analytical column and a conductivity detector. The eluent solution composition was 2.2 *mM* Na_2CO_3 and 2.8 *mM* $NaHCO_3$ and the regenerant solution was 0.075 *N* H_2SO_4.

Quality assurance was conducted by spiking withdrawn samples with certified element standards.

Sequential Extraction Procedures

Water-soluble phase (F0): Samples from the acid waste suspensions were taken weekly throughout the 3 months of incubation. The suspension samples were centrifuged and the supernatant filtered through a 0.45 µm membrane filter. This supernatant was assumed to be water-soluble.

The first and last weeks, following removal of the water-soluble phase (F0), the waste was sequentially extracted into five fractions (F1 to F5) using procedures described below (*11, 12*). The waste samples were kept under an O_2-free N_2 atmosphere during all extraction reactions.

- *F1-Exchangeable phase:* The solid phase after separation of the water soluble fraction was extracted at room temperature for 30 min with 8 mL 0.5 M $Mg(NO_3)_2$ (g^{-1} dry weight waste), adjusted to pH 7.0 with HNO_3. The samples were agitated continuously.
- *F2-Bound to carbonate phase:* The waste residue from F1 was leached at room temperature for 5 h with 8 mL, 1M NaOAc, adjusted to pH 5.0 with acetic acid for 1 g dry weight waste. These samples were also agitated continuously.
- *F3-Bound to Fe and Mn-oxides phase:* The waste residue from F2 was extracted at 96°C for 6 h with 20 mL 0.08 M $NH_2OH \cdot HCl$ in 250 mL L^{-1} acetic acid for 1 g dry weight waste. These samples were occasionally agitated.
- *F4-Bound to organic matter and sulfides phase:* For 1 g of dry weight waste, the waste residue from F3 was extracted at 85 °C for 2 h with 3 mL 0.02 M HNO_3 and 5 mL 300 mL L^{-1} H_2O_2 (adjusted to pH 2.0 with HNO_3) were added, and extraction continued at 85 °C for another 3 h.

The sample was then cooled, 5 mL 3.2 M NH_4OAc in 20 mL L^{-1} HNO_3 was added, and the sample was diluted to 20 mL with deionized water. The samples were agitated continuously for 30 min. NH_4OAc was added to prevent adsorption of extracted metals onto the oxidized waste.

- *F5-Mineral matrix phase:* The waste residue from F4 was extracted with 25 mL concentrated HNO_3 for 1 g of dry weight waste at 105 °C, the waste was digested until 5 mL solution was left, and the sample was diluted to 25 mL with deionized water.

The above sequential extractions were conducted in 50 mL centrifuge tubes, which prevented any loss of waste between the successive extractions. Separation was conducted by centrifuging at 5000 rpm for 30 min. Supernatants were filtered using 0.45 μm Millipore filters and then analyzed for As. The residues were rinsed with 8 mL deionized water for 1 g dry weight waste and centrifuged at 5000 rpm for 30 min. These second supernatants were discarded.

Statistical analyses were performed using Statgraphics® Plus version 5.0.

Results and Discussion

Effect of Eh on Total Soluble Arsenic, Iron and Zinc

Figure 3 summarizes the effect of Eh and time on As, Fe, Zn and SO_4^{2-} solubility (total soluble concentrations) in acidic waste suspensions. Figures 3a and 3b show a similar trend in both elements As and Fe. Therefore, the large increase in As solubility in reducing environments was probably linked to the reductive dissolution of hydrated iron oxides. Arsenic solubility in soils and sediments is believed to be controlled by sorption-desorption mechanisms (*13, 14*); pH (*14, 15*); the amount and type of clay (*14, 16*); and iron oxides (*17*). In our experiment, water-soluble Fe concentrations were highly correlated ($P <$ 0.01) with dissolved total As, suggesting that Fe hydrous oxides are also important in controlling As adsorption-desorption reactions in waste environments. Dissolution of Fe oxy-hydroxides upon reduction and subsequent release of adsorbed As led to increased dissolved As concentrations. The absence of correlation between soluble Mn and As (data not shown) indicated that Mn-oxides were less important in controlling As solubility.

On the other hand, zinc shows a different solubility pattern (Figure 3c). Under aerobic conditions, soluble Zn significantly increased with incubation time and reached concentrations as high as 800 mg kg^{-1} waste. Perhaps, zinc was initially present as insoluble zinc sulfides (zinc blende and/or spharelite), however, after further oxidation (Eh $\cong$ -150 mV), sulfide was transform to sulfate and the Zn^{2+} cations were then released into the waste solution (Eh > 200 mV). The high soluble concentrations reached under aerobic conditions at the end of the experiment will indeed represent a serious threat to the environment.

Figure 3d shows increments of soluble sulfate independently of the redox conditions. These increases seem to indicate that most of the sulfur comes from initial insoluble sulfides which with time become more unstable even under moderately and highly anaerobic conditions.

Figure 4 summarizes the effect of Eh on As, Fe and Zn solubility (total soluble concentrations) in acidic waste suspensions. Figures 4a and 4b are clear examples of how As and Fe biogeochemistry are related. Under aerobic and moderately anaerobic conditions [+400, +100] mV, the mobility and solubility of As in waste solution was closely related to the presence and behavior of Fe. As will be eventually co-precipitated and became immobilized by the formation of insoluble, hydrated iron oxides (18). Immobilization (to a lesser extent) will also occur with other high surface-area compounds such as organic matter (19). In general, no significant water soluble As or Fe were detected until an Eh below – 50 mV was reached. At this point, both pentavalent As and ferric iron should have started to reduce to trivalent arsenic and ferrous iron, respectively (Figure 5). Maximum soluble concentrations of both elements were found at the interval [–50, -150] mV; however, soluble As started to decrease at –200 mV, redox potential at which sulfate could have started to reduce to sulfide (Figure 5), and this new formed sulfide could have reacted with arsenite to formed insoluble compounds (although this comment is only based on one point).

The influence of redox on As solubility in soils (14) was found to be governed by: 1) reduction of arsenate to arsenite followed by desorption, and 2) the dissolution of Fe-oxyhydroxides and concurrent release of coprecipitated arsenate. In recent experiments, Carbonell-Barrachina et al. (12) found that besides these two controlling factors, in sewage sludge suspensions there was a new and crucial factor, not important in soil environments, i.e. the strong affinity of arsenite for S and the potential formation of insoluble As sulfide minerals. In this way, the waste geochemistry seems to resemble more to the sewage sludge than to the soils.

Sulfide reacts with many divalent transition metal ions (e.g. Fe^{2+}, Cd^{2+}, Cu^{2+}, Hg^{2+}, Ni^{2+}, Pb^{2+}, and Zn^{2+}) to form very insoluble precipitates (20). Arsenic has been reported to have a strong affinity for S (18); the observed decrease in dissolved As concentrations upon reduction from approximately –150 mV to -200 mV (Fig. 4a) indicated that As solubility was perhaps limited by the formation of insoluble As sulfide minerals. Therefore, after arsenate is reduced to arsenite in reducing waste and, if sulfur is abundant, most of the As reacts with sulfides to form insoluble As sulfide minerals [realgar (AsS), orpiment (As_2S_3), and/or inclusions in copper and zinc sulfides]. Further research is,

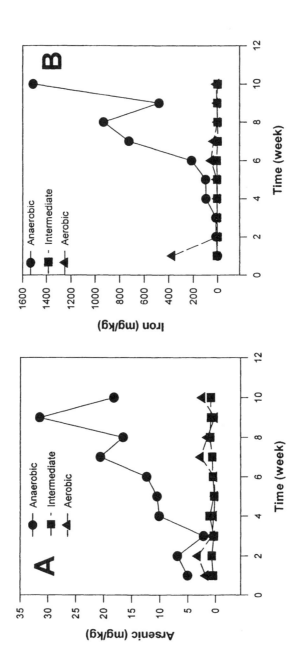

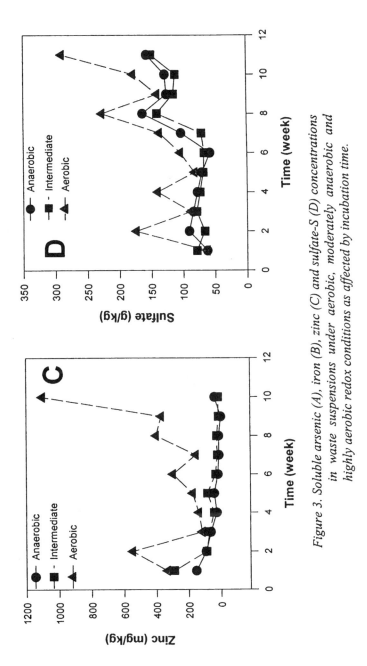

Figure 3. Soluble arsenic (A), iron (B), zinc (C) and sulfate-S (D) concentrations in waste suspensions under aerobic, moderately anaerobic and highly aerobic redox conditions as affected by incubation time.

192

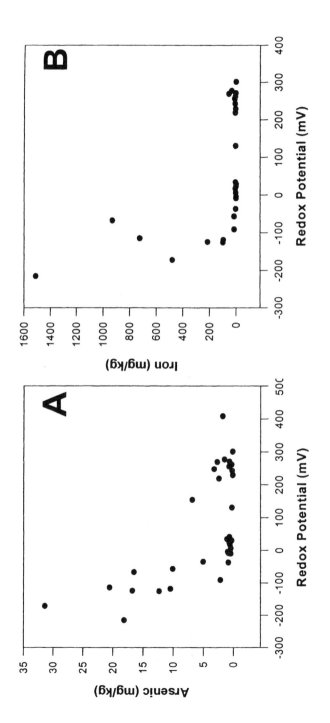

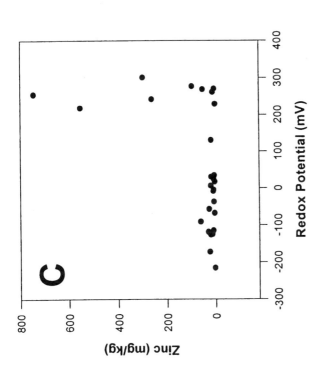

Figure 4. Soluble arsenic (A), iron (B) and zinc (C) concentrations in waste suspensions as affected by redox conditions.

194

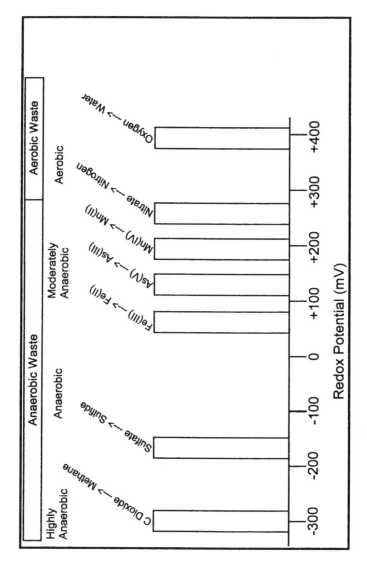

Figure 5. Main redox reactions controlling the biogeochemistry of heavy metals and trace elements in any environment.

however, needed to isolate these minerals and to establish the conditions of formation of each one of them.

Arsenic sulfides are reported to be easily oxidized under aerobic conditions (21). Therefore, in areas where redox fluctuations are possible or where removal of the wastes from the reducing environment and their exposure to oxygen occur, the sulfides would undergo oxidization and release As in large quantities into the environment.

The role of As sulfides and hydrous Fe oxides as important sinks and modes of transport for As in the environment is therefore of great environmental significance, because changes in redox potential and acidity can bring about the formation and dissolution of these sulfides and colloidal hydrous oxides, hence directly controlling the mobilization of As in the waste environment.

Effect of Eh on Arsenic Partitioning

The general pattern for the different fractions is summarized in the following equation: F5 >> F3 $\cong$ F4 >> F2 > F0 > F1 (Figure 6). Therefore, results from arsenic partitioning show that, in general, the two dominant active As fractions were As bound to insoluble large molecular humic compounds and/or sulfides (F4) and As bound to Fe(III) and Mn(IV) oxides (F3), considering that the fraction with the highest concentration of As is the mineral matrix phase (F5). A similar situation was previously described by Carbonell-Barrachina et al. (4) in sewage sludge.

The As strongly associated to the mineral matrix phase (F5) changed with time, increasing and decreasing under highly aerobic and anaerobic conditions, respectively. The increase in the F5 fraction was reached from decreases in all the other fractions under highly aerobic conditions, indicating that under these redox circumstances, the As somehow became more stable and insoluble. It is possible that As was co-precipitated or occluded within Fe oxy-hydroxides crystals in the mineral matrix and became unavailable even for our extraction processes. This could be a reason why the F3 fraction in the second partitioning was maximum under anaerobic conditions when the expected situation would be to find maximum levels of As bound to iron oxides under highly aerobic conditions, when these adsorbent compounds are stable. On the other hand, at the beginning of the incubation period, the F3 fraction was maximum under highly aerobic conditions, as expected.

At the second partitioning, though under anaerobic conditions (low Eh values), Mn(IV) and Fe(III) oxides are usually reduced to soluble Mn(II) and Fe(II), respectively, there is still a relatively high percentage of As bound to these oxy-hydroxides (F3). A similar situation was found for the F4 fraction (As bound to insoluble sulfides and organic matter) under aerobic conditions, where a relatively high percentage of As was present in this fraction, when sulfide was

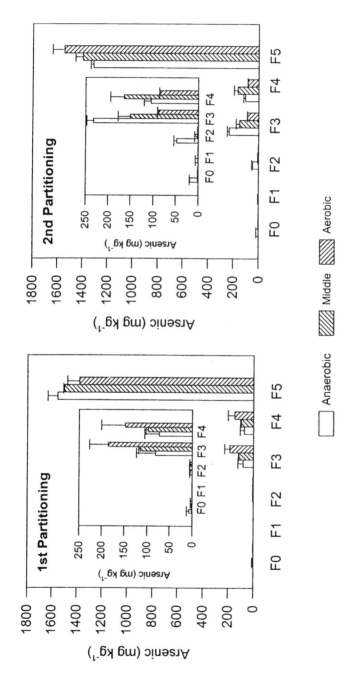

Figure 6. Distribution of arsenic present in pyrite mine waste in the various chemical fractions at the beginning (1st) and at the end (2nd) of the incubation time under highly aerobic, moderately anaerobic and anaerobic conditions.

expected to have been oxidized to sulfate and the elements initially adsorbed or precipitated onto them were supposed to be already soluble. Authors have no other explanations for these experimental findings than the postulation that the redox reactions are not as fast as theoretically expected and kinetics may play an important role in this type of transformation.

Some of the As was bound to carbonates (F2). In our experiment, As in the fraction F2 was maximum at the end of the incubation period (second partitioning) and under anaerobic conditions. Guo et al. (11) reported an increase in the As bound to carbonates at Eh levels below –130 mV, however, at Eh levels from 430 mV to –130 mV, this fraction decreased.

Conclusions

In summary, As has a complex biogeochemistry in waste ecosystems. Some of the processes are chemical, some may occur only through microbial mediation, and others may occur either chemically or with microbial mediation.

Redox potential was shown to control solubility and chemical partitioning of As, Fe and Zn in pyrite mine acid waste suspensions. This parameter, Eh, is of utmost importance in assessing the fate of As and Zn in As-containing wastes.

Under highly aerobic conditions, As solubility was low and mainly associated with adsorption or co-precipitation onto amorphous Fe oxides. Upon reduction, As mobilization increased significantly and solubility was maximum. Under moderately reduced conditions (0-100 mV), As solubility seemed to be controlled by the dissolution of iron oxy-hydroxides; however, at –200 mV, As chemistry was dominated by the formation of insoluble sulfides.

A similar solubility pattern was found for soluble Fe, showing that As and Fe biogeochemistry are positively related.

Sulfides also play an important role in retaining and remobilizing As. An organic amendment rich in S compounds would drastically reduce the potential environmental risks derived from As pollution under highly anaerobic conditions by precipitating this toxic metalloid as insoluble and immobile sulfides.

Zinc showed, however, a completely different behavior than As and Fe, and its soluble concentration only increased under aerobic conditions, reaching concentrations up to 800 mg kg^{-1} waste, which are a serious risk to the environment. Zinc was initially present as insoluble zinc sulfides, however, after further oxidation, sulfide was transformed to sulfate and Zn^{2+} become soluble.

When disposal of As-containing wastes is planned, consideration should be given to maintaining high redox conditions necessary for minimum As solubility and mobilization. If, however, the waste contains high amounts of Zn, the situation is completely different and highly anaerobic conditions are recommended.

In the view of these conclusions, the authors concluded that the Spaniard authorities took the right decisions in the Aznalcóllar environmental accident (this was an example of the worst possible scenario when remediation of an arsenic- and zinc-polluted environment was confronted), first trying to phyto-remediation with plants accumulating large amounts of Zn in their tissues, and simultaneous adding amorphous iron oxy-hydroxides under aerobic conditions and before more anaerobic conditions could be reached.

Acknowledgements

Funding for this research was provided by the *Conselleria de Cultura, Educació i Ciència (Generalitat Valenciana)*, through Project GV99-141-1-13, for which the authors are deeply indebted.

References

1. Pain, D. J.; Sánchez, A.; Meharg, A. A. *Sci. Total Environ.* **1998**, *222*, 45-54.
2. Nandak, P. B. A.; Dushenkov, V.; Motto, H.; Raskin, I. *Environ. Sci. Technol.* **1995**, *29*, 1232-1238.
3. *Terwelle, H. F.; Slater, E. C. Biophys. Acta. 1967, 143, 1-17.*
4. Carbonell-Barrachina, A. A.; Jugsujinda, A.; Sirisukhodom, S.; Anurakpongsatorn, P.; Burló, F.; DeLaune, R. D.; Patrick, W.H., Jr. *Envir. Int.* **1999**, *25*, 613-618.
5. Masscheleyn, P. H.; DeLaune, R. D.; Patrick, W. H.; Jr. *Environ. Sci. Technol.* **1991**, *25*, 1414-1419.
6. Masscheleyn, P. H.; DeLaune, R. D.; Patrick, W. H., Jr. *Environ. Sci. Technol.* **1990**, *24*, 91-96.
7. Patrick, W. H., Jr.; Williams, B. G.; Moraghan, J.T. *Soil Sci. Soc. Am. Proc.* **1973**, *37*, 331-332.
8. C.S.I.C. (Centro Superior de Investigaciones Científicas). **1998**, http://www.csic.es/hispano/coto/aznalco.htm
9. Bhumbla, D. K.; Keefer, R. F. In *Arsenic in the Environment. Part I: Cycling and Characterization*; Nriagu, J. O., Ed.; John Wiley & Sons, Inc.: New York, NY, 1994; pp 51-82.
10. Patrick, W. H., Jr.; Henderson, R. E. *Soil Sci. Soc. Am. J.* **1981**, *45*, 855-859.

11. Guo, T.; DeLaune, R. D.; Patrick, W.H., Jr. *Environ. Int.* **1997**, *23*, 305-316.
12. Carbonell-Barrachina, A. A.; Jugsujinda, A.; Burló, F.; DeLaune, R. D.; Patrick, W. H., Jr. *Water Res.* **2000**, *34*, 216- 224.
13. Livesey, N. T.; Huang, P. M. *Soil Sci.* **1981**, *131*, 88-94.
14. Masscheleyn, P. H.; DeLaune, R. D.; Patrick, W. H., Jr. *J. Environ. Qual.* **1991b**, *20*, 522-527.
15. Goldberg, S.; Glaubig, R. A. *Soil Sci. Soc. Am. J.* **1988**, *52*, 1297-1300.
16. Bar-Yosef, B.; Meek, D. *Soil Sci.* **1987**, *144*, 11-19.
17. Pierce, M. L.; Moore, C. B. *Water Res.* **1982**, *16*, 1247-1253.
18. Ferguson, J. F.; Gavis, J. *Water Res.* **1972**, *6*, 1259-1274.
19. Mitchell, P.; Barr, D. *Environ. Geochem. Health.* **1995**, *17*, 57-82.
20. Allen, H. E.; Fu, G.; Deng, B. *Environ. Toxicol. Chem.* **1993**, *12*, 1441-1453.
21. Mok, W. M.; Wai, C. M. In *Arsenic in the Environment. Part I: Cycling and Characterization;* Nriagu, J. O., Ed.; John Wiley & Sons, Inc.: New York, NY, 1994; pp 99-118.

Chapter 15

Total Arsenic in a Fishless Desert Spring: Montezuma Well, Arizona

Anne-Marie Compton-O'Brien[1], Richard D. Foust Jr.[1,*],
Michael E. Ketterer[1], and Dean W. Blinn[2]

[1]Department of Chemistry and Merriam-Powell Center for Environmental
Research and [2]Department of Biological Sciences, Northern Arizona
University, Flagstaff, AZ 86001–5698

Total arsenic was measured in various matrices from
Montezuma Well, a unique ecosystem with many endemic
species located in north-central Arizona, U.S.A. Montezuma
Well water contains natural arsenic levels of 100 µg/L.
Analysis was performed by nitric acid/hydrogen peroxide
microwave digestion followed by quadrupole inductively
coupled argon plasma mass spectrometry. Sediment, soil,
Potamogeton illinoiensis roots, *P. illinoiensis* leaves, *Berula
erecta* roots, *B. erecta* leaves, *Fissidens grandifrons*, *Hyallela
montezuma*, *H. azteca*, *Ranatra montezuma*, *Telebasis salva*,
Belostoma bakeri, *Motobdella montezuma* and DW-rinsed *M.
montezuma* were collected and analyzed. Total arsenic values
ranged from 1.0 to 2,810 µg/g (dw). Biodiminution of total
arsenic in the littoral zone food web was observed. Total root
arsenic levels were higher than leaf arsenic levels. Organisms
show different mechanisms for coping with the elevated levels
of arsenic found in the Well. The plants, inhibit vertical
transport, others exclude arsenic and some absorb arsenic onto
their surface in higher quantities than in their cells.

Introduction

Few studies have tracked arsenic in freshwater food webs. A compilation of studies by Eisler (*1*) of field collected species of flora and fauna provides crucial information on dry weight (dw) and fresh weight (fw) of total arsenic in freshwater and marine organisms. However, the majority of the studies are for marine ecosystems; freshwater data mainly include fish and plants. The data frequently lack food web interactions. To date, there are only a few comprehensive freshwater food web studies of arsenic (*2-4*). These studies show a biodiminution of arsenic in freshwater. This study is an attempt to provide information on a freshwater food web in the southwestern U.S.A.

Montezuma Well is part of Montezuma Castle National Monument, located in north-central Arizona. It is a thermally constant, collapsed travertine spring, measuring 112 m in diameter with an average depth of 6.7 m (*5*). The Well contains naturally high concentrations of dissolved carbon dioxide (>450 mg/L), and is thus fishless (*6*). Montezuma Well has very high levels of productivity. Productivity of phytoplankton is on the order of 602 $gCm^{-2}yr^{-1}$(*6*).

The aquatic plant *Potamogeton illinoiensis* may grow up to 6 m in Montezuma Well, as compared to <1 m in other locations. Studies have shown that moderate levels of arsenic tend to increase the growth of certain plants. Arsenate and phosphate have similar uptake mechanisms, however, arsenate does not provide an energy source as phosphate does. Therefore, when arsenate is competing for phosphate in a plant, it acts as if it is P-deficient and increases nutrient uptake, resulting in faster growth rates (*8*). This theory, lack of herbivory and optimal physio-chemical conditions, may explain why *P. illinoiensis* thrives in Montezuma Well.

Montezuma Well contains elevated concentrations of geogenic arsenic, a known human carcinogen and a biogeochemically important element (*9*). The arsenic in Montezuma feed water presumably originates in the Verde formation and enters the well through several fissures in the well bottom as an equilibrium mixture of As(III)/As(V) (*10*). Upon mixing with the highly oxygenated water of Montezuma Well the arsenic is converted to 100% arsenate, the form of arsenic found in the water column and in surface water (*11*).

The trophic interactions of Montezuma Well have been thoroughly studied (*5,7,12*), thus providing excellent groundwork for an investigation of total arsenic in freshwater trophic levels. In addition to being a well-studied system, Montezuma Well has the advantage of being a closed ecosystem with many endemic species. Of the many endemic species present, those included in this study are: *Hyalella montezuma* (a freshwater amphipod) (*13*), *Ranatra montezuma* (water scorpion) (*12*), and *Motobdella montezuma* (a new genus of a non-bloodsucking leech) (*14*).

Materials and Methods

Samples were collected from Montezuma Well in May and September 2001. This corresponds to the peak productivity season (*15*). The following samples were collected for analysis: sediment, soil, *Hyalella azteca*, *H. montezuma*, *Telebasis salva* (damselfly nymphs), *Ranatra montezuma*, *Belostoma bakeri* (waterbug), *Motobdella montezuma*, *Fissidens grandifrons (moss)* and *Berula erecta* and *Potamogeton illinoiensis* (aquatic plants). The plant samples were hand picked and stored in Ziplock bags. Roots and leaves were separated in the field. *B. bakeri*, *T. salva*, and *R. montezuma* were also collected individually by hand and stored in separate centrifuge tubes. The amphipods were collected with a 254 µm pore-size plankton tow net. *H. azteca*, which have no dorsal spines and live near shore, were gathered with a surface plankton tow net. *H. montezuma* were collected by boat in the center of the Well with a weighted plankton tow net at a depth of approximately 10 m. *M. montezuma* are nocturnal predators and were collected at dusk as they begin their ascent to the surface for nightly foraging. Sediment samples were obtained with a plastic scoop next to where the plant roots were collected. Soil samples were collected below the rim of the Well, about 15 m from the shoreline. Sediment and soil samples were stored in plastic Ziplock bags and kept cold until returned to the laboratory, where they were immediately frozen in liquid nitrogen and stored in a freezer at -80°C until analysis.

Chemical analysis of biological samples for total arsenic is difficult because 1) arsenic is present at trace amounts and 2) because both inorganic and organic forms of arsenic are often present (*16-19*). Investigators routinely use Standard Reference Materials (SRMs) to identify and compensate for the analytical problems (*18,19*), but 75% arsenic recoveries are frequently reported for biological specimens (*20*).

We used SRMs that matched the sample matrix as closely as possible. Our percent recoveries ranged from 90-113 percent. The SRMs used are listed in Table I. A closed vessel microwave digestion system was used to digest all the samples reported in this study. GF-AAS and ICP-MS were the analytical techniques used in this work. Therefore, all sample digestions were done with an HNO_3/H_2O_2 digest at 180°C in an Ethos closed-vessel, temperature controlled microwave digestion system (*21*).

GF-AAS was used to characterize total arsenic in dried plant samples from Montezuma Well. A Perkin-Elmer transversely heated graphite furnace for electrothermal atomization (THGA) was used with a Perkin Elmer 600 AAS. The instrumentation incorporated a Zeeman background correction. A Perkin Elmer EDL arsenic lamp was employed. Instrument settings were chosen based on the instrument guidebook specifications. A wavelength of 193.7 nm and a slit width of 0.7 nm were selected. The high char temperature (1100°C) required for

Table I. Standard Reference Materials Used for Total Arsenic Analysis

Standard Reference Material	Sample Matrix	Certified Arsenic Value (mg/kg)	Percent Recovery Results
Estuarine Sediment NIST 1646	Montezuma Well Sediment	11.6±1.3	94.0
Montana Soil NIST 2711	Montezuma Well Soil	105.0±8	113
Orchard Leaves NIST 1571	*Potamogeton illinoiensis, Berula erecta, Fissidens grandifrons*	14.0±2	103
Dogfish tissue DORM-2 NCR	*Hyalella montezuma, Hyalella azteca, Ranatra montezuma, Telebasis salva, Belostoma bakeri, Motobdella montezuma*	18.0±1.1	99.3

GF-AAS has a negative impact on percent recovery of SRMs with high organic arsenic content such as DORM-2, which is primarily arsenobetaine. ICP-MS provided an elegant solution to this problem and was used to determine total arsenic in extracts from plants.

A Perkin-Elmer Elan 500 ICP-MS with a glass nebulizer and water-cooled spray chamber was used to determine total arsenic in plants, sediment, soil and organisms. A peristaltic pump with a flow rate of 1.5 mL/min pumped sample to the nebulizer. Gallium was used as an internal standard and mass to charge (m/z) ratios were measured by peak-hopping at 75 and 69. All standards were made up using Spex plasma standards in 2% nitric acid. Two milliliters of concentrated nitric acid were added to each calibration standard to provide matrix matching with the digested samples. Serial dilutions of the standards were performed to make calibration standards ranging from 0-1000ppb. Analysis was based on EPA method 6020 (21).

Results and Discussion

All data were analyzed in conjunction with the appropriate matrix-matched SRMs, calibration standards and digest blanks. A population of 3 to 6 replicates of plant, organism, soil and SRM were sampled, digested and analyzed. The results of the percent recoveries of the SRMs are shown in Table I. Total arsenic for *Potamogeton illinoiensis* roots that were dried prior to analysis was 8.87 µg/g. The value for the fresh weight root arsenic was significantly higher, 26.84 µg/g. Given the high volatility of many organic species of arsenic commonly found in plant samples, it is believed that some of the arsenic was volatilized in the drying process (*19*). All data were therefore taken on samples that were not dried, and converted to dry weight by correcting for percent moisture.

Burló et al. (*8*) show consistently higher concentrations of arsenic in arsenic-treated tomato plant roots than in shoots and leaves and explain this phenomenon with a theory of upward transport limitation as a way for plants to tolerate high arsenic environments. The trend of higher plant root arsenic is clearly replicated in this study (See Figure 1).

While there was no difference between fresh-weight total arsenic in the sediment and soil samples, the percent solids were much different (15.5% and 79.8%, respectively). Fresh-weight arsenic was converted to dry-weight arsenic based on the percent solids data. This gave a value for sediment arsenic that was 4 times higher than that for the soil arsenic (see Table II). Both sediment and soil samples are derived from the same arsenic-bearing geologic formation, the Verde Formation (*9*). This suggests that the arsenic is being transported to Montezuma Well through groundwater. This is supported by hydrogeological data collected by the U.S. Geological Survey (*10*).

There are two distinct food chains in Montezuma Well, one in the littoral zone and one in the open-water zone near the center of the Well (7). The open-water food web is much less complex than that of the littoral zone, consisting of phytoplankton, *Hyalella montezuma*, and *Motobdella montezuma*, in order of ascending trophic level (7). Unfortunately, the total arsenic data for this food web lacks the phytoplankton, so it cannot be determined if bioaccumulation or biodiminution is occurring in the open water. However, it is interesting to note that the leech, which lives near the deep, anaerobic sediment, sometimes even burrowing into it, has very high total arsenic compared to the other trophic levels. In addition, it is important to note that a t-test showed no statistically significant difference between the two species of amphipods ($\alpha=0.05$), which occupy the same trophic level in this system.

Mason et al. (*3*) point out the importance of a phosphate wash to remove any arsenic that may be adhering to the surface (skin or exoskeleton) of organisms. It should be noted that the high value of 2,810 µg/g (dw) for the leech was determined without a phosphate rinse. The leeches did not appear to have a thick layer of sediment on their skins and were collected at the surface and rinsed in Montezuma Well water. However, the leeches do have an outside layer of mucilage, which can only be seen by a microscope. This may be a factor in the elevated value of total arsenic determined for the leeches. A distilled water rinse brought the value for total arsenic in *M. montezuma* to a value of only 45.5 µg/g (dw) (See Table II). This suggests that surface absorption is indeed a real problem and phosphate washes should be performed.

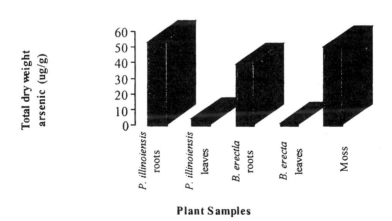

Total dry weight arsenic in Plant Samples

Figure 1. Total dry weight arsenic in aquatic plant samples, Montezuma Well.

The littoral zone food web is more complex than the open-water food chain. *H. azteca* will feed on phytoplankton and epiphyton or the roots of *Berula erecta* (*22*). *Telebasis salva* is both predator and prey. It will consume *H. azteca* and will be consumed by the top predators of the littoral zone, *Belostoma bakeri* and *Ranatra montezuma* (*7*). Many of these trophic interactions occur in the littoral zone of the well where *P. illinoiensis* grows prolifically. A more detailed description of the food web in Montezuma Well can be found in Runck and Blinn (*7*). However, there is no known invertebrate herbivory on *P. illinoiensis*. The total arsenic (dw) in the littoral zone food web is shown in Figure 2 in order of ascending trophic level. Damselfly nymph total arsenic was below the detection limit, suggesting that they are somehow excluding arsenic. It is clear from the data that arsenic is indeed biodiminishing in this system. This is consistent with the freshwater food web arsenic studies performed by Chen (*4*) and Mason (*3*).

Conclusions

Arsenic enters Montezuma Well dissolved in groundwater and is deposited in the sediment, 427 µg/g (dw). Arsenic in the water column remains high at a concentration of ~100 µg/L. Two species of aquatic plants in Montezuma Well accumulate arsenic in their roots at concentrations significantly higher, 39-53 µg/g (dw), than in their shoots and leaves, 1.3-4.1 µg/g (dw), indicating a relative paucity of vertical transport of arsenic. Two species of amphipods and an endemic water scorpion contain arsenic at levels of approximately 1 µg/g (dw) and damselfly nymphs are able to completely exclude arsenic. Total arsenic in the littoral zone food web of Montezuma Well therefore decreases with increasing trophic level. *Motobdella Montezuma,* an endemic leech and the top predetator in Montezuma Well, is an exception to this trend with a total arsenic concentration of 2,810 µg/g (dw). Most of *M. Montezuma's* arsenic, however, is contained in a mucilage layer on the leech's exterior.

Acknowledgements

We wish to acknowledge the financial support of the U.S. Department of Energy through contract No. DE-FC04-90AL66158, administered through the HBCU/MI ETC Consortium. We would like to thank the Merriam-Powell Center for Sustainable Environments and the Verde Watershed Education Program for their financial assistance. We acknowledge the help of Dr. Molly S. Costanza-Robinson. Finally, we thank the staff of the National Park Service and Montezuma Well.

Table II. Total Arsenic in Samples by ICP-MS

Sample Analyzed for Total Arsenic	Avg. Total Arsenic (μg/g) fw	Avg. Total Arsenic (μg/g) dw
Montezuma Well Sediment	67±5 (n=5)	427
Montezuma Well Soil	76±10 (n=5)	95
Potamogeton illinoiensis roots	27±5 (n=4)	53
Potamogeton illinoiensis leaves	2.0±0.3 (n=4)	4.1
Moss	27±3 (n=2)	50
Berula erecta roots	20±3 (n=4)	39
Berula erecta leaves	0.7±.2 (n=4)	1.3
Endemic amphipod	1.6±0.3 (n=3)	3.4
Non-endemic amphipod	1.7±0.3 (n=3)	3.2
Endemic water scorpion	0.7±0.07 (n=3)	1.5
Damselfly nymphs	<0.5 (n=3)	NA
Aquatic insect	0.6±0.2 (n=3)	1.0
Endemic leech	926±100 (n=2)	2,810
Endemic leech, rinsed	15±1 (n=2)	45.5

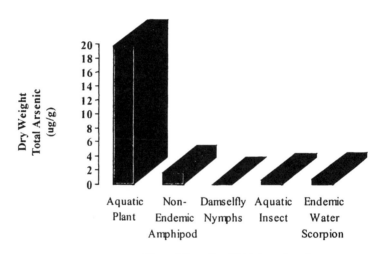

Figure 2. Dry Weight Total Arsenic in Littoral Zone Food Web, Montezuma Well, AZ.

References

1. Eisler, R. In *Arsenic in the Environment Part II: Human Health and Ecosystem Effects*; Nriagu, J. O.; IBSN 0-471-30436-0; Wiley: New York, 1994, Chap. 11, pp 185-259.
2. Muñoz, O.; Devesa, V.; Suñer, M. A.; Vélez, D.; Montoro, R.; Urieta, I.; Macho, M. L.; Jalón, M. *J. Agric. Food Chem.* **2000**, *48*, 4369-76.
3. Mason, R. P.; Laporte, J. M.; Andres, S. *Arch. Environ. Contam. Toxicol.* **2000**, *38*, 283-97.
4. Chen, C.; Folt, C. L. *Environ. Sci. Technol.* **2000**, *34*, 3878-84.
5. Runck, C.; Blinn, D. W. *J. N. Am. Benthol. Soc.* **1993**, *12*, 136-47.
6. O'Brien, C.; Blinn, D.W. *Freshwater Biology* **1999**, *42*, 225-34.
7. Runck, C.; Blinn, D. W. *Limnol. Oceanogr.* **1994**, *39*, 1800-12.
8. Burló, F.; Guijarro, I.; Carbonell-Barrachina, A. A.; Valero, D.; Martínez-Sánchez, F. *J. Agric. Food Chem.* **1999**, *47*, 1247-53.
9. Foust Jr., R. D.; Mohapatra, P., Compton, A.-M., Reifel, J. **2002**, *Appl. Geochem.* in press.
10. Konieczki, A. D.; Leake, S. A. *Hydrogeology and Water Chemistry of Montezuma Well in Montezuma Castle National Monument and Surrounding Area, Arizona;* U.S. Geological Survey: Tucson, AZ 1997.
11. Compton-O'Brien, A. M.; Costnaza-Robinson, M., Ketterer, M. E., Foust, Jr., R. D., Arsenic Investigations in Montezuma Well, Central Arizona, *TheScientific World*, **2002**, submitted for publication.
12. Runck, C.; Blinn, D. W. *Ecoscience* **1995**, *2*, 280-85.
13. Blinn, D. W.; Davies, R.W. *Freshwater Biology* **1990**, *24*, 401-07.
14. Govedich, F. R.; Blinn, D. W., Hevly, R. H., Keim, P. S. *Can. J. Zool.* **1999**, *77*, 52-57.
15. Boucher, P.; Blinn, D. W., Johnson, D. B. *Hydrobiologia* **1984**, *119*, 149-60.
16. Davis, A.; Ruby, M. V.; Bloom, M.; Schoof, R.; Freeman, G.; Bergstrom, P. D. *Environ. Sci. Technol.* **1996**, *30*, 392-99.
17. Wang, X.; Zhang, Z.; Sun, D.; Hong, J.; Wu, X.; Lĕe, F. S.-C.; Yang, M. S.; Leung, H.W., *Atom. Spectr.* **1999**, *20*, 86-91.
18. Koch, I.; Wang, L.; Ollson, C. A.; Cullen, W. R.; Reimer, K.J. *Environ. Sci. Technol.* **2000**, *34*, 22-26.
19. Geiszinger, A.; Goessler, W.; Keuhnelt, D.; Francesconi, K.; Kosmus, W. *Environ. Sci. Technol.* **1998**, *32*, 2238-43.
20. Roelandts, I.; Gladney, E. S. *Fresnius J. Anal. Chem.* **1998**, *360*, 327-38.
21. *Test Method for Evaluating Solid Waste: Physical/Chemical Methods (SW-846) through update III;* U.S. Environmental Protection Agency, CD-ROM version available through the National Technical Information Service, Washington, DC, 1998.
22. Rowe, K.; Blinn, D. W. *Bull. N. Amer. Benthol. Soc.* **2001**, 18, 223-224.

Chapter 16

Arsenic Contamination of Groundwater, Blackfoot Disease, and Other Related Health Problems

Chien M. Wai[1], Joanna Shaofen Wang[1], and M. H. Yang[2]

[1]Department of Chemistry, University of Idaho, Moscow, ID 83844
[2]Department of Nuclear Sciences, National Tsing Hua University, Hsinchu, Taiwan 300, Republic of China

Arsenic contamination of groundwater can occur by natural leaching of minerals and by human activities. In aquatic environments, arsenic usually exists in +3 and +5 oxidation states, both as inorganic and organometallic species. Inorganic arsenite is more toxic than arsenate which in turn is more toxic than monomethylarsonic acid and dimethylarsinic acid. Deep well waters often have arsenic concentrations far greater than the current maximum contaminant level of 10 ppb and with arsenite/arsenate ratios >1. The Blackfoot disease found in southwest Taiwan nearly half a century ago was related to the drinking of deep well waters containing high concentrations of arsenic with high fractions of arsenite by local villagers. Similar arsenic poisoning problems were later found in Inner Mongolia, Bangladesh and India, all related to the drinking of groundwaters contaminated with arsenic. This global arsenic contamination problem is perhaps one of the most serious environmental problems facing human beings today.

Introduction

Arsenic is a naturally occurring and ubiquitous element found in the earth's crust. It is classically considered as a metalloid and shares many of its toxic attributes with the other heavy metals such as lead and mercury. Contamination of groundwater with arsenic is a global environmental problem because arsenic can enter groundwater systems from weathering of minerals in rocks and soil. The arsenic standard set by World Health Organization (WHO) is 50 ppb (0.05 μg/mL) and WHO's guideline is 10 ppb in drinking water. In the U.S.A. the current maximum contaminant level (MCL) of arsenic in water is 50 ppb, a level established in 1942 by the U.S. Public Health Service (*1*). This is also the permissible level of arsenic in bottled water according to the U.S. Code of Federal Regulations (CFR) (*2*). However, analyses (*3,4*) suggest that the current standard of 50 ppb has a substantial increased risk of cancer and is not sufficiently protective of public health. The U.S. Environmental Protection Agency (EPA) is required by the Safe Drinking Water Act Amendments of 1996 to propose a new standard of arsenic by January 2000 and to finalize that regulation by January 2001(*5*), but in 2000 Congress extended it to June 22, 2001. Based on the accumulating scientific information and data on the health effects of arsenic, EPA in January 2001 issued regulations that set a MCL level of 10 ppb arsenic standard for drinking water (*6*). The EPA estimated that the new standard would affect around 13 million people, mainly in the West, Midwest, and New England where arsenic levels in many well waters are greater than 10 ppb. The current Bush Administration withdrew the 10-ppb standard in March 2001, three days before it was to take effect. In October of 2001, EPA affirmed the appropriateness of the MCL and reinstated 10 ppb as the new MCL for arsenic in drinking water. Water systems must meet this standard by January 2006.

The chemical form of arsenic in drinking water is not specified by the CFR, although it is well established that the toxicity of arsenic depends on its chemical form. Arsenic exists in natural waters in different oxidation states depending on the redox environment. The trivalent inorganic species arsenite is more toxic to the biological systems than the pentavalent species arsenate (*7,8*). Organoarsenicals such as monomethylarsonic acid (MMA) and dimethylarsinic acid (DMA) also exist in the natural environments, but their toxicities are lower than the inorganic arsenic species pathophysiologically (*7,8*). The trivalent arsenic is more toxic because it can bind to thio groups in biological systems. Information on the distribution of arsenic species and speciation of arsenic is therefore important to assess its toxicity in drinking water.

Since arsenic is an ubiquitous element in the earth's crust and it can be concentrated in well waters, arsenicosis has become an emerging epidemic in many areas of the world. Blackfoot disease(BFD), an arsenic related disease, was first observed in Taiwan in the 1930s and peaked in the 1950s (9).   The disease was found to correlate to the high arsenic contents in the groundwaters consumed by local inhabitants in several villages in southwest Taiwan (10-12). The symptoms of BFD start with spotted discoloration on the skin of the extremities, especially on the feet. The spots change from white to brown and eventually to black, hence the name BFD.   The affected skin gradually thickens, cracks and ulcerates. Amputation of the affected extremities is often the final resort to save the BFD victims. After 20-30 years of exposure to high levels of arsenic, internal cancers may also appear. Arsenic related diseases were later reported to occur in other areas of Asia including Inner Mongolia of China (13), Bangladesh (14), and India (15). A large number of populations in Bangladesh and India are currently affected by arsenic in drinking water obtained from subsurface sources.   Other countries such as Vietnam (16,17), Chile (18,19), Argentina (20), Finland (21), and the United States (22) also have groundwater arsenic contamination problems. This global arsenic contamination problem is getting considerable attention from the scientific communities today. This paper summarizes current information concerning arsenic contamination in groundwaters of some selected areas and related environmental and health problems.

Environmental Arsenic Distribution and Toxicity

In order to understand the global arsenic problem, it is necessary to understand the chemistry of arsenic. The toxicity, bioavailability, bioaccumulation, and transport of arsenic are often dependent upon its species in the system under investigation. Because each arsenic species possesses unique physical and chemical properties and causes specific effects in living systems, measurement of the total concentration of arsenic provides little information about its toxicity or bioavailability. Toxicity tests have shown that the most toxic form of arsenic is arsenite, which is as much as 60 times more toxic than arsenate, due to its ability to react with enzymes in human metabolism, and several hundred times more toxic than MMA or DMA (23). Arsenite is also significantly more mobile in groundwaters than arsenate.   Penrose (23) compiled the approximate toxicity order of various arsenic compounds, which, in decreasing order, is: arsines > arsenite > arsenoxides > arsenate > pentavalent arsenicals > arsonium compounds > metallic arsenic. The chemical formulas of different arsenic species are shown in Table I.

The natural abundance of arsenic in soil is of great importance both for the assessment of environmental quality and for devising countermeasures against soil pollution. The levels of arsenic may be much higher in soils contaminated by human activities. The levels of arsenic in soil of various countries have been reported in 1979 to range from 0.1 to 40 ppm (mean 6 ppm) by Bowen (24), and from 1 to 50 ppm (mean 6 ppm) by Backer and Chesnin in 1975 (25). The total concentrations of arsenic in soils from a number of countries were also described by Huang in 1994 (26).

Table I. Chemical Forms of Arsenic Species

1.	Arsine	AsH_3
2.	Arsenoxide	$As(O)(OH)$
3.	Cacodylic Acid	$(CH_3)_2HAsO$
4.	Dimethylarsine	$(CH_3)_2AsH$
5.	Dimethylarsinic Acid (DMA)	$(CH_3)_2As(O)(OH)$
6.	Methylarsine	$(CH_3)AsH_2$
7.	Monomethylarsonic Acid (MMA)	$CH_3As(O)(OH)_2$
8.	P-Arsanilic Acid	$NH_2C_6H_4As(O)(OH)_2$
9.	Phenylarsine Oxide	C_6H_5AsO
10.	Phenylarsonic Acid	$C_6H_5As(O)(OH)_2$
11.	Arsenite	AsO_2^-
12.	Arsenate	AsO_4^{3-}
13.	Trimethylarsine	$(CH_3)_3As$
14.	Triphenylarsine Oxide	$(C_6H_5)_3AsO$

In aquatic environments, arsenic can exist in several oxidation states, both as inorganic and organometallic species (27). Arsenic in surface river waters is present primarily as an inorganic ion, arsenate. Reduced arsenic (arsenite) and methyl arsenicals (MMA and DMA) are also occasionally present (28,29). In contaminated rivers, sediments can contain substantial amounts (100-300 ppm or higher) of arsenic that are potentially mobile during water-sediment interactions. The mobility of arsenic in these sediments is affected by the physical and chemical forms of the arsenic species as well as by environmental conditions. Dissolved arsenic species can be adsorbed on or co-precipitated with suspended solids and carried down to the river sediments. On the other hand, a build-up of arsenic compounds in the bottom sediments of a river may subsequently be released to the overlying water (30).

For water with arsenic levels less than 0.2 ppm (31), the major health concern is an increased chance of getting some types of cancer such as skin, bladder, lung and possible liver and kidney cancers. As arsenic levels in water become greater than 0.2 ppm and length of water use becomes longer than a year, the following health effects may occur on the skin including: a "pins and needles" sensation in your hands and feet, skin changes in color appearing as a fine freckled or "raindrop" pattern in the trunk, hands, and feet, and unusual skin growth (wart-like) on the palms and soles. Several years of low level arsenic exposure can also cause various skin lesions. Hyperpigmentation (dark spots), hypopigmentation (white spots) and keratoses of the hands and feet will appear. After a dozen or so years, skin cancers are expected (*31*).

Arsenic is present in most foodstuffs in concentrations of less than 1 μg/g (1ppm). However, marine fish may contain arsenic concentrations up to 5 ppm wet weight and concentrations in some crustacean and bottom-feeding fish may reach several tens of ppm, predominantly in the form of organic arsenic (*31*).

In both animal and man, organic arsenic compounds ingested via fish and crustacean are readily absorbed from the gastrointestinal tract and 70 % - 80 % are eliminated within a week, mainly in the urine. Urine is a suitable indicator for assessment of exposure to inorganic arsenic, since most studies show that the elimination of arsenic, in both animals and man, takes place mainly via the kidneys. Arsenic levels in the hair of unexposed human adults are usually below 1 ppm. Levels up to about 80 ppm have been recorded in subject with chronic arsenic poisoning caused by ingestion of contaminated well waters (*31*).

In humans, the highest arsenic concentrations are found in skin, hair, and nail, all tissues rich in keratin. Data from mice and rabbits show the highest arsenic concentrations are found in liver, kidney, lung, and intestinal mucosa at short times after a single exposure to trivalent or pentavalent inorganic arsenic. However, arsenic in the trivalent form generally causes higher concentrations in tissue levels than the pantavalent form. Inorganic arsenic is methylated in the body mainly to MMA and DMA. In humans exposed to low doses of trivalent or penatavalent inorganic arsenic, the urinary excretion consists of about 20 % MMA and 60 % DMA, the rest being inorganic arsenic.

Arsenic Species in Groundwater and Analytical Methods

In aquatic environments, arsenite can be converted to arsenate under oxidizing conditions (e.g. aerated surface water). Likewise, arsenate can also become arsenite under reducing conditions (e.g. anaerobic groundwater). However, the conversion in either direction is quite slow, so the reduced species can be found in oxidized environments and vice versa. Microbes, plants and animals can also convert these inorganic arsenic species into organic compounds involving carbon and hydrogen atoms, such as MMA and DMA. These compounds are much less commonly found in natural waters (*32*).

The major species of interest for most analytical work are inorganic arsenic, MMA and DMA. In most environmental water samples these will be the predominant forms of arsenic. Other arsenic species that have been assayed include arsine and methylarsines, triphenylarsine oxide (TPAO), arsanilic acid,

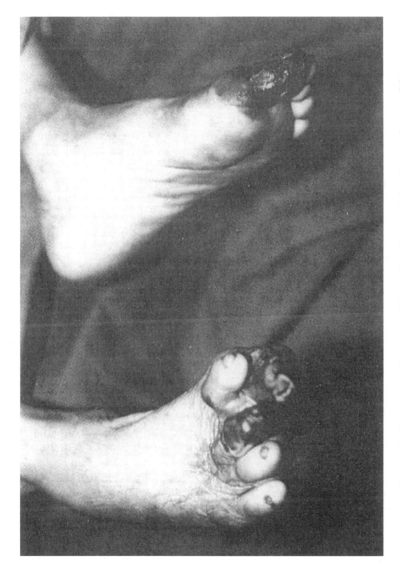

Figure 1. Feet of a BFD victim. Adapted with permission from reference 86. Copyright 1993 Kaohsiung Medical College Press.

phenylarsonic acid (PAA), and arsenate mononucleotide complexes (Table I). Most of the speciation techniques determine different arsenic species rather than valence states. Total arsenic in water can be measured directly by flame or graphite furnace atomic absorption spectrophotometry (AAS or GFAAS). The most commonly used techniques for the preconcentration of arsenic involve its transformation into arsine. Subsequent measurements of arsine can be carried out using spectrophotometry, flames and electrothermal devices for AAS *(33)*, atomic fluorescence (AFS)*(34)*, or atomic emission spectroscopy (AES)*(35)*. Other separation methods include solvent extraction*(36)*, ion exchange chromatography (IEC)*(37)*, liquid chromatography (LC)*(33)*, gas chromatography (GC) *(38)* and eletrochemistry*(39)*. Neutron activation analysis (NAA) using radiochemical separation is a very sensitive method for the determination of arsenic, with detection limits near 1 *ng (40,41)*.

Current speciation methods for arsenic rely mainly on separations based upon the principles of selective hydride generation, chromatography, solvent extraction, and electrochemistry. Often, one separation method is used to isolate or concentrate some arsenic species before applying the second technique for final separation and quantification as in the common employment of a solvent extraction step prior to separation by chromatography.

Hydride generation (HG) techniques for the speciation of arsenic involve selective reduction of the hydride-forming arsenic species to the corresponding arsines *(42)*. The arsines are generated in a reaction chamber by reduction with sodium tetrahydroborate ($NaBH_4$) at different pH and separated either by GC, HPLC, or by sequential volatizaion. At pH 5-7, arsenite can be reduced to arsine and at pH <1, both arsenite and arsenate are reduced to arsine. MMA and DMA are reduced to methylarsine and dimethylarsine at pH<1. After generation, the arsines may either passed directly into the detection device or accumulated prior to detection. Separation is achieved by controlling and varying the pH between sample aliquots. When a hydride collection step is incorporated prior to detection, it is usually accomplished by freezing the arsines out in a liquid nitrogen-cooled trap *(43)*. After collection, the volatile arsines can then be separated in the order of their boiling points (e.g. arsine, b.p. −55 °C; methylarsine, b.p. 2 °C; and dimethylarsine, 35.6 °C) by slow warming of the trap, and subsequently transferred to the detection device. Because the various arsines pass through the detector at different times, the analysis readout is similar in appearance to a chromatogram and analogous to a thermal volatilization curve. A comparison of observed volatilization times for arsenic compounds from samples to those of standard solutions can be used to identify the arsenic species *(44)*.

Preconcentration and separation of arsenic species by liquid extraction prior to NAA, AAS, and GFAAS have been applied to several sample systems, especially environmental waters *(45)*. Most of the methods are based on the selective liquid extraction of As(III) and differential determination of As(III) and As(V). The dithiocarbmate extraction method is one of the techniques used for preconcentration of arsenic and other trace metals. Dithiocarbamate reagents form water insoluble complexes with arsenite that can be extracted into an organic solvent such as methyl isobutyl ketone (MIBK) or chloroform. In a second aliquot, a reducing agent is used to convert arsenate to arsenite followed by solvent extraction with dithiocarbamate reagents. The organic arsenic compounds, MMA and DMA, in aqueous solutions, can be converted to arsenite such as CH_3AsI_2 and $(CH_3)_2AsI$, using reducing agents consisting of a mixture of potassium iodide, sodium thiosulfate, and sulfuric acid *(40,41)*.

GC*(38)*, LC*(33)*, IC*(46)*, SFC*(40,41)*, and open-column IEC*(37)* have been successfully employed to separate various arsenic species. General chromatographic concerns such as choice of eluent and flow rate, stationary phase, and retention time must be optimized for the species of interest and analytical objectives.

Both capillary and packed columns have been utilized with a variety of detection methods. The choice of detector depends on the specific derivatization procedure and degree of selectivity required. Detection methods that have been coupled to GLC for arsenic determination include inductively coupled plasma (ICP), AAS, AFS, microwave emission spectroscopy (MES), electron capture detection (ECD), and flame ionization detection (FID) *(47,48)*.

In addition to the volatile species that can be determined by GC, ions and non-volatile organometallic species can be separated with LC. When sensitive and selective detectors are used for measurement, the determination of nanogram amounts of eluants can be obtained with HPLC. Speciation of As(III) and As(V) in sediment extracts by HPLC-HG atomic absorption spectrophotometry was discussed recently *(33)*. Several detectors have been coupled to HPLC including ultraviolet (UV), AAS, GFAAS, ICP, and conductivity (CD). HPLC methods permit the use of a large variety of separation columns and both the stationary and mobile phases can be varied to enhance separation, whereas in GC the stationary phase is the only variable.

Electrochemical speciation methods are usually based on their ability to distinguish inorganic As(III) and As(V), since the latter is not electroactive. Using stripping voltammetric methods for speciation provides the advantage of *in situ* preconcentration capacity. One of the most widely applied electrochemical techniques is anodic stripping voltammetry (ASV). The presence of dissolved oxygen causes significant interference, so it must be removed from the ASV cell. Interference can also occur in ASV from the formation of interelemental species at the electrode surface (39).

Arsenic Related Diseases

Blackfoot Disease in Taiwan

Arsenic contamination of deep well waters and its correlation to the BFD was found in Taiwan nearly half a century ago (9-12, 49-52). Outbreaks of the BFD increased rapidly around 1950 when the number of deep artesian wells drilled by local villagers for drinking reached a maximum. Geological descriptions of the BFD area and early investigations of arsenic in the artesian wells of this area were reported by Tseng et al. (9). Both arsenicosis and BFD were limited to people drinking artesian well waters with a high concentration of arsenic (0.10-1.8 ppm). The number of patients suffering from this disease has been decreasing since 1956 after purified tap water was made available to the local dwellers. The population of this endemic area in southwest Taiwan is about 100,000 and the prevalence was 8.9 per 1000 in 1968 (12). Inhabitants of the endemic areas are very prone to chronic arsenism. Figure 1 shows the feet of a BFD victim. Figure 2 shows arsenic contaminated areas in southwest Taiwan.

In a 1977 survey of more than 40,000 local inhabitants in a BFD area in Taiwan (11), a positive dose-response relationship between the contents of arsenic in well water and the prevalence rate for skin cancer was established. Medical examinations, with special attention to skin lesions and peripheral vascular disorders, were performed on a population of 40,421 in the endemic area of Taiwan. A total of 428 cases of skin cancer and 370 of BFD disease were recorded. The overall prevalence rates for skin cancer, keratosis and hyperpigmentation were 10.6, 71.0 and 183.5 per 1000, respectively. The male to female ratio in skin cancer was 2.9(11). In a survey of artesian well waters in this area, the arsenic concentrations ranged from 0.01-1.82 ppm and most of the wells had 0.4-0.6 ppm of arsenic. No information was given on the chemical form and valence state of the arsenic in the water. The causes of death were recorded for 528 patients with BFD symptom and 244 cases of skin cancer from this area from 1961 to 1977.

A systematic study of arsenic species and other trace elements in the well waters of a representative village (Putai) in the BFD area in southwest Taiwan was conducted in 1992 (50). The metal contents in the well waters of the BFD area were compared with those obtained from another city in Taiwan where no

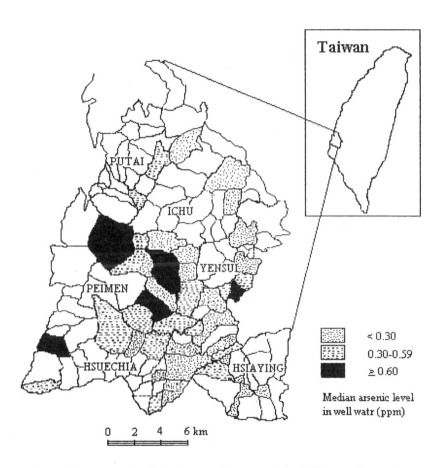

Figure 2. Arsenic levels in deep well waters of the BFD areas in southwest Taiwan.

BFD case had ever been reported.   Averaging over the three wells studied, the total dissolved arsenic concentration was 671±149 ppb, with a range of 470-897 ppb for all the well waters with 54 samples collected from Putai area.   The average value of the dissolved arsenic in all the well waters analyzed by this study was about 13 times greater than the 50 ppb MCL.   The main arsenic species found in in the well waters of the BFD area were inorganic arsenic species As(III) and As (V), with an average ratio of As(III)/As(V) about 2.6. The individual wells showed a variation of As(III)/As(V) ratio from 1.1 to 5.2 . The MMA and DMA were below detection in the well waters.

In 1985, Chen et al. (52) evaluated the relationship between arsenic containing well waters and cancers in endemic areas of BFD in comparison with the general population of Taiwan.   Both the standardized mortality ratio(SMR) and cumulative mortality rate were significantly high in the BFD-endemic areas for cancers of bladder, kidney, skin, lung, and liver.   A dose-response relationship was observed between SMR of the cancers and BFD prevalence rate of the villages in the endemic areas.   This positive dose-response relationship was also observed between the exposure to artesian well waters and cancers of bladder, lung and liver (53).   SMRs of cancers were greater in villages where only artesian wells were used as drinking water supply.   Chen et al. (54) gave another report regarding a seven-year follow-up study in the endemic areas of arsenic.   A total of 263 patients with BFD and 2293 healthy residents were recruited and followed up for seven years.   The results suggest that BFD patients have significantly increased cancer incidence for cumulative arsenic exposure.

The medical records of a specific BFD endemic community were investigated recently (55).   The local inhabitants were divided into four groups according to age (under or over 40 year old) and gender.   The results suggested a significantly declining trend for mortality rate ratios of all malignant tumors from 1971 to 1973 in the study areas, especially in females.   A decrease of mortality rate ratios from malignant cancers, compared with the local and national references, was found in those aged over 40 for both male and female. The decrease was mainly due to a fall in internal organ and skin cancer mortality rates.   Overall, the results suggested that the improvement of drinking water supply to eliminate arsenic exposure from artesian well water could reduce the mortality incidence of arsenic related disease.

Chiou et al. (56) recently reported another arsenicosis-endemic area in northeast Taiwan in which each household had its own well for obtaining drinking water.   This situation is different from southwest Taiwan described above, in which many households share only a few wells in their villages.   Risk of transitional cell carcinoma in relation to ingested arsenic in drinking water during 1991-1994 in 8,102 residents was investigated (56).   This study of evaluation of each individual exposure to arsenic was based upon the arsenic concentration in well water of each household that was determined by hydride generation combined with AAS. There was a significantly increased incidence of urinary cancers for the study cohort compared with the general population in Taiwan.

Arsenic Related Diseases in Other Areas of Asia

Inner Mongolia of China

Since the first patient was diagnosed as suffering from arsenic poisoning in Inner Mongolia of China in 1990, newly affected areas and victims have been found one after another (57). In 1996, a joint survey team was organized by Asia Arsenic Network experts and members of the Institute for Control and Treatment of Endemic Disease in Inner Mongolia, and an extensive field investigation was carried out among 15 villages in 3 counties from August to December (57).

Most arsenic affected areas in Inner Mongolia are located in the arid region of the Hetao Plain between the Yellow River on the south and the Inshan Mountains on the north (Figure 3). Arsenic was not found both in the air and in the soil in the range of concern during the environmental survey. However, arsenic concentrations in 96 percent of the tested well waters exceeded 50 ppb, with the highest concentration of 1080 ppb. Meanwhile fluorine in drinking water was also found in high concentrations (57).

During the health survey, 1,728 people including 126 people in the control group were examined for skin lesions. 612 people, or 35.4 %, showed skin lesions related to arsenicosis such as pigmentation, depigmentation, keratosis, or skin cancer, on the abdomen, palm, and sole. Male was easier to be affected by arsenic contaminated well waters, according to the data of percentage of people with symptoms. No one was found to have skin lesions in the control group (57) (Table II).

According to some of the research data compiled by the end of 1995, arsenic contamination in Inner Mongolia spread to 655 villages in 11 counties, and 1,774 patients with arsenic caused diseases were confirmed. During the field survey from August to December 1996, 35.4 percent of the population examined showed skin lesions characteristic of arsenic poisoning, suggesting that there were a lot of arsenicosis patients in the 15 villages investigated(50).

Another report (58) described the arsenic pollution of ground waters and contamination of well waters in the Hetao Plain, Inner Mongolia of China. In particular, a heavily arsenic-polluted area was distributed along the old channels formed by faulting movement since the Quaternary age. Several field

222

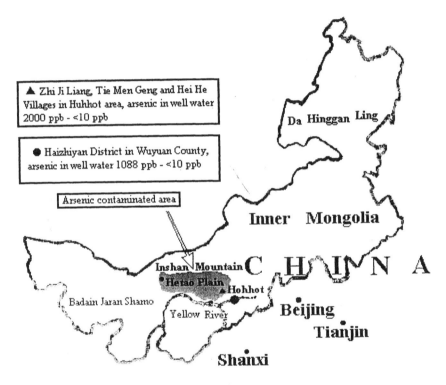

Figure 3. Arsenic contaminated areas in Hetao Plain, Inner Mongolia, China

investigations and lab analyses for the arsenic concentration in water and sediments revealed where the arsenic came from and how it was accumulated in the groundwater and the sediments. The mechanisms of migration, accumulation, and remigration of the arsenic from the hydrogeochemistry and geology viewpoints and geological setting were discussed. This study suggested that the arsenic pollution of ground water in the Hetao Plain of Inner Mongolia was brought about by two main causes: (1) geological and hydrogeochemical processes and (2) the influence of human activity, such as penetration of irrigation water. In the polluted area, the arsenic content is highly concentrated in clay and silty clay layers located at less than 40 meter in depth. They are probably the major sediment source for arsenic pollution of ground water at the present time. Accumulation of the arsenic in the sediments is caused by oxidation followed by biological action and colloid formation. The sediments containing arsenic are derived from weathering and erosion of iron sulfide deposits and igneous and metamorphic rocks in the Rangsan Mountains that contain natural arsenic. Dissolution of arsenic from the sediments is caused by changes in oxidation-reduction potential, as well as acid and alkaline conditions of the groundwater.

Table II. Examinations for Skin Lesions in Inner Mongolia of China (57)

Area	No. of People Examined			People with Skin Lesions		
	Male	Female	Total	Male	Female	Total
Haizhiyan	212	221	433	116(54.7 %)	82 (37.1%)	198(45.7%)
Bayinmaodao	602	567	1169	220(36.5%)	194 (34.2%)	414 (35.4%)
Control	63	63	126	0	0	0
Total	877	851	1728	336(38.3%)	276 (32.4%)	612 (35.4%)

Bangladesh

Bangladesh has the largest arsenic affected population in the world. This area has always had a problem of getting clean water. As many as a million wells drilled into Ganges alluvial deposits in Bangladesh and West Bengal, India may be contaminated with arsenic (59-66). Measured concentrations of arsenic in the well waters range up to 1 ppm. In the southwestern and some parts of eastern Bangladesh, arsenic content in groundwater has even higher concentrations. The experts at Bangladesh Council for Scientific and Industrial Research (BCSIR) found the highest concentration (14 ppm) in a shallow tube water in Pabna, Bangladesh (63), which is far above the restricted level for drinking water in Bangladesh (50 ppb). Approximately 44 percent of the total area of Bangladesh (34 districts) and 53 million rural people may be suffering from arsenic poisoning (63).

In Bangladesh, there are three types of water resources: surface water, rain water and ground water. Surface water is obviously the most convenient water resource. But on a closer inspection, there are many problems with this resource of water. During the Monsoon season, waters from the melting snows from the Himalayas, and rainwater from other regions of the Indian sub-continent, accumulate in the low-lying areas of Bangladesh before eventually passing into the Bay of Bengal. Flooding, therefore, becomes a yearly occurrence. In the dry season of the year, water scarcity results. Existing surface water is already rendered useless by extreme contamination. People in Bangladesh rely on surface water for their daily requirements. The water is directly taken from ponds and from shallow hand pumped wells. However, this water has become increasingly polluted. The pollution stems from poor sewage systems in Bangladesh. Newly established industrial plants frequently dump their waste in the water. This ultimately makes its way into the low-lying plains of Bangladesh. This contamination leads to various health problems such as cholera and the extensive pollution of the environment. The rainwater from the Monsoon rains cannot be effectively harnessed, since the country lacks the technology to functionally collect and store the rainwater for the future use. The remaining alternative is using groundwater since there are many fresh water aquifers in Bangladesh.

In the middle of 1980s, with the assistance of the World Bank, many shallow tube wells were dug to meet the daily water requirements of the local people. In 1992, it was discovered that the well water, which had provided a solution to the country's water supply problems, came with a hidden poison. The well water was laced with naturally occurring arsenic (66). No one had thought to check for arsenic contamination of well water when the wells were being dug. According to the latest report (61), perhaps as many as half of the 4 million wells drilled are contaminated with arsenic with concentrations exceeding 50 ppb. The catastrophe of arsenic contamination in well water occurred in 1950s in Taiwan apparently did not alert public attention in other parts of the world. Today, arsenic crisis in Bangladesh is considered the worst massive environmental health catastrophe.

Figure 4 is a map prepared in early 1998 showing high levels of arsenic primarily in the western part of the country (61). The arsenic-rich groundwater is mostly restricted to the alluvial aquifers of the Ganges Delta (59), therefore the source of arsenic-rich oxyhydroxides must lie in the Ganges source region upstream of Bangladesh. Weathered base-metal deposits occur in the Ganges basin, so that weathering of these arsenic-rich base metal sulfides must have supplied arsenic-rich iron oxyhydroxide to downstream Ganges sediments during

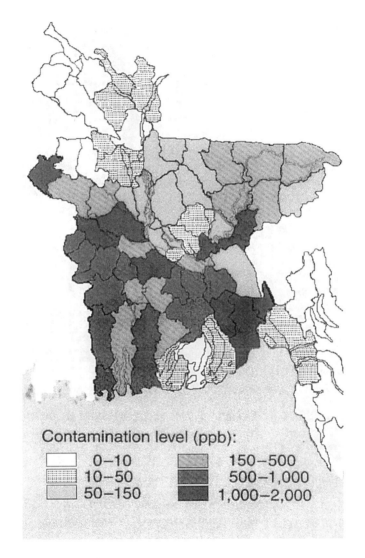

Figure 4. Arsenic levels in groundwaters of Bangladesh. Adapted with permission from reference 61. Copyright 1998 ACS.

Late Pleistocene-Present times *(59)*. The arsenic-rich iron oxyhydroxides are now being reduced, causing the current problem *(60-62)*.

India

The duration of arsenic exposure in West Bengal, India is uncertain, but it is thought that the problem began in the late 1960s when digging of tubewells commenced as part of a state-wide irrigation plan *(67)*. Because groundwater was cleaner than water from ponds and the polluted Ganges River, people prefer to use tubewell water. Since 1983 when the first group of patients was identified to be suffering from arsenic poisoning, the numbers of victims and affected areas have been increasing every year *(68, 69)*. Areas affected by arsenic contamination are all located in the upper delta plain of Ganges River. More than 800,000 people from 312 villages are drinking arsenic contaminated water *(70, 71)*. More than 175, 000 people exhibit arsenical skin lesions. Hair, nails, scales, urine and liver tissue analyses show that the residents have drunk this arsenic contaminated water for years. Most of the three stages of clinical manifestation of arsenicosis of skin lesions are observed in the inhabitants of the affected villages in six districts *(70, 72)*: dermatitis, keratosis, conjunctivitis, bronchitis and gastroenteritis in the first stage; peripheral neuropathies, depigmentation and hyperkeratosis in the second stage; and gangrene in the limbs in the last stage. Continual poisoning by arsenic results in the enlargement of the liver, kidneys, and spleen which often develops into malignant tumors, lung, skin and bladder cancer.

The analyses suggested there was no MMA and DMA in well water and arsenite and arsenate present in well water in these six districts were in the ratio of 1:1 *(73, 74)* according to the chemical analysis of the real samples.

Nickson et al. *(60)* suggested that the arsenic in the alluvial sediments was derived from sulphide deposits in the Ganges basin. However, the copper belt of Bihar, which contains small amounts of arsenopyrite, and the coal basins of the Damodar Valley, which contain moderate concentrations of arsenic, are drained by rivers that flow far to the south of the Ganges tributary system. Alternatively, arsenic may occur as contaminants from past and present mining and smelting activities.

Vietnam

A report in 2001 shows a serious arsenic contamination of Red River alluvial tract in the city of Hanoi and in the surrounding rural district *(16,17)*. Due to naturally occurring organic matter in the sediments, the groundwaters are anoxic and rich in iron. Arsenic levels in some Vietnam groundwater wells sampled exceed 3000 ppb. In a highly affected remote rural area, the groundwater used directly as drinking water had an average concentration of 430 ppb. Analysis of groundwater from the lower aquifer for the Hanoi water supply showed arsenic levels of around 280 ppb in three of eight water-treatment plants and 37-82 ppb in another five plants. The arsenic in the sediments may be associated with iron oxyhydroxides and released to the groundwater by reductive dissolution of iron. Oxidation of sulfide phases could also release arsenic to the groundwaters.

Other Parts of the World

Some areas in the United States including Minnesota (75), Oregon (76), California (77), Alaska (78), and Utah (79, 80) were reported a couple of decades ago to have arsenic in drinking water higher than the MCL level of 50 ppb. Some new contaminated areas including New Hampshire (22) and Wisconsin (81) were reported in 1999. Randomly selected 992 drinking water samples were collected from New Hampshire local households and analyzed using hydride generation ICP-MS. In these randomly selected household water samples, concentrations of arsenic range from <0.0003 to 180 ppb, with water from domestic wells containing higher concentration of arsenic than the one from municipal sources. Water samples from drilled bedrock wells had the highest arsenic concentrations. The spatial distribution of elevated arsenic concentrations (50 ppb) correlates with Late-Devonian Concord-type granitic bedrock. Analysis of rock digests indicates arsenic concentrations are up to 60 mg/kg (ppm) in pegmatites.

Allan H. Smith of UC Berkeley (82) emphasized that over 2.5 million people in the U.S might be supplied with water containing more than 25 ppb arsenic and 350,000 people supplied with water of 50 ppb arsenic or more. The arsenic contamination of water is probably a hidden global problem that is gradually being unfolded in recent years. Countries like Canada (83), Chile (18,19), Mexico (84,85), Argentina (20), and Finland (21) all appear to have a potential threat from arsenic contamination of groundwater.

References

1. *Code of federal Regulations,* 40 CFR 141.11, revised July 1992, U.S. Government Printing Office: Washington, DC, 1992.
2. *Code of federal Regulations,* 21 CFR 103.35, revised April 1992, U.S. Government Printing Office: Washington, DC, 1992.

228

3. Morales, K. H.; Ryan, L.; Kuo, T. L.; Wu, M. M.; Chen, C. J. *Environ. Health Perspective* **2000**, *108(7)* , 655.
4. Karagas, M. R.; Stukel, T. A.; Morris, J. S.; Tosteson, T. D.; Weiss, J. E.; Spencer, S. K.; Greenberg, E. R. *American J. Epidemiology* **2001**, *153(6)*, 559.
5. *Safe Drinking Water Act Amendments of 1996*; Arsenic Amendments added in Section 1412(b)(12)(A), Aug. 6, 1996.
6. U.S. Environmental Protection Agency, *National Primary Drinking Water Regulations, Arsenic and Clarifications and New Source Contaminants Monitoring;* Final Rule, Federal Register Part VIII, 2001; 66, pp 6976-7066.
7. *Handbook on the Toxicology of Metals*; Friberg, L.; Nordberg, G.F.; Vouk, V. B. 2nd Ed. , 1986; Vol. II, pp 43-83.
8. Le, X. C.; Ma, M.; Lu, X.; Cullen, W. R.; Aposhian, H. V.; Zheng, B. *Environ. Health Perspectives* **2000**, *108(11)*, 1015.
9. Tseng, W. P.; Chu, H. M.; How, S. W.; Fong, J. M.; Lin, C. S.; Yeh, S. *J. Natl. Cancer Inst.* **1968**, *40*, 453.
10. Yeh, S; How, S. W. *Rep. Inst. Pathol. Natl. Univ. Taiwan*, **1963**, *14*, 25.
11. Tseng, W. P. *Environ. Health Perspectives* **1977**, *19*, 109.
12. Lou, F. J. *The Lancet* **1990**, *336*, 115.
13. Luo, Z. D., et al., *Arsenic: Exposure and Health Effects*, Abernathy, C. O.; Calderon, R. L.; Chappell, W. R., Eds.; Chapman and Hall, London, 1997; pp 55-68.
14. Tondel, M.; Rahman, M.; Magnuson, A.; Chowdhury, I. A.; Faruquee, M. H.; Ahmad, S. A. *Environ. Health Perspectives* **1999**, *107(9)*, 727.
15. Pandey, P. K.; Khare, R. N.; Sharma, R.; Sar, S. K.; Pandey, M.; Binayake, P. *Current. Sci.* **1999**, *77(5)*, 686.
16. Christen, K. *Environ. Sci. Technol.* July 1, **2001**, 286A.
17. Berg, M.; Tran, H. C.; Nguyen, T. C.; Pham, H. V.; Schertenleib, R.; Giger, W. *Environ. Sci. Technol.* **2001**, *35*, 2621.
18. Biggs, M. L.; Haque, R.; Moore L.; Smith, A. H. *Science* **1998**, *281*, 785.
19. Smith, A. H.; Arroyo, A. P.; Mazumder, D. N. G.; Kosnett, M. J.; Hernandez, A. L.; Beeris, M.; Smith, M. M; Moore, L. E. *Environ. Health Perspectives* **2000**, *108(7)*, 617.
20. Concha, G.; Nermell, B.; Vahter M. *Environ. Health Perspectives* **1998**, *106(6)*, 355.
21. Kurttio, P.; Pukkala, E.; Kahelin, H.; Auvinen, A.; Pekkanen, J. *Environ. Health Perspectives* **1999**, *107(9)*, 705.
22. Peters, S. C.; Blum, J. D.; Klaue, B.; Karagas, M. R. *Environ. Sci. Technol.* **1999**, *33*, 1328.

23. Penrose, W. R. *Arsenic in the Marine and Aquatic Environments: Analysis, Occurrence, and Significance*, CRC Crit. Rev. Environ. Control, 1974; Vol. 4, pp 465-482.

24. Bowen, H. J. M. *Elemental Chemistry of the Elements.* Academic Press, London and New York, 1979; pp 60-61.

25. Backer, D. E.; Chesnin, L. *Adv. Agron.* **1975**, *27*, 305.

26. Huang, Y. C. *Arsenic in the Environment*, Nriagu, J. O. Ed.; John Wiley & Sons, Inc., 1994; Ch. 2, pp 17-49.

27. Cullen, W. R.; Reimer, K. J. *Chem. Review* **1989**, 89, 713.

28. Anderson, L. C. D.; Bruland, K. *Environ. Sci. Technol.* **1991**, *25*, 420.

29. Sanders, J. G. *Mar.Chem.* **1985**, *17*, 320.

30. Mok, W. M.; Wai, C. M. *Arsenic in the Environment*, Nriagu, J. O. Ed.; John Wiley & Sons, Inc., 1994; Ch. 5, pp 99-117.

31. *Environmental Health Criteria 18*, Arsenic, World Health Organization: Geneva, 1981; pp 50-146.

32. Vahter, M. Ch. 5, *Metabolism of Arsenic*; Fowler, B. A. Editor, in *Biological and Environmental Effects of Arsenic;* Elsevier Science Publishers *B.V.*, Amsterdam, New York, Oxford, 1983.

33. Manning, B. A; Martens, D. A. *Environ. Sci. Technol.* **1997**, *31(1)*, 171.

34. Woller, A.; Mester, Z.; Fodor, P. J. *Anal. At. Spectrom.* **1995**, *10*, 609.

35. Roychowdhury, S. B.; Koropchak, J. A. *Anal. Chem.* **1990**, *62*, 484.

36. Mok, W. M.; Shah, N. K.; Wai, C. M. *Anal. Chem.* **1986**, *58*, 110.

37. Faix, W. G.; Caletka, R.; Krivan, V. *Anal. Chem.* **1981**, *53*, 1719.

38. Yu, J. J.; Wai, C. M. *Anal. Chem.* **1991**, *63*, 842.

39. Lown, J. A.; Johnson, D. C. *Anal. Chim. Acta* **1980**, *116*, 41.

40. Laintz, K. E.; Yu, J. J; Wai, C. M. *Anal. Chem.* **1992**, *64*, 311.

41. Laintz, K. E.; Shieh, G. M.; Wai, C. M. *J. Chromatographic Sci.* **1992**, *30*, 120.

42. Didina, J.; Tsalev, D. L. *Hydride Generation Atomic Spectroscopy*; Wiley: New York , 1995.

43. Holak, W. *Anal. Chem.*, **1969**, *41(12)*, 1712.

44. Bermejo-Barrera, P; Moreda-Pineiro, J.; Moreda-Pineiro, A.; Bermejo-Barrera, A. *Anal. Chim. Acta* **1998**, *374*, 231.

45. Lajunen, L. H. J. *Spectrochemical Analysis by Atomic Absorption and Emission*; Royal Society of Chemistry: Cambridge, 1992.

46. Frenzel, W.; Titzenthaler, F.; Elbel, S. *Talanta* **1994**, *41*, 1965.

47. Suzuki, N.; Satoh, K.; Shoji, H.; Imura, H. *Anal. Chim. Acta* **1986**, *185*, 239.

48. Ebdon, L.; Hill, S.; Ward, R. W. *Analyst*, **1986**, *111*, 1113.

49. Chen, C. J.; Chuang, Y. C.; Lin, T. M.; Wu, H. Y. *Cancer Res.* **1985**, *45*, 5895.

50. Chen, S. L.; Dzeng, S. R.; Yang, M. H.; Chiu, K. H.; Shieh, G. M.; Wai, C. M. *Environ. Sci. Technol.* **1994,** *28,* 877.
51. Kuo, T. *Rep. Inst. Pathol. Natl. Taiwan Univ.* **1968,** *20,* 7.
52. Chen, C. J.; Chuang, Y. C.; Lin, T. M.; Wu, H. Y. *Cancer Res.* **1985,** *45,* 5895.
53. Chen, C. J.; Chuang, Y. C.; You, S. L.; Lin, T. M.; Wu, H. Y. *Br. J. Cancer* **1986,** *53,* 399.
54. Chiou, H. Y.; Hsueh, Y. M.; Liaw, K. F.; Horng, S. F.; Chiang, M. H.; Pu, Y. S.; Lin, J. S. N.; Huang, C. H.; Chen, C. J. *Cancer Res.* **1995,** *55,* 1296.
55. Tsai, S. M.; Wang, T. N.; Ko, Y. C. *J. Toxicol. Environ. Health* **1998,** *55(6),* 389.
56. Chiou, H. Y.; Chiou, S. T.; Hsu; Y. H.; Chou; Y. L.; Tseng, C. H.; Wei, M. L.; Chen, C. J. *American J. of Epidemiology* **2001,** *153(5),* 411.
57. Guo, X. J. http://phys4.harvard.edu/~wilson/inner_mongolia_paper2.html, *96% of Well Water is Undrinkable,* the Institute for Control and Treatment of Endemic Diseases in Inner Mongolia A. R., 1999.
58. Gao, C. R. *Arsenic Pollution of Groundwater in the Hetao Plain of Inner Mongolia, China* (Grad. Sch. Sci. Technol., Niigata Univ., Igarashi 2-8050, Niigata, Japan 950~2181), *Chikyu Kagaku (Chigaku Dantai Kenkyukai)* **1999,** *53(6),* 434, (Japan), Chigaku Dantai Kenkyukai, CA 132:127189h.
59. Acharyya, S. K.; Chakraborty, P.; Lahiri, S.; Raymahashay, B. C.; Guha, S.; Bhowmik, A. *Nature* **1999,** *401,* 545.
60. Nickson, R.; McArthur, J.; Burgess, W.; Ahmed, K.M.; Ravenscroft, P.; Rahman, M. *Nature* **1998,** *395,* 338
61. Lepkowski, W. *Chem. Eng. NEWS,* Nov. 16, 1998; p 27.
62. Dhar, R. K. *Current Sci.* **1997,** *73,* 48.
63. Karim, M.; Komori, Y.; Alam, M. *J. Environ. Chem.* **1997,** *7(4),* 783.
64. Das, D.; Basu, G.; Chowdhury, T. R.; Chakraborty, D. *in Proc. Intl. Conf. on Arsenic in Groundwater, Calcutta,* 1995; pp 44-45.
65. Bhattacharya, P.; Chattargee, D.; Jacks, G. *Water Resources Development* **1997,** *13,* 79.
66. Lepkowski, W. *Chem. Eng. NEWS,* May 17, 1999; p 45.
67. Bagla P.; Kaiser J. *Science* **1996,** *274,* 174.
68. Chakraborty A. K.; Saha, K. C. *Indian J. Med. Res.* **1987,** *85,* 326.
69. Mazumder, D. N. G.; Haque, R.; Ghosh, N.; De, B. K.; Santra, A.; Chakraborty, D.; Smith, A. H. *International J. Epidemology* **1998,** *27,* 871.
70. Das, D.; Samanta, G.; Mandal, B. K.; Chowdhury, T. R.; Chanda, C. R.; Chowdhury, P. P.; Basu, G. K.; Chakraborti, D. *Environ. Geochem. and Health,* **1996,** *18,* 5.
71. Chen, C. J.; Chen, C. W.; Wu, M. M.; Kuo, T. L. *Br. J. Cancer* **1992,** *66,* 888.

72. Das, D.; Chatterjee, A.; Mandal, B. K.; Samanta, G.; Chakraborti, D. *Analyst* **1995,** *120,* 917.
73. Samanta, G.; Chowhury, T. R.; Mandal, B. K.; Biswas, B. K.; Chowdhury, U. K.; Basu, G. K.; Chanda, C. R.; Lodh, D.; Chakraborti, D. *Microchem. J.* **1999,** *62(1),* 174.
74. Chatterjee, A.; Das, D.; Mandal, B. K.; Chowdhury, T. R.; Samanta, G.; Chakraborti, D. *Analyst* **1995,** *120,* 643.
75. Feinglass, E. J. *The New England J. Medicine* **1993,** *288(16),* 828.
76. Morton, W.; Starr, G.; Pohl, D.; Stoner, J.; Wagner, S.; Weswig, P. *Cancer* **1976,** *37,* 2523.
77. Goldsmith, J. R.; Deane, M.; Thom, J.; Gentry, G. *Water Res.* **1972,** *6,* 1133.
78. Harrington, J. M.; Middaugh, J. P.; Morse, D. L; Housworth, J. *American J. Epidemiology* **1978,** *108(5),* 377.
79. Lewis, D. R.; Southwick, J. W. *Environ. Health Perspectives* **1999,** *107(5),* 359.
80. Southwick, J. W. *Arsenic, Industrial, Biochem. Environ. Perspectives,* Lederer, W. L.; Fensterheim, R. J., Eds.; New York, Van Nostrand Reinhold, 1983; pp 210-225.
81. Burkel, R. S.; Stoll, R. C. *Ground Water Monit. Rem.* **1999,** *19(2),* 114.
82. Smith, A. H.; Hopenhayn-Rich, C.; Bates, M. N.; Goeden, H. M.; Hertz-Picciotto, I.; Duggan, H. M.; Wood, R.; Kosnett, M. J.; Smith, M. T. *Environ. Health Perspectives* **1992,** *97,* 259.
83. Boyle, D. R.; Turner, R. J. W.; Hall, G. E. M.; *Environ. Geochem. Health* **1998,** *20(4),* 199.
84. Rosas, I.; Belmont, R.; Armienta, A.; Baez, A. *Water, Air, Soil Pollut.* **1999,** *112 (1-2),* 133.
85. Rodriguez, C. R.; Armienta H. A.; Longley, L. *Groundwater Urban Environ. Proc. IAH Congr., 27th* 1, **1997,** pp 527-530.
86. Chang, C. H.; Yu, H. S.; Chen, G. S.; Wang, M. T.; Ko, S. S. *Kaohsiung J. Med. Sci.* **1993,** *9,* 559.

Chapter 17

Dissolved Gaseous Mercury Profiles in Freshwaters

Steven D. Siciliano, Nelson O'Driscoll, and D. R. S. Lean

Ecotoxicology Laboratory, Department of Biology, University of Ottawa, 30 Marie Curie, P.O. Box 450, Station A, Ottawa, Ontario K1N 6H6, Canada

The importance of microbial processes for the regulation of dissolved gaseous mercury (DGM) in deep freshwaters has not been previously investigated. In this study, we evaluated microbial mercury reductase and oxidase activities in depth profiles from Jack's Lake. In addition, detailed DGM depth profiles were determined for four sampling stations on Lake Ontario. Our results illustrate that microbial processes are an important factor regulating DGM in the hypolimnion. Levels of DGM in William's Bay in Jack's Lake were 6 times higher than that observed at Brooke's Bay. This was accompanied by a 10 fold reduction in mercury oxidase activity in William's Bay. When DGM concentrations are expressed on an aerial basis, DGM concentrations above the thermocline in Lake Ontario average 1.5 ng DGM m^{-2} and in small freshwater lakes it ranged between 0.1 and 0.8 ng DGM m^{-2}. Further, it was demonstrated that the majority of DGM in large freshwater lakes such as Lake Ontario exists below the thermocline where photochemical oxidation and reduction processes cannot occur. The importance of this DGM to atmospheric flux rates is discussed.

Introduction

The speciation of mercury in freshwaters plays a critical role in determining the environmental fate of this toxicant (1). Divalent mercury (Hg^{2+}) rapidly associates with biological and mineralogical colloids and as a result, is mobilized in the environment (2). Methylmercury (CH_3Hg^{1+}) is a potent neurotoxin that bioaccumulates in aquatic and terrestrial food chains (3). In contrast to these toxic forms of mercury, elemental mercury (Hg^0) is volatile and relatively non-toxic. The emission of Hg^0 from freshwater bodies is postulated to be a major route of mercury removal from lake water columns (4). Modeling this evasion for an entire lake ecosystem is complicated because it is not known if the atmospheric-water boundary is the principal controller of Hg^0 or if instead the aerial extent of dissolved gaseous mercury (DGM) is more important (5).

The aerial extent of DGM in freshwater has not yet been determined but it is known that DGM profiles in lake water can vary significantly. For example, DGM in the water column of Ranger Lake varied from 240 fM to 6 fM (6) and in Lake Ontario it varied from 1290 fM to 696 fM (7). Elemental mercury in freshwater lakes is regulated by a combination of photochemical and biological processes. The photochemical process is postulated to be mediated by reactive iron in the water column, which initiates a free radical reaction scheme that reduces Hg^{2+} to Hg^0 (8). Other parameters such as Cl and dissolved organic carbon compounds also mediate the photochemical transformations of Hg^0 (9). Recently our laboratory group demonstrated that microorganisms play an important role in DGM cycling in freshwater systems. The photochemical production of hydrogen peroxide initiates the microbial oxidation of Hg^0 to Hg^{2+} and mercury reductase activity reduces this Hg^{2+} back to Hg^0 (10). The majority of DGM research has taken place in the uppermost portion of the water column. Little is known about deep water DGM processes but these deep waters may contain the majority of DGM and thus play an important role in modulating DGM evasion from lake surfaces.

In addition to the atmosphere-water boundary, there are other boundaries in lake water that are known to influences the distribution of contaminants in freshwaters. At the thermocline sharp gradients of NO_3, NH_4, SO_4, H_2S, Fe^{3+}, Fe^{2+}, CH_4, N_2O, and H_2O_2 have been observed (11-14). Changes in the redox state of water might also influence biologically mediated transformations of DGM because as metal's move from one redox state to another, H_2O_2 is produced (14) and this may influence the H_2O_2 dependent mercury oxidase enzyme. If there were limits to mercury flux through the water column, these may limit DGM evasion by reducing the amount of DGM capable of reaching the lake surface. In this work, we investigate DGM concentrations and associated enzyme activities in detailed depth profiles under a variety of conditions. The purpose of this study was to characterize DGM distributions outside of the influence of photochemical processes and thereby estimate the total DGM load of a freshwater lake.

Materials and Methods

Samples from two bays, Brookes and Williams, in Jack's Lake, (44°,41',20" N, 72°,02',54" W), were collected from a fiberglass boat using a Go-Flo sampler on July 21, 2000. Jack's Lake is a mesotrophic lake near the Canadian Shield with an average of 14 mg of Ca^{2+} L^{-1}, 12 µg P L^{-1} and pH of 7.2 (15). Brookes Bay has a dissolved organic carbon (DOC) concentration of 7.8 mg C L^{-1} and Williams Bay has a DOC of 6.0 mg C L^{-1}. Samples from Lake Ontario were collected using a Go-Flo sampler on September 12, 2000 at 10:26 from Station 29 (43°,49',51" N, 78°,52',08" W) and at 18:30 from station 743 (43°,31',13" N, 78°,11',16" W). On September 14, 2000, samples were collected at 09:08 from Station 73 (43°,38',01" N, 76°,17',12" W) and at 17:49 from station 586 (43°,29',07" N, 77°,02',48" W).

DGM was analyzed by bubbling approx. 20L (1 L min^{-1} for 20 minutes) of mercury free air produced by a Tekran 1100 Zero Air Generator through a 1 L water sample contained in a closed glass graduated cylinder. The bubbled gas was analyzed for dissolved gaseous mercury using a Tekran 2537A with pre-cleaned Teflon lines and connections. This analytical system had a daily detection limit of 5-25 fM. Daily detection limit was determined as three times the standard deviation of the baseline. The average percent difference between duplicates was 32 % (n=24). After analysis of DGM in Jack's Lake, 500 ml of lake water was combined with 100 mL of glycerol and the samples were frozen in amber glass bottles for microbial analysis.

Microbial mercuric reductase and oxidase activity was assessed on protein extracts of 500 ml of unfiltered lake water. Microbial cells were concentrated and lysed as previously described (16) and assessed for mercury reductase activity (17). Mercuric reductase consumes NADPH to reduce mercury. To assess mercuric reductase, NADPH consumption over a 20 minute time period is compared between samples with or without 20 nmoles of Hg^{2+}. One unit of enzyme activity (U) was defined as the equivalent to 1 µmole of NADPH consumed in response to the Hg^{2+} aliquot, i.e. NADPH consumption in the presence of mercury – NADPH consumption without mercury. Microbial oxidation of elemental mercury was measured using 1 mL additions of water saturated with Hg^0 to 200 µL enzyme extracts of lake water (18). Enzyme extracts were incubated at 22°C for 1 hour and a U designated as 10 fmoles of inorganic mercury formed. Boiled controls were prepared by heating enzyme samples (100°C) for 10 minutes and background mercury oxidation is subtracted from the reported value. Protein levels were quantified using the Lowry Protein Assay (*19*).

Aerial concentrations of DGM were calculated by estimating the amount of DGM present in a square meter water column that extends to the bottom of that section of lake. For surface water samples, the thermocline was assumed to be the bottom. Hence the amount of DGM present in a m^2 that extends from the surface to the thermocline was calculated. Similarly, aerial concentrations below the thermocline were calculated by extending a m^2 column to the lake bottom and calculating DGM present in that column.

Results and Discussion

Dissolved gaseous mercury concentrations in Brookes Bay increased with depth throughout the epilimnion and decreased immediately above thermocline (Figure 1). In the hypolimnion, DGM concentrations were highest just below the thermocline and then decreased with depth. Levels of mercury reductase activity in the epilimnion followed a similar pattern with maximal mercury reductase activity co-inciding with the maximum DGM concentrations. However, in the hypolimnion this trend did not continue with mercury reductase activity remaining relatively constant despite decreasing DGM concentrations with depth. Mercury oxidase activity was highest in the surface waters, corresponding with a low DGM concentration and then steadily decreased until 7 meters where a sharp increase in activity was evident. This is consistent with the observation that mercury oxidase activity is closely linked to H_2O_2 (*10*) The increase at 7 meters coincided with the lowest level of DGM observed in the hypolimnion.

In William's Bay (Figure 2), similar trends were observed with DGM concentrations reaching a peak in the epilimnion at 2.5 meters followed by a sharp decrease within the thermocline and oxocline and then an increase in the hypolimnion just below the thermocline. In contrast to Brookes Bay, mercury reductase activity was higher in the hypolimnion compared to the epilimnion and bore little relation to observed DGM concentrations. Mercury oxidase activity in William's Bay was 10 times less than that observed at Brooke's bay but it followed a similar pattern with maximal activity observed near the surface and a rapid decrease with depth. Perhaps reflecting low mercury oxidase activity, the concentrations of DGM are 4 times greater in William's compared to Brooke's bay and there is a correspondingly large differential between mercury reductase and oxidase activity in William's bay. This illustrates that both DGM formation and transformation by microorganisms are important in regulating DGM concentrations in freshwaters.

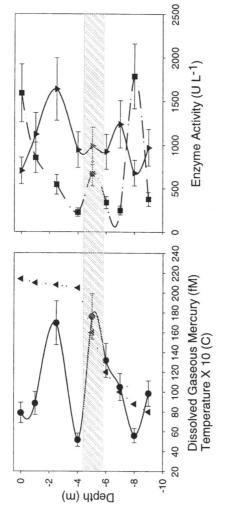

Figure 1. Depth profiles of dissolved gaseous mercury (●), temperature (▲), mercury reductase activity (▼) and mercury oxidase activity (■) in Brookes Bay, Jack's Lake. Each data point is the average of duplicate samples taken at each depth with error bars indicating the range. The shaded box indicates the water depth at which the maximum change in water temperature was observed.

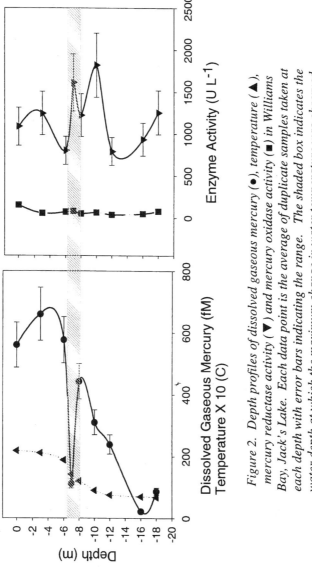

Figure 2. Depth profiles of dissolved gaseous mercury (●), temperature (▲), mercury reductase activity (▼) and mercury oxidase activity (■) in Williams Bay, Jack's Lake. Each data point is the average of duplicate samples taken at each depth with error bars indicating the range. The shaded box indicates the water depth at which the maximum change in water temperature was observed.

Our results in Jack's Lake are similar to that observed in Pettaquamscutt (20) and Kejimikujik (N. O'Driscoll, Personal Communication) in which DGM and temperature were closely linked but differ from results obtained at Ranger Lake (6). Dissolved gaseous mercury is formed by a combination of photochemical and biological processes (10). It is likely that basis for the differences in DGM concentrations observed in different lakes is the result of a complex interaction between iron cycling, Cl levels, organic matter and microbial activity (8,9). Figure 5 illustrates the complexity of reactions regulating DGM concentrations in freshwaters. This figure suggests redox based processes may be responsible for deep water DGM transformations. However, as of yet, few investigations have assessed the interaction between abiotic and biotic transformations of DGM in deep water.

Profiles of DGM in the shallow stations of Lake Ontario were similar to that observed at Jack's Lake (Figure 3). DGM concentrations were their highest at the lake surface and then rapidly decreased with depth. Just above lake bottom, DGM concentrations increased again, in the case of Lake Ontario from approximately 200 fM to 400 fM and in the case of Jack's Lake from 30 fM to 100 fM. DGM concentrations in Lake Ontario shallow stations bore little relation to changes in temperature. At Station 73, DGM concentrations remained at approximately 200 fM despite a 10° drop in temperature from 21°C to 9°C. Similarly, at Station 29, DGM concentrations dropped from 1000 fM to less than 200 fM with no change in water temperature. Similar results were obtained in 1998 at a depth profile for Station 983, 35 m deep, in which DGM concentration increased from 400 fM to 500 fM despite a 20° decrease in temperature (7). Similarly, at station 988, 27 m deep, DGM concentrations increased from 700 fM to 1300 fM and a corresponding 10° decrease in water temperature. It appears that changes in temperature with depth is not related to changes in DGM concentrations with depth.

In contrast to the shallow stations, deep stations of Lake Ontario displayed high concentrations of DGM in the hypolimnion (Figure 4). Similar DGM patterns were observed in the surface waters with a high concentration at the surface followed by a rapid decline but underneath the thermocline, DGM concentrations increased to double that seen in the surface waters. This is now the second study that has observed significant concentrations of DGM near the lake bottom (7). As of yet there is no explanation for this increase in DGM. It is unlikely that it is a H_2O_2 mediated pathway. However, up-core enrichment of mercury in lake sediments has been repeatedly observed. It is possible, that microorganisms exposed to inorganic mercury are reducing it to Hg^0. The mer gene has been observed in sediments (21) but the prevalence of this pathway in Lake Ontario is not known. Alternatively, sediment redox processes may be contributing to Hg^0 production. The importance of sediment processes on DGM in the water column is an area that warrants further investigation.

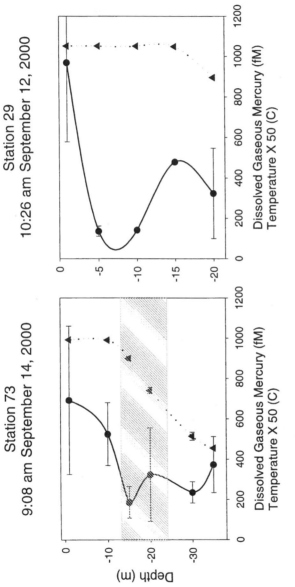

Figure 3. Depth profiles of dissolved gaseous mercury (●) and temperature (▲) in shallow stations of Lake Ontario. Each DGM data point is the average of duplicate samples taken at each depth with error bars indicating the range. The shaded box indicates the water depth at which the maximum change in water temperature was observed.

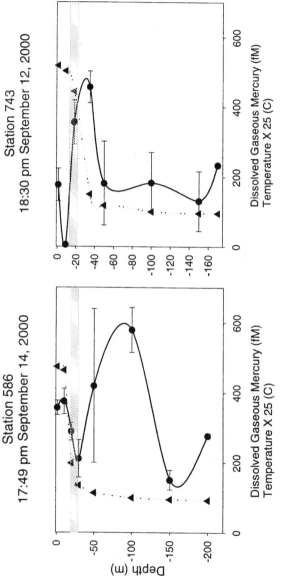

Figure 4. Depth profiles of dissolved gaseous mercury (•) and temperature (▲) in deep stations of Lake Ontario. Each DGM data point is the average of duplicate samples taken at each depth with error bars indicating the range. The shaded box indicates the water depth at which the maximum change in water temperature was observed.

241

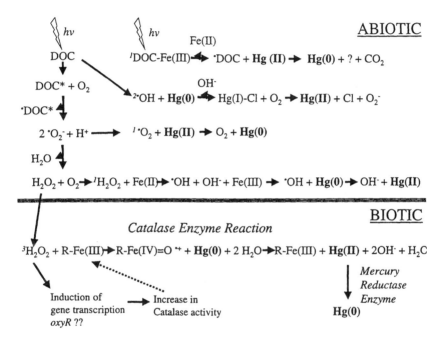

Figure 5. Conceptual diagram outlining the importance of sunlight for the two known biological and four known abiotic transformations of divalent and elemental mercury in freshwaters. [1]-reactions described by Zhang and Lindberg(2001), [2]-reactions described by Lalonde et al. (2000) and [3]-reactions described by Siciliano et al. (Submitted). The relative importance of each reaction pathway has yet to be determined. Credit: Siciliano et al. Environmental Science and Technology. Submitted. Microbial reduction and oxidation of mercury in freshwater lakes.

Aerial concentrations of the surface of Lake Ontario remained relatively constant at approximately 1.5 mg m^{-2} (Table I) and are very similar to those obtained previously for lake Ontario (7). The high concentration of DGM near Pickering may be related to higher total Hg (0.99 ng L^{-1}) in the area compared to the remainder of Lake Ontario (0.33 ng L^{-1}) and underlines the potential importance of DGM for the reclamation of polluted lakes. Jack's Lake had a much more variable aerial concentration that was at least 50% lower than that observed in Lake Ontario. Previous aerial values obtained for Ranger Lake (6) are between the aerial DGM concentrations of Brookes and William's bays in Jack's Lake. Average total gaseous mercury (TGM) flux rates recorded for a three day period for two shallow freshwater lakes, North Cranberry and Big Dam West, were 1.1 ng TGM m^{-2} h^{-1} and 5.4 ng TGM m^{-2} h^{-1} respectively (22). The values for Big Dam West are skewed by a one day period during the day in which TGM flux was 12.5 ng TGM m^{-2} h^{-1}, trimming this value gives an average flux rate of 2.5 ng TGM m^{-2} h^{-1}. The following year flux rates at Big Dam West over a 48 h period were 1.2 ng TGM m^{-2} h^{-1} and for another small lake, Puzzle lake, they were 0.9 ng TGM m^{-2} h^{-1} (unpublished observations). In contrast, flux rates over the deep water of Lake Ontario were found to be between 2 and 4 ng TGM m^{-2} h^{-15}.

DGM flux to the atmosphere is partially controlled by the concentration of DGM is the uppermost region of the epilimnion. In turn, these DGM concentrations are the result of two processes, net production/destruction in the irradiated layer and the diffusion of DGM from the remainder of the epilimnion. Our comparison of aerial DGM concentrations to observed TGM flux suggests that diffusion from the remainder of the hypolimnion may be partially responsible for the increased TGM flux observed in larger lakes. It is possible that due to the large reservoir of DGM in the hypolimnion of large lakes that diffusion from this readily available pool of DGM may offset periods when DGM is being destroyed by photochemical or biological processes in the uppermost layer of the epilimnion. We acknowledge that this hypothesis has not been rigorously tested in this manuscript. Rather our intention is to draw investigators attention to this possibility which awaits further experimental verification.

Aerial concentrations beneath thermocline are much higher than the aerial DGM concentration for surface water with the majority of DGM in the deep stations being below the thermocline where photochemical processes are not occurring. This hyplimnotic DGM is likely not available for flux to the atmosphere while the thermocline is in existence. But come fall turnover, stations such as Station 586 have 15 mg DGM m^{-2} whereas Station 73 has only 2.8 mg DGM m^{-2}. The influence this has upon DGM flux to the atmosphere is not known.

Table I. Aerial Concentrations[1] of Dissolved Gaseous Mercury at the Surface and at the Thermocline in Freshwater Lakes

Portion of Water Column[2]	Lake Ontario						Jack's Lake		Ranger Lake[4]
	Station 586	Station 743	Station 73	Station 29	Station 983[3]	Station 988[3]	Brooke's Bay	William's Bay	
Above Thermocline	1.4	1.5	2.0	1.5	1.3	4.9	0.10	0.85	0.24
Below Thermocline	13	5.6	0.89	0	1.5	0	0.094	0.53	0.10
Water Column Depth (m)	210	171.5	38	21.5	35	27	9	18	12

[1]Areal dissolved gaseous mercury concentrations are expressed as total mg of DGM m^{-2} of water at the boundary. This is calculated by determining the amount of DGM in a one m^3 column of lake water that extends from the surface to the thermocline and for the thermocline value, from the thermocline to the lake bottom.

[2]The water column was divided into that region above and below the thermocline. Stations 29 and 988 did not have a detectable thermocline.

[3]Value computed from Figure 2, Amyot et al. 2000. Distribution and transformation of elemental mercury in the St. Lawrence River and Lake Ontario. Can. J. Fish. Aquat. Sci. 57 (Suppl. 1): 155-163. Station 988 is near Pickering, Nuclear Powerplant and has concentrations on inorganic mercury three times that (0.99 ng L^{-1}) that found elsewhere in Lake Ontario.

[4]Value computed from Figure 6. Amyot et al. 1994. Sunlight-induced formation of dissolved gaseous mercury in lake waters. Environ. Sci. Technol. 28 (13): 2365-2371.

This study has illustrated that microbial processes are an important regulator of DGM concentrations at depth in freshwater lakes. Differences in these processes may be partially responsible for the large differences observed between William's and Brooke's Bay and illustrates the dynamic nature of DGM. From our short-term data it appears that there is a relationship between lake physical characteristics and DGM distributions. It appears that a substantial amount of DGM is present below the thermocline. The consequence of this DGM to seasonal mercury fluxes is an area that deserves further research.

Acknowledgements

We would like to thank the staff and crew of the Canadian Coast Guard vessel, Limnos for their assistance in conducting this study as well as Dr. Jeff Ridal for stimulating discussions on the influence of thermocline on trace contaminant fate in freshwater systems. Thanks also to Dr. Steven Beauchamp and Dr. Robert Tordon of Environment Canada for allowing us access to their TGM flux data. We also thank S. Winch, A. Khan and M. Russell for their technical assistance. This work was supported by a NSERC operating grant to D. Lean.

Reference

1. Morel, F. M. M.; Kraepiel, A. M. L.; Amyot, M. *Annu. Rev. Ecol. Syst.* **1998**, *29*, 543-566.
2. Balogh, S. J.; Meyer, M. L.; Hansen, N. C.; Moncrief, J. F.; Gupta, S. C. *Journal of Environmental Quality* **2000**, *29*, 871-874.
3. Lawson, N. M.; Mason, R. P. *Biogeochemistry* **1998**, *40*, 235-247.
4. Loux, N. T. *Environ. Toxicol. Chem.* **2000**, *19*, 1191-1198.
5. Poissant, L.; Amyot, M.; Pilote, M.; Lean, D. R. S. *Environ. Sci. Technol.* **2000**, *34*, 3069-3078.
6. Amyot, M.; Mierle, G.; Lean, D. R. S.; McQueen, D. J. *Environ. Sci. Technol.* **1994**, *28*, 2366-2371.
7. Amyot, M.; Lean, D. R. S.; Poissant, L.; Doyon, M.-R. *Can. J. Fish. Aquat. Sci.* **2000**, *57*, 155-163.
8. Zhang, H.; Lindberg, S. E. *Environ. Sci. Technol.* **2001**, *35*, 928-935.
9. Lalonde, J. D.; Amyot, M.; Kraepiel, A. M. L.; Morel, F. M. M. *Environ. Sci. Technol.* **2001**, *35*, 1367-1372.
10. Siciliano, S. D.; O'Driscoll, N.; Lean, D. R. S. *Environ. Sci. Technol.* Submitted.
11. Knowles, R.; Lean, D. R. S. *Can. J. Fish. Aquat. Sci.* **1987**, *44*, 743-749.

12. Lean, D. R. S.; Knowles, R. *Can. J. Fish. Aquat. Sci.* **1987**, *24*, 2133-2143.
13. Taylor, W. B.; Fricker, H.-J.; Lean, D. R. S. *Can. J. Fish. Aquat. Sci.* **1987**, *44*, 2178-2184.
14. Cooper, W. J.; Lean, D. R. S.; Carey, J. H. *Can. J. Fish. Aquat. Sci.* **1989**, *46*, 1177-1231.
15. Scully, N. M.; Vincent, W. F.; Lean, D. R. S.; MacIntyre, S. *Aquatic Science* **1998**, *60*, 169-186.
16. Ogunseitan, O. A. *Journal of Microbiological Methods* **1997**, *28*, 55-63.
17. Ogunseitan, O. A. *Applied and Environmental Microbiology* **1998**, *64*, 695-702.
18. Smith, T.; Pitts, K.; McGarvey, J. A.; Summers, A. O. *Applied and Environmental Microbiology* **1998**, *64*, 1328-1332.
19. Koch, A. L. In *Methods for general and molecular bacteriology*. Gerhardt, P., Ed.; American Society of Microbiology: Washington, DC, 1994; pp 248-277.
20. Mason, R. P.; Fitzgerald, W. F.; Hurley, J. P.; Hanson, A. K.; Donaghay, P. L.; Sieburth, J. M. *Limnology and Oceanography* **1993**, *38*, 1227-1241.
21. Hobman, J. L.; Wilson, J. R.; Brown, N. L. In *Environmental Microbe-metal Interactions*; Lovley, D. R., Ed.; ASM Press: Washington, DC., 2000; pp 177-197.
22. Boudala, F. S.; Folkins, I.; Beauchamp, S.; Tordon, R.; Neima, J.; Johnson, B. *Water, Air and Soil Pollution* **2000**, *122*, 183-202.

Chapter 18

Toward a Better Understanding of Mercury Emissions from Soils

Hong Zhang[1], Steve Lindberg[2], Mae Gustin[3], and Xiaohong Xu[4]

[1]Department of Environmental Health Sciences,
University of Michigan, Ann Arbor, MI 48109
[2]Environmental Sciences Division, Oak Ridge National Laboratory,
Oak Ridge, TN 37831–6038
[3]Department of Environmental and Resource Sciences,
University of Nevada, Reno, NV 89557–0013
[4]Department of Atmospheric, Oceanic, and Space Sciences,
University of Michigan, Ann Arbor, MI 48109

Despite progress in the last decade in the research of Hg emission from soils, considerable knowledge gaps still exist and point to the need for further extensive and elaborate research in this field. In this chapter, we discuss these uncertainties and identify some future priorities for the research of soil Hg emission. Our intension is to promote a more extensive and critical assessment of our current knowledge and understanding on soil Hg emissions and their effect on global Hg cycling.

Introduction

Mercury (Hg), a unique volatile metal at normal temperatures, is a well-known persistent, bioaccumulative toxic pollutant (*1*). Its persistence is enhanced by its long-distance transport in the atmosphere, with a residence time of ~6-12 months (*2*). Mercury emitted from anthropogenic and natural sources will sooner or later deposit to the earth surface, thus entering terrestrial as well as aquatic systems, in which it may be biotically transformed to methyl-mercury [CH_3Hg^+, $(CH_3)_2Hg$)], the major toxic species of Hg. The persistence of Hg is reinforced by its capability of reemitting from soils and waters to the atmosphere and thus returning to its global journey. Understanding this bi-directional transport of Hg on a global scale is critical for modeling and predicting the transport and fate of Hg in the environment (*2*). Hence, Hg emission from soils plays an important role in regional and global cycling of Hg (*3*). Mercury can be emitted from soils naturally enriched with Hg as a result of geologic processes. Mercury can also reemit from soils with Hg deposited from the atmosphere. An accurate assessment of Hg emissions and reemissions is crucial to evaluation of any regulation and control measures for anthropogenic Hg emissions.

Recognition of the importance of Hg emissions from soils stimulated much research in this regard. During the past decade, steady progress has been made in field quantification of Hg emissions from both enriched and background soils, in improvement of the methods for Hg emission flux measurements, in characterization of soil Hg emission trends and affecting factors, and in understanding of soil Hg emission processes. This research has reinforced our view on the significant role of soil Hg emissions in global Hg cycle, improved the projection of the global Hg budget, and updated the picture of global Hg cycling. The growth of research in soil Hg emissions has also made us recognize a large number of uncertainties associated with our estimation of the overall contribution of soil Hg emissions to the atmospheric Hg pool and the effect of the emissions on global Hg cycling. In this chapter, we attempt to address these uncertainties, discuss our knowledge gaps, and identify some future research priorities in the research of soil Hg emission. Presented here are the authors' preliminary thoughts, which represent neither a comprehensive scrutiny, nor an exhaustive review in this field. The major purposes of this effort are to stimulate more extensive and critical assessment of our current knowledge and understanding of Hg emissions from soils and their effects on global Hg cycling, to search means for resolving the related uncertainties, and to promote more advanced and elaborate studies on soil Hg emissions.

Some Recent Developments in Research on Mercury Emission from Soils

It would be useful to first briefly review recent developments in the research on Hg emission from soils. These developments have significantly improved our knowledge and understanding of Hg emissions from soils and their effects on global Hg cycling. Here we attempt to highlight some of the important progress with respect to field observation, emission flux measurement methodology, emission scaling, processes study, application of new analytical techniques, and new possibly significant Hg emission sources.

Trends in soil Hg emissions and affecting factors

Field observations in the last decade have established that soil Hg emissions exhibit diel trends closely following solar radiation variation (e.g., see *4, 5, 6*). Moreover, it was found that the effect of solar radiation on soil Hg emissions appeared to be an independent phenomenon. In other words, it was not a consequence of soil temperature change induced by solar radiation. This perception is suggested by rapid increases in Hg emissions nearly immediately after the soil received solar radiation while the soil temperature changed little (*4, 7*). Similar observations obtained in controlled laboratory studies also support this notion (e.g., see *8*)

Another important characteristic of soil Hg emissions is the "watering" effect (e.g., during precipitation). It was observed in a study over naturally enriched soil at Steamboat Springs, NV, that Hg emissions increased significantly immediately after a storm started (*9*). It was later revealed that this effect was also an independent phenomenon. In a study at Ivanhoe, NV, we observed immediate increases in Hg emissions from a soil artificially watered in the nighttime. The emission fluxes increased in the dark while the soil temperature changed little (Zhang *et al.*, unpublished data). Soil-watering tests conducted in our laboratory confirmed the independent nature of the "watering" effect (Zhang and Lindberg, unpublished data).

In the attempt to study soil Hg emission processes under controlled conditions, i.e., constant temperature and in the dark or constant weak radiation (e.g., room fluorescent light), we placed soil samples in a dynamic flux chamber using outside ambient air as flushing flow to monitor Hg emissions under controlled conditions inside our laboratory at Oak Ridge National Laboratory. Contrary to our expectation, we observed repeatable, seasonally variable diel trends of soil Hg emissions under the controlled conditions (*7*). This mysterious phenomenon observed in Oak Ridge remains to be fully understood. More similar tests need to be conducted in other places to independently confirm this phenomenon or determine if this is local environment-specific.

Intercomparison of Hg emission measurement methods

The first intercomparison study in soil Hg emission measurement methods was accomplished in September of 1997 at Steamboat Springs, NV, at a naturally enriched soil site. An international group of scientists participated in this project, The Nevada Study and Tests of the Release of Mercury From Soils (Nevada STORMS). Four micrometeorological methods and seven different field flux chambers were employed to compare the results of the soil Hg emission measurements. It was found that generally, the mean Hg fluxes measured with micrometeorological methods during daytime periods were nearly three times higher than the mean fluxes measured with field dynamic flux chambers (10). This study revealed for the first time the significant discrepancy in soil Hg emission flux measurements between micrometeorological methods and dynamic flux chamber methods. The disagreement pointed out the existence of various uncertainties regarding soil Hg emission flux estimation through field measurements and raised questions as for which methodology is more reliable.

Flushing flow effect on soil Hg emission fluxes measured by dynamic flux chamber operations

After the Nevada Study, a follow-up workshop was organized in May of 1998 to assess the accomplishments of this project and discuss the emerging issues. One of the critical research needs identified during the workshop was the examination of the effect of flushing flow rate on soil Hg emission fluxes measured by dynamic flux chamber operations. It was suspected that low flushing flow rates could be in part responsible for the observed underestimation of soil Hg emission fluxes measured with dynamic flux chambers (10). This promoted further research devoted to investigation into the effect of flushing flow rate.

Our field tests over enriched soils in Nevada in October of 1998 showed that high flushing flow rates indeed led to higher emission fluxes measured with dynamic flux chambers. Subsequent laboratory tests under controlled conditions carried out in Oak Ridge National Laboratory and University of Nevada, Reno, all confirmed the effect of flushing flow rate (11). These tests stimulated our interest in developing a physical model to explain and quantify the flushing flow rate effect. A two-resistance exchange interface model was developed to simulate soil Hg emissions measured using dynamic flux chamber operations.

We found that two major mechanisms are responsible for underestimation of the emission fluxes at low flushing flow rates: (a) internal accumulation of emitted Hg as a result of insufficient removal of the Hg from the chamber at low flushing rates, and (b) higher exchange resistance of the soil/air boundary layer at low flushing rates (12). The findings from the intercomparision project and subsequent study on the effect of flushing flow rate suggest that many previous soil Hg emission flux measurements obtained by dynamic flux chamber methods could have underestimated real emission fluxes, especially for the enriched soils. Therefore, higher flushing flow rates were recommended in order to obtain accurate soil Hg emission flux estimation.

Application of new analytical technique in soil Hg emission research

A modern analytical technique has been employed recently in environmental Hg research. Synchrotron X-ray Absorption Fine Structure (XAFS) spectroscopy, a non-destructive element-specific structural method that requires no special sample preparation, has been used to probe the chemical speciation of Hg in mine wastes from a variety of abandoned Hg mine sites in California and Nevada. With this technique, it has been revealed that the main Hg-bearing phases in calcines are cinnabar and metacinnabar. Also identified were several relatively soluble Hg-bearing phases that were not previously detectable by XRD, including montroydite (HgO), schuetteite ($HgSO_4$), and several Hg-Cl phases. XAFS has also been applied to examine the sorption of Hg(II) on model mineral surfaces to determine the effects of inorganic ligands (e.g., SO_4^{2-}, Cl^-) on Hg(II) sorption, and to determine the chemical form(s) of Hg associated with colloids generated in laboratory column experiments on mine wastes (13).

Fire-induced Hg emissions

Although the effect of fire on the atmosphere and global climate change has been actively studied for nearly a decade, its effect on Hg emissions from soils was brought to attention very recently. Brunke et al. (14) published the first data ever known on gaseous Hg emissions from biomass burning, obtained from a savanna fire in the Cape Peninsula, South Africa, in January 2000. Measurements of total gaseous Hg during this episode provided Hg/CO and Hg/CO_2 emission ratios of 2.1×10^{-7} and 1.2×10^{-8} mol/mol, respectively, suggesting the global Hg emissions from biomass being 930 and 590 t/yr, respectively. This estimate appeared to point to a major Hg emission source. The Hg emitted from biomass burning probably comes mostly from atmospheric deposition. Recent research on the source of Hg in foliage has demonstrated that the atmosphere is the primary source of Hg in foliar tissue (*15*). Plants thus represent an unaccounted for source and sink of atmospheric Hg. Hence, the fire-induced Hg emissions represent a long neglected recycling of atmospheric Hg. In consideration of the large areas affected by fires, the severe heating generated by fires, and the frequency at which fires occur globally, the potential contribution of Hg emissions from global biomass burning to atmospheric Hg burden could be significant. Recent measurements of Hg emissions from foliage and ground litter from some forests across the United States, conducted at the U.S. Forest Service Fire Science Laboratory (*16*), re-emphasized the importance of fire-induced Hg emissions and the consequent impacts on global Hg cycling. Forest fires may account for up to 8% of total global Hg emissions from all sources, and up to 18% of global natural emissions, according to scientists from the National Center for Atmospheric Research and the Meteorological Service of Canada (*17*).

Towards a Better Database for Mercury Emissions from Soils

In the past decade or so, the database for soil Hg emission fluxes has steadily increased (e.g., see *3, 18*), with a considerable growth in the database for Hg emissions from naturally enriched and mining contaminated soils in the past 5 years (*3, 6, 19*). These soils are found primarily in the western United States in association with recent tectonic activities. Many western areas with widespread naturally enriched soils are in areas of rugged terrain, so are not suitable for micrometeorological methods for measuring soil Hg emission fluxes. As a result, dynamic flux chambers are the primary tools for in situ Hg flux measurements. Recognition of the flushing flow rate effect as described previously make it imperative that high flushing flow rates be adopted in new measurements of soil Hg emission fluxes using dynamic flux chambers in order to avoid underestimation of soil Hg emission fluxes. In addition, the flux data previously obtained with low flushing flow rates need to be corrected for the flow rate effect upon employment of these data for scaling.

The data characterizing Hg fluxes associated with naturally enriched areas and background soils remain very limited. For the latter, the published data are only for a few places, such as Sweden forest, Quebec field, Tennessee forest, Florida field, and northern Michigan forest and field (*2, 18*, and refs. within). The lack of necessary spatial and temporal coverage is significant. Contrary to enriched soils, which are concentrated mostly in the western regions, natural background soils are widely distributed all over the terrestrial surfaces. Considering the large magnitude of the actual background areas in which Hg emissions occur, a comprehensive and representative database for background soil Hg emissions is definitely needed. In selecting soil types and areas for sampling, the type of vegetation cover and climate should be considered. Vegetation cover should include forest, agricultural land, grassland, and desert. As for geographic distribution, consideration should be given to the soils in tropical, subtropical, temperate, and boreal regions. Seasonal data are also needed for each soil type covering various geographic zones.

Apparently, more data are needed for Hg emissions from soils over the world for different types of substrates, and the data from tropical and subtropical areas are especially insufficient. Lack of sufficient data for representative substrate types has hindered the attempt to construct global Hg cycling models.

Fire-induced Hg emissions could represent a significant fraction in the total terrestrial Hg emission pool. Quantifying fire-induced Hg emissions is only at its beginning. Two sources could contribute to the fire-induced emissions: (a) biomass-burning and (b) thermal release of Hg from soils induced by fire. It is unclear which source is more important. These questions need to be answered so that a reasonable model could be constructed to predict fire-induced Hg emissions from terrestrial surfaces. The work conducted at the University of Nevada on losses of Hg from burned soil plots (Johnson *et al.*, unpublished data) as well as the experiments of soil pyrolysis carried out at the University of Michigan (Nriagu *et al.*, unpublished data) seem to indicate the importance of fire-induced Hg emissions from soils.

Towards Improved Methodologies for Estimating Soil Mercury Emissions

As discussed above, recent studies demonstrated a significant difference between soil Hg emission fluxes measured with micrometeorological methods and dynamic flux chamber methods. The finding that underestimation of soil Hg emission fluxes occurs at low flushing flow rates used for dynamic flux chamber operations indicates the necessity of adoption of optimum operation conditions (especially high flushing flow rates) for flux chamber measurements. Therefore, it is necessary to conduct intercomparisons between micrometeorological methods and dynamic flux chamber operations at high flushing flow rates to reassess the agreement between the two methodologies. More field tests are also needed to assess the flushing flow rate effect on dynamic flux chamber measurements of Hg emissions from background soils. The accuracy of the micrometeorological methods also needs to be critically assessed. In addition, introduction of new micrometeorological approaches [e.g., Relaxed Eddy Accumulation (REA), see 20] may be of interest.

Towards Better Scaling of Soil Mercury Emissions

Estimation of annual Hg emission over a large area or region requires integration of measured local or discrete fluxes over space and time. This is a complex process involving a large number of parameters that affect the emissions. Factors to be considered include substrate type, surface type, vegetation type, Hg concentration and speciation of the surface, anthropogenic interference, and environmental conditions such as solar radiation, temperature, precipitation, wind, and atmospheric chemistry (*3*). Development of a hierarchy of parameters in terms of their role in controlling emissions is also necessary. The variations associated with flux, space, and time, the three basic parameters involving scaling operations, are enormous. This is further complicated by the inter-dependence of the three parameters upon each other. One special difficulty in scaling appears in situations of large areas with small fluxes, which involves large scaling uncertainties, because very small flux variations can lead to big differences in scaled total emissions over a large region (e.g., see *19*). Two basic approaches have been used: (a) direct scaling using flux data and (b) indirect scaling using correlations between emission fluxes and soil Hg concentrations (*3, 6, 19*). Good linear correlations have been found between Hg emission fluxes and total Hg concentrations of soils for a large number of enriched soil sites (*3*).

Initial scaling of Hg emissions appears promising (*3*), but many issues remain to be resolved. First, more flux data are needed covering different surface types and different seasons; the validation of correlations between emission fluxes and Hg concentrations of soils need to be tested over more regions and covering different seasons and surface types, especially over background soils; effects of environmental factors on emission fluxes need to be quantified. Another important issue is whether Hg emissions from enriched soils behave the same way as the emissions from background soils. This largely determines whether the same scaling techniques can be used for both types of soils.

Several other issues are also of concern. Are the factors controlling Hg emissions from enriched soils versus Hg emissions from background soils the same? Is there a soil classification system that is specifically suitable for assessing soil Hg emissions? What are those environmental, geologic and land management parameters that are dominant in influencing emissions? Equally important is development of a sound scaling model (i.e., the approach to scaling). This is directly related to our understanding of the soil types representing different Hg emission characteristics.

Towards Better Understanding of Soil Mercury Emission Processes

Hg emissions from soils are controlled by physical and chemical processes, and perhaps also partly by biological processes. Identification and understanding of these processes are important for elucidating the effect of soil Hg emission on global Hg cycling and modeling soil Hg emissions. Although our knowledge of soil Hg emission processes has been growing in the last decade (21), many gaps still exist. More efforts are needed to quantify various effects on soil Hg emissions and to elucidate the emission mechanisms so that soil Hg emissions can be modeled based on emission processes. The process-based models may generate more realistic predictions of soil Hg emissions. In general, the major physical processes controlling soil Hg emissions are sorption of Hg(0) on soil and desorption of Hg(0) from soil, sorption of Hg(II) on soil and desorption of Hg(II) from soil, and diffusion of Hg(0) in soil in some cases (e.g., wet soil). The major chemical processes are decomposition of Hg-containing minerals and reduction of Hg(II). How fast does each process undergo in soil? Which process dominantly contributes to soil Hg emission? Do different types of soils have the same dominant emission controlling process(es)? What are the factors affecting each emission process? These are some questions of large concern for understanding soil Hg emission processes.

Quantification of the effects of solar radiation and precipitation

It is clear that solar radiation and precipitation strongly affect soil Hg emissions. However, quantification of these effects is insufficient. More tests, both in the field and laboratory, are needed to establish the empirical correlations between solar radiation input and between precipitation input and increases in soil Hg emissions and increases in soil Hg emissions, for different types of soils. The wavelengths responsible for the solar radiation effect should be elaborately identified. The sensitivity of background soils to the solar radiation effect needs to be assessed. The nature of the solar radiation effect is still not well understood, but it has been hypothesized to result from desorption caused by absorption of solar energy by Hg atoms (18, 22).

We have found that the precipitation effect was intensive immediately after real or simulated precipitation reached the soils, but the effect decreased with time after the precipitation stopped. This effect has been mechanistically explained by exchange between incoming water molecules with Hg(0) adsorbed on soil particle surfaces (9). However, a quantitative model remains to be constructed to describe the mathematical relationship between the amount of precipitation entering a soil and the increases in Hg emission from the soil

Reduction of Hg(II) on background soils

Availability of Hg(0) is important for Hg emission from background soils, and thus reduction of Hg(II) could be the key process controlling Hg emissions from background soils. Both thermal and photochemical reduction of Hg(II) may occur on soils. But, it is unclear which kind of reduction is more important on surface soils. We hypothesize that photochemical reduction is the major controlling emission process for background soils (*4, 7*). Yet, this hypothesis needs to be tested. The factors controlling photoreduction of Hg(II) on surface soils should be identified and quantified. The kinetics of reduction of Hg(II) in background soils, thermal or photochemical, needs to be measured for different types of soils covering temperate regions and tropical and subtropical regions.

Reduction of Hg(II) on enriched soils

Reduction of Hg(II) is also important for naturally enriched soils (Gustin *et al.*, unpublished data). More likely, decomposition of cinnabar and subsequent reduction of Hg(II) or other natural weathering processes determine the production of Hg(0) on enriched soils. However, these soils are probably already enriched with Hg(0) in the long course of natural weathering. As a result, desorption of Hg(0) may significantly determine Hg emissions. It needs to be clarified whether Hg(II) reduction or Hg(0) desorption dominantly contribute to Hg emissions from enriched soils. The knowledge of kinetics of reduction of Hg(II) in enriched soils is critical to identifying the primary emission process. Furthermore, assessment is needed to reveal the role of photochemical reduction of Hg(II) in overall reduction of Hg(II) in enriched soils. The kinetics of reduction of Hg(II) in enriched soils, thermal or photochemical, needs to be measured for different types of soils covering soils in temperate regions and in tropical and subtropical regions.

One important question of high interest regarding reduction of Hg(II) in soils is where the reduction occurs; in other words, which is more important, the reduction on soil particles (surface reaction) or the reduction in soil solution? Little data are available to clearly answer these questions. The mechanisms for reduction of Hg(II) in soils and related affecting factors are little known. Hence, mechanistic studies on soil Hg(II) reduction is an important research need.

Kinetics of Hg(0) sorption/desorption on soils

There is little doubt that desorption of Hg(0) from soil particles plays an important or even dominant role in Hg emissions from soils. It is hypothesized that the desorption is a fast process. However, current kinetic data for Hg(0) desorption is insufficient to convincingly verify this hypothesis. More kinetic information is needed and it should be obtained for various types of soils for different regimes of seasons.

Kinetics of sorption/desorption of Hg(II) on soils

Reduction of Hg(II) in soils may be a major or even dominant contributing process for soil Hg emission. The sorption/desorption process controls the distribution of Hg(II) between soil particle surfaces and soil solution. Behavior of soil Hg(II) directly affects production of Hg(0) in soils. The kinetics of Hg(II) sorption/desorption in soils determines the availability of Hg(II) for its reduction, in soil solution or on soil particle surfaces. More kinetic studies on sorption/desorption of Hg(II) are needed, especially for soils of different mineralogical compositions (*21*).

Diffusion of Hg(0) in soil

Because of high irregularity of soil pore structures, diffusion of Hg(0) in soil air differs from in ambient air. It is expected that the diffusion coefficient of Hg(0) for soil air deviates from the value in ambient air. Hg(0) diffusion coefficient for soil air is an important parameter, which is required for modeling soil Hg emissions. Some models and empirical relationships have been available for estimating the diffusion coefficients for gas transporting in soil (*12, 23*). However, applications of these models and relationships to Hg(0) transporting in soil are limited and remain to be verified experimentally. This poses an uncertainty regarding process-based modeling for soil Hg emissions.

Supply of Hg(0) from deep soil

This could be an important emission process for enriched soils. It is unclear if Hg emissions from enriched soils are mainly controlled by desorption of Hg(0) from surface soils, or sustained by the supply of Hg(0) from lower column of the soil. Understanding the role of the supply of Hg(0) from deep soil should benefit modeling soil Hg emissions, especially for enriched soils.

Soil gas Hg(0) concentrations

Hg concentration of soil gas, which directly determines the emission gradient, is an important parameter necessary for understanding and estimating soil Hg emissions. It is also directly related to the equilibrium of soil Hg(0) sorption and desorption (*21, 23*). It is not easy to directly measure soil gas Hg concentrations, especially for background soils, because a large amount of soil gas may need to be sampled in order to obtain sufficient amount of Hg for analysis. Determination of Hg concentrations of soil gas in enriched soils is achievable. The latest measurements of soil gas in controlled mesocosm studies using Hg enriched soils suggest that diffusion cannot solely explain the observed emissions from the soils (*24*). Lack of soil gas Hg concentration data has left a barrier for application of processed based modeling for soil Hg emissions. It is also important to establish the empirical correlations between soil air Hg concentration and total Hg concentration of a soil. A physical model describing the correlations would be valuable.

In summary, we propose the following general hypotheses for soil Hg emission processes: For enriched soils, the emission is controlled mainly by three processes, desorption of Hg (0) from surface soil, decomposition of Hg-compounds and reduction of Hg(II) (e.g., cinnabar), and Hg(0) supply from lower soil; for background soils, the emission is mainly controlled by reduction of Hg(II) on surface soils; the Hg sources for background soils are probably primarily wet and dry atmospheric depositions; reduction of deposited Hg(II) mainly contributes to Hg emissions from background soils. Studies to test these hypotheses should be of high interest.

Towards a Process-Based Modeling of Mercury Emissions from Soils

Given its significant magnitude, Hg emission from soils has to be considered in air quality models, which are useful tools for understanding various processes associated with Hg cycling and evaluating effectiveness of control measures. Because of lack of sufficient and reliable data, however, Hg emissions from soils were often included as background concentration or varied only with latitude and season (*25, 26*). Therefore, the emission strength as a function of surface cover type, historic Hg deposition, and environmental conditions has been oversimplified. With increasing of field measurements, empirical relationships have been developed to link soil Hg emissions to environmental variables such as soil temperature and solar radiation (e.g., see *4, 22*). Efforts have also been made to implement such empirical relationships in modeling work, allowing dynamic calculation of soil Hg emission fluxes (e.g., see *27*). While being useful in estimating Hg emissions, these empirical relationships have limitations. In order to model the effect of surface cover, soil type, soil water, reduction and sorption/desorption of Hg(II) the empirical relationships must be translated into mathematical ones that may be used for modeling. Therefore, soil Hg emissions remain a large source of uncertainty in the overall assessment and modeling of Hg pollution.

The need to improve representation of soil Hg emissions calls for the development of process-based modeling. The parameters can be used in such models are listed in Table I. As can be seen, Hg concentrations may need to be collected, but all other parameters are either readily available from database such as that of USGS or can be predicted by meteorological models such as MM5. The process-based models would use parameters such as those listed in Table 1 and solve mathematical equations describing processes controlling the air-surface exchange to estimate soil Hg emissions.

The advantages of process-based modeling of Hg emission from soils are two-fold. First, the scaling will be done more easily because the representation of in situ conditions by mathematical relationships. Second, process-based modeling provides means to track Hg pool in terrestrial ecosystems over a long period, and therefore enables the Hg emission fluxes to respond to changing anthropogenic Hg emissions and atmospheric depositions.

Process-based modeling of soil Hg emissions may find its use first in urban/regional scale simulations of atmospheric Hg, because of availability of data and better presentation of spatial variation by smaller grid sizes. However, the process-based modeling on global scales may raise the issues concerning the lack of soil Hg concentration data as well as the mixing of many soil types, surface covers, Hg concentrations, and meteorological conditions within large grids. Nevertheless, regional scale modeling studies may provide spatial and temporal Hg emission data that can be used in global scale simulations.

In order to develop process-based modeling, more research is needed to obtain a solid understanding of the processes driving Hg air/soil exchange. For example, observed diel patterns of Hg emission in the dark seem to suggest an emission induced by something in the ambient air other than solar radiation. Supply of Hg from deep soil remains largely unknown, but it is important, considering Hg in topsoil is unable to support continuous emissions as observed. The effect of vegetation cover on Hg emissions also needs to be characterized.

Table I. Parameters Possibly Used in Process-Based Modeling of Mercury Emissions from Soils

Modeling Parameter	Possible Data Source
land cover	USGS database
soil type	USGS database
soil properties related to reduction, sorption/desorption of Hg(0) and Hg(II)	USGS database
soil moisture	meteorological models with soil layer
soil temperature	meteorological models
solar radiation	meteorological models
wind speed and direction	meteorological models
precipitation	meteorological models
ambient Hg concentration	measurements or model prediction
soil Hg concentration	measurements or USGS database

Effect of Global Climate Change on Soil Mercury Emissions

Recent work at the arctic on Hg cycling around polar sunrise showed how climate change may affect global Hg cycle (e.g., see 28, and refs. within). The findings of the effects of solar radiation and watering as well as soil temperature point to the hypothesis that global climate change could vary soil Hg emissions and its effect on global Hg cycling. Global climate change will change solar radiation and the frequency, intensity, and extent of precipitation. More research is needed to assess the possible impact of global change on soil Hg emissions and the consequences in terms of global Hg cycling.

Concluding Remarks

In spite of rapid growth of research in soil Hg emission in the last decade, considerable gaps still exist concerning our knowledge and understanding. Various uncertainties need to be resolved before we could achieve accurate scaling of soil Hg emissions and sound prediction of the dynamic emissions of

260

Hg from soils generated by process-based models. The enormous spatial and temporal variations and intrinsic complexity of natural systems associated with soil Hg emissions pose compelling challenges. Nevertheless, more extensive and elaborate efforts devoted to the research will bring about better knowledge and understanding of Hg emissions from soils.

Acknowledgements

This work was funded by the US EPA through US EPA STAR Grants, and also in part by the Electrical Power Research Institute and the US Department of Energy.

References

1. *Mercury Research Strategy.* EPA, Washington DC, **2000**.
2. Schroeder, W.H.; Munthe, J. *Atmos.. Environ.* **1998**, *32*, 809.
3. Gustin, M.; Lindberg, S.; Austin, K.; Coolbaugh, M.; Vette, A.; Zhang, H. *Sci. Total Environ.* **2000**, *259*, 61.
4. Carpi, A.; Lindberg, S. *Environ. Sci. Technol.* **1997**, *31*, 2085.
5. Poissant, L.; Casimir, A. *Atmos. Environ.* **1998**, *32*, 883.
6. Engle, M.; Gustin, M.; Zhang, H. *Atmos. Environ.* **2001**, *35*, 3987.
7. Zhang, H.; Lindberg, S.E.; Gustin, M.S. American Chemical Society 222nd Annual Meeting. Chicago, IL, August, **2001**.
8. Gustin, M. S.; Rasmussen, P.; Edwares, G.; Schroeder, W.; Kemp, J. *J. Geophys. Res.* **1999**, *104*, 21873.
9. Lindberg, S.; Zhang, H.; Gustin, M.; Vette, A.; Marsik, F.; Owens, J.; Casimir, A.; Ebinghaus, R.; Edwards, G.; Fitzgerald, C.; Kemp, J.; Kock, H.; London, J.; Majewski, M.; Poissant, L.; Pilote, M.; Rasmussen, P.; Schaedlich, F.; Schneeberger, D.; Sommar, J.; Turner, R.; Wallschlaeger, D.; Xiao, Z. *J. Geophys. Res.* **1999**, *104*, 21879.
10. Gustin, M.; Lindberg, S.; Marsik, F.; Casimir, A.; Ebinghaus, R.; Edwards, G.; Fitzgerald C.; Kemp, R.; Kock, H.; Leonard, T.; London, J.; Majewski, M.; Montecinos, C.; Owens, J.; Pilote, M.; Poissant, L.; Rasmussen, P.; Schaedlich, F.; Schneeberger, D.; Schroeder, W.; Sommar, J.; Turner, R.; Vette, A.; Wallschlaeger, D.; Xiao, Z.; Zhang, H. *J. Geophys. Res.*, **1999**, *104*, 21831.
11. Lindberg, S.; Zhang, H.; Vette, A.; Gustin, M.; Barnett, M.; Kuiken, T. *Atmos. Environ.* **2002**, *36*, 847.
12. Zhang, H.; Lindberg, S.; Barnett, M.; Vette, A.; Gustin, M. *Atmos. Environ.* **2002**, *36*, 835.

13. Brown, G.E.; Kim, C.S.; Shew, S.S.; Lowry, G.V. Rytuba, J.J.; Gustin, M.S. US EPA Workshop on the Fate, Transport, and Transformation of Mercury in Aquatic and Terrestrial Environments. West Palm Beach, FL, May, **2001**.
14. Brunke, E-G.; Labuschagne, C.; Slemr, F. *Geophys. Res. Lett.* **2001**, *28*, 1483.
15. Benesch, J.A.; Gustin, M.A.; Schorran, D.E.; Coleman, J.; Johnson, D.A.; Lindberg, S.E. *6th International Conference on Mercury as a Global Pollutant.* Minamata, Japan, October, **2001**.
16. Friedli, H.; Radke, L. F.; Lu, J. Y. *Geophys. Res. Lett.* **2001**, *28*, 3223.
17. Renner, R. *Environ. Sci. Technol.* **2001**, *35*, 439A.
18. Zhang, H.; Lindberg, S.; Marsik, F.; Keeler, G. *Water Air Soil Pollut*, **2001**, *126*, 151.
19. Engle, M.A.; Gustin, M. S. *Sci. Total Environ.* **2002** (in press).
20. Goodsite, M.; Dong, W.; Lindberg, S.; Meyers, T.; Brooks, S.; Skov, H. *6th International Conference on Mercury as a Global Pollutant.* Minamata, Japan, October, **2001**.
21. Zhang, H.; Lindberg, S. *J. Geophys. Res.* **1999**, *104*, 21889.
22. Gustin, M. S.; Maxey, R.A.; Rasmussen, P.; Biester, H. Symposium volume, Measurement of Toxic and Related Air Pollutants, Air and Waste Management Association, Cary, NC, **1998**, p224-234.
23. Johnson, D. W.; Lindberg, S. E. *Water Air Soil Pollut.,* **1995**, *80*, 1069.
24. Johnson, D. W.; Benesch, J. A.; Gustin, M.S.; Schorran, D.E.; Coleman, J.; Lindberg, S.E. *6th International Conference on Mercury as a Global Pollutant.* Minamata, Japan, October, **2001**.
25. Shannon J. D.; Voldner E. C. *Atmos. Environ.* **1995**, *29*, 1649.
26. Pai P.; Karamchandani P.; Seigneur C. *Atmos. Environ.* **1997**, *31*, 2717.
27. Xu, X.; Yang, X.; Miller, D. R.; Helble, J.J. *Atmos. Environ.* **1999**, *33*, 4345.
28. Lindberg, S. E.; Brooks, S.; Lin, C-J.; Scott, K.; Meyers, T.; Chambers, L.; Landis, M.; Stevens, R. *Water Air Soil Pollut.* Focus, **2001**,*1*, 295.

Chapter 19

Geochemical and Biological Controls over Methylmercury Production and Degradation in Aquatic Ecosystems

J. M. Benoit[1], C. C. Gilmour[2], A. Heyes[3], R. P. Mason[3], and C. L. Miller[3]

[1]Chemistry Department, Wheaton College, Norton, MA 02766
[2]The Academy of Natural Sciences, Benedict Estuarine Research Center, St. Leonard, MD 20685
[3]University of Maryland Center for Environmental Science, Chesapeake Biological Laboratory, Solomons, MD 20688

It is the goal of this paper to discuss the more salient recent advances in the understanding of the controls of net CH_3Hg formation in natural systems. The discussion highlights the gaps in knowledge and the areas where progress in understanding has occurred. In particular, this chapter focuses on recent developments in Hg bioavailability and uptake by methylating bacteria, on the competing roles of sulfate and sulfide in the control of methylation, and in pathways for demethylation. The role of sulfide in influencing methylation is discussed in detail. In addition, the impact of other environmental variables such as pH, dissolved organic carbon and temperature on mercury methylation are discussed. Lastly, we provide a synthesis of the variability in the methylation response to Hg inputs across ecosystems. We suggest that although methylation is a function of Hg concentration, the range of methylation rates across ecosystems is larger than the range in Hg deposition rates. Overall, we conclude that factors in addition to the amount Hg deposition play a large role in controlling CH_3Hg production and bioaccumulation in aquatic ecosystems.

Introduction

Mercury (Hg) inputs to the environment have been increased dramatically since industrialization and anthropogenic sources of Hg to the atmosphere now dominate the input (*1-3*). While inorganic Hg is the major source of Hg to most aquatic systems, it is methylmercury (CH_3Hg) that bioconcentrates in aquatic food webs and is the source of health advisories worldwide that caution against the consumption of fish containing elevated CH_3Hg (*4-7*). Although a small fraction of the Hg in atmospheric deposition is CH_3Hg, the dominant source of CH_3Hg to most aquatic systems is *in situ* formation, or formation within the watershed (*8-12*). The current consensus, based mainly on temperature-dependency of Hg methylation and its response to biological substrates (*13-16*), is that biological methylation of inorganic Hg to CH_3Hg is more important than abiotic processes in natural systems. Biological methylation was first demonstrated in the late 1960's (*17*) and it is now generally accepted that sulfate reducing bacteria (SRB) are the key Hg methylators (*13,18-20*) although a number of organisms besides SRBs have been shown to produce CH_3Hg in pure culture from added Hg(II) (*21*).

The role of SRBs in methylation has been demonstrated through studies using specific metabolic inhibitors, addition of sulfate, and coincident measurement of sulfate reduction rate and CH_3Hg production. The addition of BES, a specific inhibitor of methanogens, was shown to increase Hg methylation while molybdate, a specific inhibitor of sulfate reduction, dramatically decreased CH_3Hg production in saltmarsh sediment (*13*). Since this early study, molybdate inhibition of mercury methylation, and coincident depth-profiles of sulfate-reduction rate and Hg methylation have demonstrated the importance of SRBs in estuarine, freshwater lake, saltmarsh, and Everglades sediments (*19,22-24*). Furthermore, addition of sulfate has been shown to stimulate mercury methylation in concert with stimulation of sulfate reduction, most notably during the whole lake sulfuric acid addition experiment in Little Rock Lake, WI (*18,25,26*) and in a series of short and long-term sulfate addition studies in freshwater ecosystems (*19,20,24,27*). However, while sulfate stimulates both sulfate reduction and mercury methylation at low sulfate concentrations, the build up of dissolved sulfide at high sulfate concentrations can inhibit Hg methylation (*28-30*). The mechanism of sulfide inhibition of Hg methylation is discussed in detail in this chapter. The sulfate addition experiments suggest that increased atmospheric sulfuric acid deposition in this century ("acid rain") may have lead to enhanced Hg methylation in remote freshwater ecosystems (*20,24*).

Overall, as many factors influence both methylation and the reverse reaction, demethylation, *in situ* CH₃Hg concentration is a complex function of its rate of formation and loss.

Community structure studies, using molecular probes and other techniques, have shown correspondence between the distribution of certain types of SRB and Hg methylation in sediments, and between sulfate reduction rate and Hg methylation rate (*31-34*). The primary site of methylation is just below the oxic/anoxic interface, which is often near the sediment surface in aquatic systems (*9,16,35-37*). It should be noted, however, that CH₃Hg can be produced in environments where sulfate reduction is low, such as upland soils, where other bacteria and fungi may be important methylators. However, little work has been done in these upland environments, as studies have rather focused on environments within aquatic ecosystems where the CH₃Hg produced has greatest likelihood of entering the aquatic food chain, and where sulfate reduction is a dominant degradation pathway for organic matter in sediments. Even though sulfate reduction and Hg methylation are linked, it should be noted that some SRB can methylate Hg while growing fermentatively (*38*).

One obvious mediator of Hg methylation rate is the concentration of inorganic Hg substrate, and its chemical form. Although there is a significant relationship between Hg and CH₃Hg across ecosystems, Hg does not appear to be largest source of variability in CH₃Hg production among ecosystems. The relationship between Hg and CH₃Hg concentrations in surface lake, river and estuary sediments and wetland soils across many ecosystems is weak but there is, on average, about 1% of the total Hg as CH₃Hg for the lower concentration (<500 ng/g) sites (Figure 1), which represent the range in Hg concentration of natural, unimpacted environments. The measured concentration at any time point is an integration of the impact of all the processes influencing CH₃Hg, such as differing loading rates (*39,40*) and methylation and demethylation rates (*16*), which vary spatially and temporarily (with season and temperature). Such variation is not accounted for in the data used in this plot, which include published and unpublished values from ongoing studies - see Figure caption for references. Only data collected and analyzed using trace-metal-free techniques were included here and the relationship is geographically biased, and favors contaminated systems. Additionally, as the data were not normally distributed, a log relationship is plotted ($r^2 = 0.41$; $p<0.01$; Fig. 1).

The data in Figure 1 appear to cluster into two sets, with Hg concentrations exceeding 500 ng g⁻¹ having little increased impact on CH₃Hg production. For ecosystem types, the relationship has been found to be significant for estuaries ($r^2 = 0.78$, $p<0.01$), lakes ($r^2 = 0.64$, $p=0.01$) and rivers ($r^2 = 0.68$, $p=0.01$) but not for wetlands ($r^2=0.29$, $p>0.05$) based on data in Figure

1. Overall, within single rivers or wetlands, or even clusters of similar ecosystems, significant relationships can exist, but the relationships currently have no predictive power, given the importance of other parameters, discussed below, in influencing methylation rate by controlling the availability of Hg to, and activity of, the methylating bacteria, and given our current level of understanding. Effective regulation of Hg pollution requires the ability to predict the relationship between Hg and CH_3Hg among ecosystems, a goal that researchers and modelers seem to be slowly approaching. Detailed investigations of the mechanisms of CH_3Hg formation, degradation, fate and transport are required, so that the factors controlling the levels of CH_3Hg in fish can be understood. Clearly, while many other factors influence Hg methylation, the supply and availability of Hg is a key parameter.

In addition to the effect of sulfide, other chemical factors influencing methylation include the supply of labile carbon (57,58) although the role of dissolved organic carbon (DOC) is complex. The distribution of methylation activity is tied to the distribution of biodegradable organic matter but complexation of Hg by DOC may influence Hg bioavailability. Maximal net methylation is often observed in surface sediments (15,16) where microbial activity is greatest due to the input of fresh organic matter. As a result, systems with high levels of organic matter production, such as wetlands, recently flooded reservoirs, or periodically flooded river plains, may exhibit extremely high rates of methylmercury production (10,42,59,60). New research on Hg complexation with DOC is highlighted below. Temperature is another important variable (61) as the temperature responses of methylation and demethylation have been reported to differ (16,62). However, seasonal changes in Hg complexation that affect methylation and demethylation differently could account for these observations.

Demethylation of CH_3Hg can occur via a number of mechanisms, including microbial demethylation and reduction by *mer* operon-mediated pathways, and by "oxidative demethylation processes" (21,63-71). In addition, photochemical CH_3Hg degradation in the water column has been demonstrated (72). The *mer*-based pathway is an inducible detoxification mechanism, while oxidative demethylation is thought to be a type of C1 metabolism. Recent research suggests that oxidative demethylation is the dominant process in uncontaminated surface sediments (65,70,71).

It is the goal of this paper to discuss the more salient recent advances in the understanding of the controls of net CH_3Hg formation in natural systems. Therefore, rather than being a complete review of the literature, this chapter will provide an in-depth examination of some of the pertinent recent papers and current developments, and will endeavor to highlight the gaps in knowledge and the areas where progress in understanding has occurred. In particular, this chapter focuses on recent developments in Hg bioavailability and uptake by

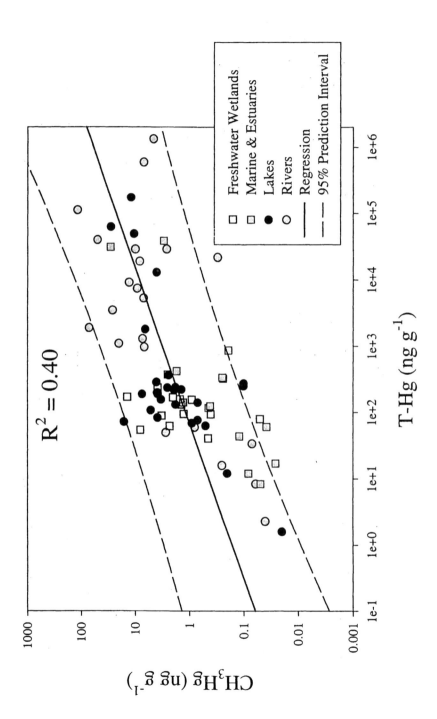

267

Figure 1. Mercury (Hg) and methylmercury (CH_3Hg) in near surface (0-4 cm) sediment in freshwater wetlands from: North and South Carolina (41), Ontario, Canada (42), Florida Everglades (37); Marine and Estuarine sediments from: coastal N. and S. Carolina (34), The Chesapeake Bay and its Estuaries (43,44), coastal Florida (45), coastal Texas (46), Slovenia coast (47), coastal Poland (48) coastal Malaysia (48), Anadyr Estuary, Russia (48); Lakes: New Jersey (41), New York State (49), Wisconsin (41,50), California (51), Finland (52), Poland (48); Rivers: S. Carolina (41), Wisconsin (53), Nevada (54), Alaska (55), Germany (56), Poland (48).

methylating bacteria, on the competing roles of sulfate and sulfide in the control of methylation, and in pathways for demethylation. Lastly, we provide a synthesis of the variability in the methylation response to Hg inputs across ecosystems. We suggest that although methylation is a function of Hg concentration, the range of methylation rates across ecosystems is larger than the range in Hg deposition rates, and that factors in addition to Hg deposition play a large role in controlling CH_3Hg production and bioaccumulation in aquatic ecosystems.

Mercury Speciation and Methylation

Although mercury resistant bacteria possessing the *mer* operon have the ability to actively transport Hg(II), this operon is not present in SRB that methylate mercury (*38*). It is generally accepted that CH_3Hg is produced in an accidental side reaction of a metabolic pathway involving methylcobalamin (*73*), although this pathway has only been demonstrated in one SRB. Therefore, it is not likely that SRB have acquired an active transport for this toxin. A limited number of experiments with SRB support this idea (*38*). For this reason, diffusion across the cell membrane has been proposed as the important uptake mechanism (*30,38,74*). This hypothesis is consistent with studies that have demonstrated diffusion of neutral mercury complexes (chloride complexes) across artificial membranes and into diatom cells (*75-77*).

The diffusion hypothesis is also supported by the relationship between methylation and the distribution of neutral Hg sulfide complexes in sediments. It has been noted in many field studies that the rate of Hg methylation, and the CH_3Hg concentration in sediments, decrease as the sediment sulfide concentration increases (*28,29,36,37,74,78,79*). A number of mercury complexes exist in solution in the presence of dissolved sulfide, including $HgS°$, $Hg(SH)_2°$, $Hg(SH)^+$, HgS_2^{2-} and $HgHS_2^-$ (*80-84*) and it is possible that inorganic Hg uptake by SRB occurs via diffusion of the dissolved neutral Hg complexes, such as $HgS°$. If so, then the bioavailability of Hg to the bacteria would be a function of sulfide levels, as this is the ligand controlling Hg speciation in solution in low oxygen zones where SRB are active. It has been shown through chemical complexation modeling that the speciation of Hg tends to shift toward charged complexes as sulfide levels increase (*74,84,85*), decreasing the fraction of Hg as uncharged complexes, and, as a result, the bioavailability of Hg to methylating bacteria.

The existence of neutrally charged Hg-S complexes, and the notion of decreasing bioavailability in the presence of sulfide, was demonstrated in the laboratory using a surrogate measure of membrane permeability, the octanol-water partition coefficient (K_{ow}; 85). These experiments showed that $HgS°$ and

$Hg(SH)_2^\circ$, the neutral complexes present under the experimental conditions, had relatively high K_{ow}'s such that they would be taken up at rates more than sufficient to account for methylation in both pure cultures and in the field, based on the estimated permeability through the cell membrane, which is a function of K_{ow} (*30,74,86*). There is the potential for polysulfide formation in porewaters and the solubility of Hg in the presence of HgS(s) has been shown to be dramatically increased by the complexation of Hg with polysulfide species (*87*). However, these interactions do not appear to significantly enhance the concentration of neutral Hg species, as measured by changes in the K_{ow}, suggesting that the dominant polysulfide complexes are charged (e.g., HgS_xOH^- and $Hg(S_x)^{2-}$; *87*) and thus unavailable for uptake and methylation by SRB.

It should be noted here that theoretical chemical calculations suggest the neutral species in solution is $HOHgSH^\circ$ rather than HgS°, as it is likely that Hg would form a more stable linear complex (*88*). The results of Benoit et al. (*30*) are not in disagreement with this notion even though $HOHgSH^\circ$ would have a lower permeability, because of its larger molecular volume, than HgS°. The relationship between permeability and K_{ow} is extremely non linear (*89*) and $HOHgSH^\circ$ would have a permeability that is a factor of five less than that of HgS°. In the pure culture studies, estimated uptake was greatly in excess of methylation rate (Figure 2a) even if the neutral complex is considered to be $HOHgSH^\circ$.

Experiments with pure cultures and other studies indicate that not all the Hg taken up by the bacteria is methylated as there are other sinks within the cell for Hg (*30*). This is shown illustratively in Figure 2b, where Hg_C represents the pool of Hg available for methylation inside the cell. Given uptake of a neutral Hg complex ($[HgL_n^\circ]$), the intracellular steady state concentration of Hg_C is given by:

$$[Hg_C] = k_D.[HgL_n^\circ]/(k_B + k_M)$$

where k_M is the methylation rate constant, k_D is the diffusion rate, HgL_n° is the concentration of neutral complex in solution and k_B is the rate constant that incorporates all the other processes that are rendering Hg unavailable for methylation within the cell. The rate of formation of CH_3Hg, assuming no loss mechanisms, is then given by:

$$d[CH_3Hg]/dt = k_M.[Hg_C] = k_D.k_M.[HgL_n^\circ]/(k_B + k_M)$$

If k_M is much greater than k_B, then the rate of CH_3Hg formation is directly related to the rate of diffusion across the membrane. However, in the opposite case, the rate of methylation is dependent on both the rate of uptake and the rate at which Hg is being sequestered within the cell. However, there would still be a

Fig. 2a

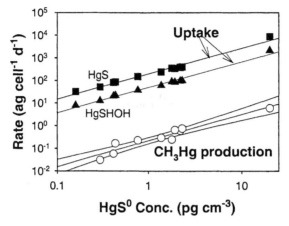

Fig. 2b

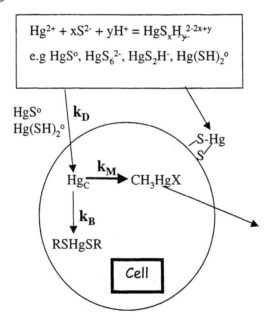

Figure 2. a) Estimated mercury (Hg) uptake rate, assuming passive diffusion of neutral Hg complexes - either modeled as HgS° or HOHgSH°, and the simultaneous Hg methylation rate in pure cultures of D. propionicus. b) Model representation of the assumed process of Hg uptake and methylation within this organism. The rate of methylation is designated as k_M; k_B is the combined rate of competing reactions that sequester Hg and make it unavailable for methylation and k_D is the uptake rate. See text for details.

linear relationship between the neutral complex uptake rate and methylation rate across a sulfide gradient, for example, if the rate constant, k_B, was relatively independent of the environmental conditions outside the cell, which is likely. At the limit, if $k_B \gg k_M$, methylation will not occur to any measurable degree.

Studies with laboratory cultures of *Desulfobulbus proprionicus* across a sulfide gradient have shown that the uptake rate, as estimated from the dissolved Hg speciation and the estimated permeability of the complexes, and methylation rate changed in a similar fashion, and both increased with increasing HgS° concentration (Figure 2a), supporting the above hypothesis (86). These results also suggest that k_B is greater than k_M in this case and that a small fraction of the Hg taken up is methylated. The relative percent methylated would depend on a number of factors that would differ between organisms, due to differences in physiology, size and membrane composition. We propose that differences in Hg partitioning within cells may partially explain the large differences in methylation rates among various strains of SRB.

The hypothesis of neutral complex bioavailability controlling methylation begs the question why methylation appears to be confined mostly to SRBs in the environment. Clearly, these organisms dominate in the region where HgS° dominates the Hg speciation. Why then, in oxic or suboxic environments, where $HgCl_2^\circ$ and other neutral Hg complexes exist, is there is little evidence of methylation? In studies with diatoms, Mason et al. (76) demonstrated that little of the $HgCl_2$ taken up (<10%) reached the cytoplasm of the organisms, with most Hg being rapidly bound within the cellular membrane. On the other hand, CH_3HgCl was less strongly bound within the membrane and a greater fraction was found in the cytoplasm of the diatom (63%). The intracellular distribution is related to the rate and degree of reaction of the accumulated complex with cellular sites. The rate of sequestration (k_B), which would determine where the Hg would become bound within a cell, depends on the exchange reaction between the neutral complex and the cellular reaction site (RH); for example, $HgS^\circ + RH = HgR^+ + SH^-$. The kinetics of this process would be to some degree determined by the reaction mechanism (adjunctive or disjunctive), but are a strong function of the relative magnitude of the equilibrium constants for the accumulated complex and HgR^+. Given that most cellular binding sites for Hg are likely thiol groups, the rate of the exchange reaction should be faster for $HgCl_2$ than for HgS°, as the associated equilibrium constant ($HgCl_2^\circ + RH = HgR^+ + H^+ + 2Cl^-$) is much greater for $HgCl_2$. Furthermore, given the high stability of HgS°, its rate of dissociation will be slower than that of $HgCl_2$. Therefore, in the presence of HgS°, a higher fraction of the Hg diffusing across the outer membrane is transferred to the site within the cytoplasm where methylation can occur compared to other organisms, because of the kinetics of the intracellular reactions.

Thus, it is not sufficient that the pathway for methylation exists within an organism but also that the Hg can be transported within the cell to the site of the reaction. As discussed below, methylation has been linked to the acetyl-CoA pathway in one bacterium (73) and a simple explanation of why some organisms which have this pathway do not methylate Hg is that the Hg is being bound to other intracellular sites before being transported to the site of methylation. Thus, there are kinetic and biochemical factors that influence the relative degree of methylation between organisms besides Hg bioavailability and Hg content in the medium. The kinetics and location of the intercellular reactions are an important modifier of the methylation rate and clearly more studies on the intracellular mechanisms of methylation are needed.

A simple model of Hg partitioning in sediments was developed to explore how Hg partitioning to solids impacts Hg concentration and bioavailability in sediment pore waters (74). The model developed used adsorption reactions as the mechanism controlling porewater concentration, since field measurements show that pure cinnabar equilibrium dissolution dramatically over predicts the concentration of Hg in porewaters. It should also be noted that there is little field evidence to support the formation of the pure HgS solid phase in environmental media (90-92). In the model, water column speciation was driven primarily by sulfide concentration. The adsorption of Hg was modeled with two types of binding sites, singly or double coordinated thiols, which could represent either an interaction with organic thiol and/or inorganic sulfide groups in the solid phase. This model was not only able to predict both the measured distribution of Hg in sediment porewaters in two ecosystems (the Patuxent River and the Florida Everglades) but the model-predicted concentration of neutral Hg-S complexes correlated well with the $in\text{-}situ$ CH_3Hg concentration (74).

The model and the laboratory culture studies (30,74,86) cover the speciation of Hg in the presence of sulfide but it is known that Hg binds strongly to organic matter, and that dissolved organic carbon (DOC) impacts Hg methylation. The impact of organic content on Hg methylation appears to be complex (16,57,58). While Hg binds strongly to DOC, laboratory complexation studies using DOC isolates from the Florida Everglades suggest that this binding is not sufficient for Hg-DOC complexation to be important in systems where sulfide is present (93,94). Thus, while DOC has been shown to be the most important complexing ligand in surface waters in the absence of sulfide (94,95), it is likely to be unimportant in Hg complexation in sediment porewaters under typical DOC concentrations and >0.01 ΦM sulfide (93,94). However, binding of Hg to organic matter is important in the solid phase. Laboratory studies suggest that in oxic regions, organic complexation is much more important than binding of Hg to metal oxide phases in all except very low organic matter sediments (96). It has been suggested that Fe-S formation scavenges Hg in anoxic regions of the sediment (97) and that Hg binds strongly to pyrite such that, even when only small amounts are present, it is the dominant solid phase binding Hg in sediments (*90*). These studies focused on regions of low organic content and, in contrast, our sediment sequential extraction studies (*96*) show that Hg is associated with the organic fraction even in the presence of significant solid sulfide phases (AVS and pyrite). Furthermore, it has been shown that the sediment particle-dissolved distribution coefficient (K_d) is a strong function of organic carbon (*46*). In the environment, concentrations of Fe, S and C typically co-vary in sediments, and all often correlate with Hg, and it is therefore difficult to ascertain from field data which is the ultimate controlling phase (*44,98*). Laboratory and field studies (*96,99,100*) suggest that the binding of Hg to organic matter involves interaction with the thiol groups of the organic molecules and thus, in a sense, the complexation of Hg to inorganic sulfide phases or to organic matter are comparable as both involve the interaction between Hg and a reduced S species.

The role of pH needs to be considered as the complexation with sulfide and thiols involves acid-base chemistry. An inverse correlation between lake water pH and mercury in fish tissues has been observed in a number of studies (*101 and references therein*) suggesting that pH influences methylation and demethylation in aquatic ecosystems. In some freshwater studies, methylation was reduced with decreasing pH (*27,35*) while the impact on demethylation was small. In other studies, increasing rates of mercury methylation in epilimnetic lake waters and at the sediment surface were found with lowered pH (*57,102,103*). Winfrey and Rudd (*35*) reviewed potential mechanisms for low pH effects on mercury methylation and suggested that changes in mercury binding could account for the seemingly conflicting results seen in all of these studies. They pointed out that lowering pH may lead to increased association of mercury with solid phases, decreased dissolved pore water mercury, and (presumably) to lower availability of Hg(II) to bacteria. The model discussed above (*74*) can be used as a simple predictor of the impact of pH on Hg methylation. Considering the reaction of Hg with the solid phase ($RSH + Hg^{2+} = RSHg^+ + H^+$), and the dissolved speciation, the following overall reaction can be postulated:

$$RSHg^+ + H_2O + HS^- = HgS^\circ + RSH + H^+$$

In the pH range of 7-10 (pK_{a1} ~7 for H_2S and assuming the pK for RSH is around 10), an increase in pH, at constant sulfide, will result in an increase in HgS° relative to $RSHg^+$ ([HS^-]/[RSH] is essentially constant) and thus methylation should increase with pH. Below a pH of 7, decreasing pH (increasing [H^+]) leads to decreasing [HS^-], and as a result, HgS° will decrease relative to $RSHg^+$ with pH i.e., methylation rate should decrease. This theoretical consideration supports the notions put forward by Winfrey and Rudd (*35*) and suggests that a decrease in pH will lead to a decrease in methylation rate in sediments because of changes in the concentration of bioavailable Hg in porewaters. The magnitude of the effect will depend on the pH range as the impact of pH is more marked at low pH. Overall, these considerations suggest that sulfide concentration will have the most significant impact on Hg bioavailability in porewater but that other factors such as organic matter content, pH, temperature and the presence or absence of inorganic sulfide phases all play a role in controlling Hg bioavailability to methylating bacteria.

The conflicting influences of sulfate and sulfide on the extent of Hg methylation are well illustrated by the studies in the Florida Everglades (*37*). Studies over four years at eight sites that cover a large gradient in sulfate and sulfide showed that the highest methylation rates, and the highest %CH$_3$Hg in the sediment, were at sites of intermediate sulfate-reduction rates and sulfide concentration (Figure 3). In the Everglades, the north-south trend in sulfate concentration leads to a similar trend in sulfate reduction rate and porewater sulfide. As the sulfide concentration decreases, the relative concentration of predicted HgS^o concentration increases. The peak in methylation rate results from the combination of the increasing availability of Hg to the SRB coupled with the decreasing sulfate reduction rate north to south. These results confirm the importance of both Hg speciation, and bacterial activity, in controlling Hg methylation rate. Overall, the sites with the highest Hg methylation are those that also have the highest fish CH$_3$Hg concentrations (*104*), confirming the direct link between the extent of Hg methylation and fish CH$_3$Hg levels.

Experiments in which Everglades sediments were incubated with additional sulfate or sulfide further demonstrated the interplay between bacterial activity and Hg speciation. In cores taken from a relatively low sulfate site, addition of sulfate stimulated methylation, and sulfate reduction, over that of unamended control treatments even though sulfide levels increased slightly (see example experiments in Figure 4). In these sulfate-limited sediments, the higher induced bacterial activity more than compensated for the slightly lower bioavailability of Hg at the higher sulfide levels. Addition of sulfide alone however resulted in inhibition of methylation. It is clear from this and other experiments that inhibition occurs at low ΦM sulfide concentrations in Everglades sediments. However, high rates of Hg methylation have been demonstrated in highly sulfidic saltmarsh sediments (*32*). Perhaps the high rates of sulfate-reduction in these sediments make up for the very low percentage of dissolved Hg that would exist as neutral species. For Everglades sites with higher ambient sulfate, addition of sulfate did not increase methylation but addition of more sulfide led to an inhibition of Hg methylation. The field data across sites show a decrease in methylation rate when concentrations of sulfide increase above 10 ΦM (Figure 3), consistent with the core incubation data. Overall, the results of the field and laboratory studies show that the balance between sulfate availability, which controls SRB activity, and sulfide production and accumulation, which control Hg bioavailablity, are critical in modeling methylation rates. Ongoing mesocosm studies in the Everglades and in a boreal ecosystem should provide more quantitative equations for Hg cycling models.

Studies in a number of sites have now demonstrated a relatively strong relationship between the concentration of CH_3Hg in sediments and the instantaneous (short-term) rate of Hg methylation. Methylation rates can be estimated by Hg spike additions, either as a radioactive or a stable isotope, preferably to microbial communities held relatively intact (*18,41,103,105*) (Figure 5). The methylation rate constant, k_M, is calculated as the amount of new isotopic CH_3Hg formed per unit time, divided by the pool size of substrate. Methylation rate is derived by multiplying k_M by the total Hg pool size. Use of either custom-synthesized, high specific-activity [203]Hg, or the use of individual Hg stable isotopes combined with analysis by ICP-MS, allows methylation measurements to be done at relatively low, near ambient levels. In addition, when using stable isotopes it is possible to track both the *in situ* Hg and the added Hg spike and compare relative methylation rates (*106*). Furthermore, methylation and demethylation can be measured simultaneously in the cores if different isotopes are used for Hg and CH_3Hg (*41,105,106*).

Figure 5 shows relationships between methylation rates and ambient CH_3Hg concentration, in sets of 1m diameter enclosures at four sites across the Florida Everglades. In these studies, [202]$HgCl_2$ was added to the surface water of the enclosures, and CH_3[202]Hg production was followed over time in surface sediments. Additionally, 5 cm sediment cores were removed from the enclosures for the estimation of instantaneous methylation rate, using [199]$HgCl_2$ injected into the cores. The figure shows the concentration of CH_3[202]Hg in sediments in the spiked enclosures after 51 days, and production of CH_3[199]Hg in cores after 2 hours, both in comparison with *in situ* sediment CH_3Hg concentrations. Short term rates, net CH_3[202]Hg production after nearly 2 months, and the *in situ* concentration of CH_3Hg in the enclosures all provided the same information about the relative degree of methylation among sites. Measurement of short-term gross methylation, from an exogenous Hg spike, appears to be a good predictor of the relative steady state CH_3Hg concentration across sites within a specific ecosystem; i.e. a good relative measure of the propensity for methylation at each site.

The relationship between new CH_3Hg production, as measured by short term incubation, and *in situ* CH_3Hg concentration remains strong within a single ecosystems over time. In the Experimental Lakes Area, Ontario, Canada (ELA) wetlands, we have observed a persistent relationship between new production and *in situ* concentration of CH_3Hg. However, the slope of the line changed seasonally. This was not a purely temperature dependent response, as CH_3Hg production and concentration peaked in the fall and not in the height of summer

276

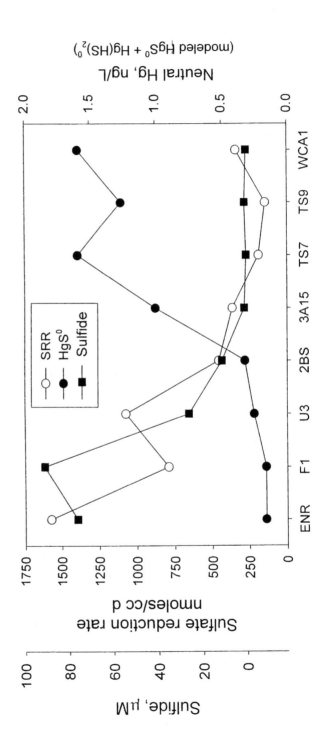

Figure 3. Measured sulfate reduction rate, porewater sulfide concentration, percent methylmercury (%CH₃Hg), mercury methylation rate and modeled porewater HgS⁰ in the upper 4 cm of Florida Everglades sediments at 8 ACME sites. Everglades sites are arranged from left to right by average surface water sulfate concentration (highest concentrations on the left). With the exception of the WCA 1 site, this represents a north to south transect, running from the Everglades Nutrient Removal Project (ENR) and Water Conservation Area 2A (F1, U3) in the north, through Water Conservation Areas 2B (2BS) and 3A (3A15), and to Taylor Slough in Everglades National Park (TS7, TS9) in the south. Data shown are averages from three years (1995-1998) of bi- to tri-annual sampling. Methods are described in ref. 37.

Continued on next page.

278

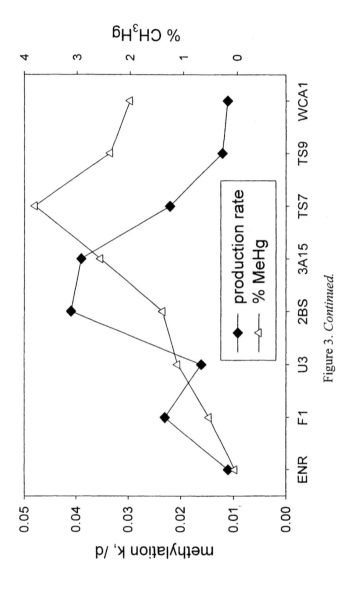

Figure 3. *Continued.*

Measured sulfide concentration

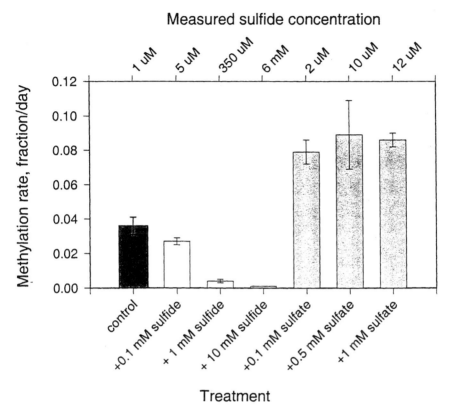

Treatment

Figure 4. Methylmercury (CH₃Hg) production in Florida Everglades sediment cores after addition of either sulfate (light grey bars) or sulfide (white bars). Sediment cores, taken from the central area of the Loxahatchee National Wildlife Refuge (LNWR), were amended with either sodium sulfide or sodium sulfate (at neutral pH), by injection into the top 4 cm of sediment. The calculated concentration of the spikes in pore water after injection is shown on the bottom axis, based on measured porosity. After 1 hr of incubation with sulfate or sulfide, mercury methylation was estimated in the cores using high specific activity [203]*Hg (18, 37). Methylation assays were conducted over 2 hr at ambient temperature. The final measured concentration of sulfide in sediment porewaters, three hours after the sulfate or sulfide spikes, is shown on the top axis. All measurements were made from triplicate cores for each treatment. Sediments spiked with sulfide sequestered much of it into the solid phase. Sediments spiked with sulfate produced measurable porewater sulfide, via sulfate reduction, within 3 hours. The LNWR is a very low-sulfate area within the Everglades. In these cores, addition of sulfate stimulated methylation even though sulfide levels increased slightly, while the addition of as little as 5 μM sulfide alone inhibited methylation.*

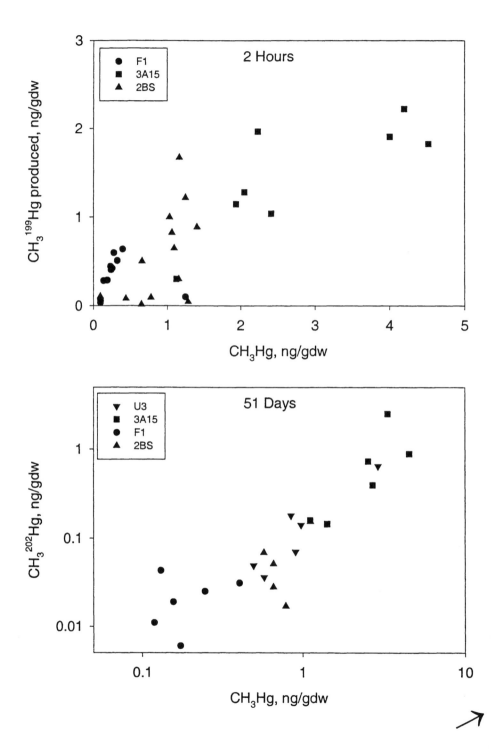

Figure 5. Relationship between in situ methylmercury (CH₃Hg) concentration and methylmercury production from stable isotope spikes in Everglades surface sediments: a) $CH_3{}^{199}Hg$ production from ^{199}Hg injected into Everglades sediment cores after an incubation period of 2 hrs. vs. in situ CH₃Hg; b) $CH_3{}^{202}Hg$ in enclosure surface sediments 51 days after the enclosures were amended with ^{202}Hg vs. in situ native CH₃Hg. These measurements were made in experimental, 1m diameter enclosures at four sites across the Everglades, as part of the ACME project. Three enclosures at each site were spiked with ^{202}Hg in May 2000, with spikes ranging from half to two times the equivalent of 1 year of atmospheric deposition. (see text for details; from Krabbenhoft et al., unpublished data).

(Figure 6). While we do not know the reason for the seasonal trend, both the measured methylation and *in situ* concentration were responding in concert to the same controlling factors.

Overall, the rate of production of CH_3Hg from exogenous spikes is generally higher than the rate of methylation from *in situ* Hg pools. This probably reflects the higher bioavailability of the added Hg compared to that in the sediment (*106*). This is illustrated in Figure 7, which shows sediment CH_3Hg levels in a lake enclosure experiment conducted as part of the METAALICUS project in the ELA. The top panel shows *in situ* CH_3Hg as a percentage of total Hg in the sediments in each of the four enclosures throughout the summer. The enclosures were spiked with ^{200}Hg either in mid June (encl.1& 2) or biweekly throughout the summer (encl. 3 & 4). The middle panel shows $CH_3{}^{200}Hg$ as percentage of ^{200}Hg accumulated in sediments. The relative percentage of the added Hg that becomes methylated is initially much higher for the added Hg. Over time, the percentage decreases and this is likely a combination of both reduction in bacterial activity in the fall and a decrease in bioavailability of the Hg over time as it is cycled through the system (methylated, demethylated and complexed to strong binding ligands). With the caveat that the method of Hg addition in our spike experiments may not truly reflect reality, these data suggest that Hg newly deposited to ecosystems is more available for methylation than existing Hg pools. However, more work is needed to further ascertain the crucial question of the relative importance of newly added versus *in situ* Hg in contributing to the Hg that is methylated and bioaccumulated in aquatic systems.

Thus, we suggest that the *in situ* CH_3Hg concentration across a series of sites within an ecosystem can be used to predict which site is likely to be more active in terms of methylation, and likely in terms of bioaccumulation, all else being equal. However, there is too little information at present to determine the degree to which these relationships can be used in a quantitatively predictive fashion between ecosystems. Overall, in comparing across systems, the greatest difficulty is in assessing the pool of Hg available for methylation, which is crucial to estimating realistic accurate methylation rates. To this point, we have not been able to measure bioavailable pools of Hg to bacteria, nor have we been able to mimic the speciation of *in situ* Hg with added Hg. Therefore, short term production remains more qualitative than quantitative. Both of these questions are the focus of ongoing research.

Biological Controls over Methylation

Different organisms clearly have different rates of Hg methylation, even among the SRB, and not all SRB methylate Hg (*24,32*). A small number of Fe-reducing bacteria, that are phylogenetically similar to methylating SRB, have

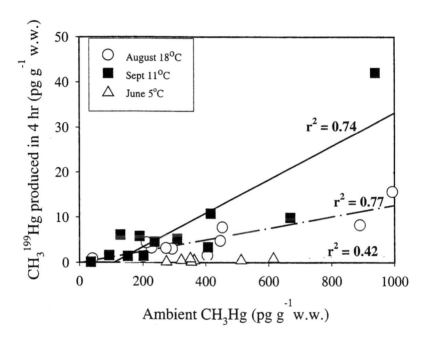

Figure 6. Native in situ methylmercury (CH₃Hg) concentration and excess CH₃¹⁹⁹Hg produced from ¹⁹⁹Hg in 4 hrs, in peat collected in June, August and September, 2000, from a lakeside, sphagnum wetland (L115) at the Experimental Lakes Area (ELA) in northwest Ontario. Work was conducted as part of the METAALICUS project.

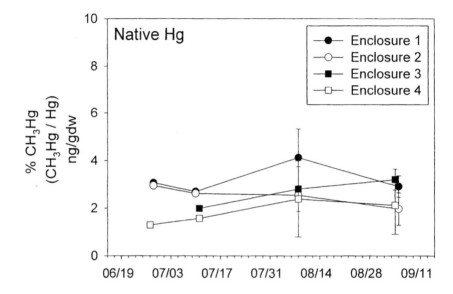

Figure 7. Sediment CH₃Hg levels in a lake enclosure experiment conducted in L329 in 2000 as part of the METAALICUS project in the Experimental Lakes Area (ELA), Ontario, Canada. The surface water of enclosures was spiked with ^{200}Hg either in mid-June (encl. 1 & 2) or biweekly throughout the summer (encl. 3 & 4). Spikes were equivalent to 1 year of atmospheric deposition. The top panel shows native in situ CH₃Hg as a percentage of total Hg in 0-4 cm depth sediments in each of four enclosures throughout the summer. The middle panel shows CH₃^{200}Hg as percentage of ^{200}Hg accumulated in sediments. The bottom panel shows water temperature in the enclosure. A much higher percentage of the spike was found methylated in sediments than native Hg, especially within 2 months of the spike.

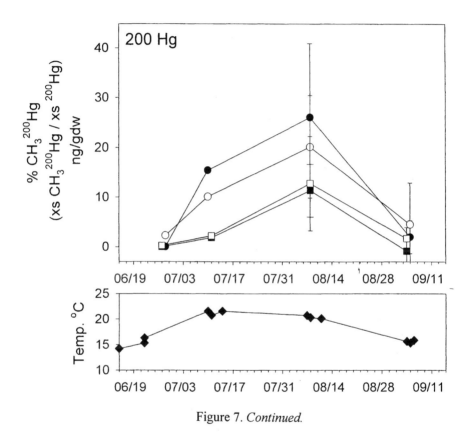

Figure 7. *Continued.*

been shown capable of methylating Hg in pure culture (*107*). While a number of organisms other than SRB have been shown to produce CH_3Hg in pure culture from added Hg(II) (see ref. *24* for a review) the relative rates of methylation by these organisms and their role in *in situ* methylation is unknown. Furthermore, while a number of SRB that are incomplete organic carbon oxidizers readily methylate Hg in culture (e.g. *Desulfobulbus propionicus*, (*30,86*) and *Desulfovibrio desulfuricans*, (*108*)), studies in the environment have suggested that they may not be the dominant Hg methylators. King et al. (*32*) showed in pure culture that, of the organisms they tested, *Desulfobacterium* methylated Hg at a substantially greater rate, under the conditions of their experiments, than the other species tested (*Desulfobacter*, *Desulfococcus*, *Desulfovibrio* and *Desulfobulbus*). *Desulfobacterium* is a complete acetate oxidizer and CH_3Hg was only produced in these cultures when sulfate-reduction was occurring. This contrasts results of others who have shown that *Desulfobulbus propionicus* can methylate Hg while growing fermentatively (*38,86*).

King et al. (*32*) also found that marine sediments amended with acetate produced more CH_3Hg than sediments amended with lactate, or unamended controls. Acetate-amended slurries were dominated by *Desulfobacterium* and *Desulfobacter*. Macalady et al. (*33*), using polar lipid fatty acid analysis, also found that *Desulfobacter*-like organisms were important Hg methylators in sediments of a Hg-contaminated freshwater system, Clear Lake, California. It appears from these results that the organisms capable of complete oxidation of acetate are potentially more efficient methylators in the environment.

However, it is clear that there is some aspect of the mechanism of Hg methylation that allows some bacteria to methylate Hg while others do not. The ability to methylate Hg is not confined to one phylogenetic group of sulfate-reducing bacteria but is scattered throughout the phylogenic tree of sulfate-reducing eubacteria (*41*). Furthermore, phylogenetically similar organisms have differing abilities to methylate Hg -e.g., *Delsulfovibrio gigas*, *D. vulgaris*, *D. salexigens* and *D. desulfuricans aestuarii* do not methylate Hg but *D. desulfuricans* LS (*109*) and ND132 (*24*) do. A pathway for methylation has been demonstrated for only one organism (*Desulfovibrio desulfuricans* LS). Berman et al. (*109*) showed that mercury methylation is an enzymatically catalyzed process *in vivo*, and suggested, based on the selective inhibition of mercury methylation in *D. desulfuricans* LS, that methylation is mediated by a cobalt porphyrin in this organism. Further work (*73,79*) led the group to propose that Hg methylation in this organism occurs via transfer of a methyl group from methyl-tetrahydrofolate to cobalamin to Hg. The methyl group may originate from serine or via the acetyl-CoA synthase pathway.

Mercury methylation by cell extracts of *D. desulfuricans* LS was 600-fold higher compared to free methylcobalamin (*73*), and thus it is not merely the presence of cobalamin that instills the ability to methylate Hg at a significant rate. Indeed, cobalamin is not unique to SRB. Cobalamin and high levels of acetyl-CoA enzymes are present in methanogens and acetogens, and indeed, cell extracts of a methanogen, have been shown to methylate Hg (*110*). However, these organisms are not thought to play a large role in environmental methylation, based on selective inhibitor studies (*13,19,22,37*). Furthermore, it is not known whether the corrinoid protein found in strain LS is always present in the SRB that do methylate Hg. It has been suggested that SRBs that methylate Hg possess a distinct or highly specific enzyme to catalyze this step. However, the identity of the enzyme responsible for methyl transfer to Hg in most methylators is not known.

Recent studies have shown that of the complete acetate oxidizers, *Desulfococcus multivorans* (1be1), *Desulfosarcina variabilis* (3be13) and *Desulfobacterium autotrophicans* all contain the acetyl-CoA pathway and methylate Hg while *Desulfobacter hydrogenophilus* does not methylate Hg but does have the pathway. Similarly, for SRBs that are not complete oxidizers, there is a correspondence between the presence of the acetyl-CoA pathway and the ability to methylate Hg for *Desulfovibrio africanus, D. sulfuricans* LS and *D. vulgaris* (Marburg). However, there are also organisms that methylate Hg that do not have this pathway (*Desulfobulbus propionicus* (1pr3) and *D. propionicus* (MUD)). Thus the acetyl-CoA pathway cannot be the only mechanism for Hg methylation *in vivo*.

Given the reactivity of Hg, it is obvious that the Hg will not be present inside cells as the free metal ion, Hg^{2+}. Thus, the transfer of the methyl group to Hg likely involves the interaction with Hg bound to a ligand, or to an enzyme. There are a number of mechanisms for methyl transfer within cells, but as the Hg is likely in the +2 state within the bacteria, the methyl group needs to be transferred as a radical or as a carbanion, and this restricts the methyl transfer process to that involving electrophilic attack by Hg(II) on cobalamin (*110*). If Hg was in the +1 state, then it could directly substitute for Ni(I) in the normal operation of the acetyl-CoA pathway i.e., be involved in a nucleophilic attack on the corrinoid methyl group. This is an intriguing but untested notion. Alternatively, it may be that the Hg is bound to a particular enzyme or thiol group in some organisms that places it in the correct location for transfer, or that steric hindrance prevents the transfer of Hg in some organisms but not in others. In the acetyl-CoA pathway, the methyl group is transferred to carbon monoxide dehydrogenase (CODH) and if Hg were bound to the active site of the CODH, the transfer of the methyl group directly to the Hg could occur. These ideas are speculative, and further studies should focus on identifying the location of Hg within the cell during methylation.

Bacterial Demethylation

Microbial degradation via the *mer* operon is the best-studied pathway of CH_3Hg degradation. The operon is widely distributed in nature, often co-existing on transposons that also contain antibiotic resistance genes (*111*). Variants on the *mer* operon that include the *mer*B gene confer "broad spectrum" resistance to a variety of organomercury compounds including methyl- and ethylmercury chloride via organomercurial lyase (*21,112*). Microbial degradation of methylmercury occurs through the cleavage of the carbon-mercury bond by the enzyme organomercurial lyase followed by reduction of Hg(II) by mercuric reductase to yield methane and Hg^o (*21*). The physiology and genetics of *mer*-mediated CH_3Hg degradation and mercury resistance have been extensively reviewed elsewhere (*69,111-115*). While the biochemistry of the commonly studied *mer* operon is fairly well understood, newer studies of the distribution of the *mer* operon in the environment are revealing unexpected polymorphism and genetic diversity (*116-121*).

As many as half of the bacteria from Hg-contaminated sites may contain the *mer* operon (*111*). However, another mechanism also appears to mediate CH_3Hg degradation. While methane and Hg^o are the primary products of *mer*-mediated Hg demethylation, CO_2 has also been observed as a major methylmercury demethylation product by Oremland and co-workers (*64,70,71*). These authors suggested that methylmercury degradation can occur through biochemical pathways used to derive energy from single carbon substrates, and they termed this process "oxidative demethylation", i.e.:

$$4CH_3Hg^+ + 2H_2O + 4H^+ = 3CH_4 + CO_2 + 4Hg^{2+} + 4H_2$$

As a presumptive C1 metabolic pathway, oxidative demethylation is not an active detoxification pathway for CH_3Hg, unlike *mer*-mediated demethylation and Hg reduction. A variety of aerobes and anaerobes (including sulfate reducers and methanogens) have been implicated in carrying out oxidative demethylation, and oxdative demethylation has been observed in freshwater, estuarine and alkaline-hypersaline sediments (*64,65,70*). However, the identity of the organisms responsible for oxidative demethylation in the environment remains poorly understood. Further, no organism has been isolated that carries out this pathway.

Pak & Bartha (*66*) confirmed the ability of two sulfate reducing bacterial strains and one methanogen strain to demethylate mercury in pure culture. They argued that the CO_2 seen in these studies resulted from oxidation of methane released from CH_3Hg after cleavage via organomercurial lyase by anaerobic methanotrophs in the sediments and that CO_2 was a secondary product and not the primary product of demethylation. However, Marvin-DiPasquale et al. (*71*) found that the rate of CO_2 production from CH_3Hg far exceeded the rate of CO_2 production from CH_4 in sediments from two of their study ecosystems, under both aerobic and anaerobic conditions.

The relative importance of *mer*-mediated versus oxidative demethylation is poorly understood (*71*). In highly contaminated environments, the *mer* operon is more prevalent among the microbial community, and Hg(II) reduction activity is enhanced (*111*). However, the rate of microbial Hg^0 production in the environment may not always be proportional to *mer* transcription (*123*). Overall rates of microbial activity, the presence of Hg-reducing genes divergent from commonly used probes, and the bioavailability of CH_3Hg to cells also play a role. In systems that are not highly contaminated, oxidative demethylation appears to dominate, under both aerobic and anaerobic conditions. The Hg concentrations that would cause a switch from one pathway to the other are only loosely defined. Most studies of Hg demethylation via oxidative demethylation have employed ^{14}C-labeled CH_3Hg and thus only the carbon products are traceable. The end-product of oxidative demethylation has been presumed to be Hg(II), but that has not been confirmed. Demethylation studies using CH_3Hg containing a specific stable Hg isotope should help resolve that issue.

Bioluminescent Sensors for Mercury

"Bioreporters" are genetically engineered microorganisms designed to rapidly assess the bioavailable concentration of contaminants, or the rate of contaminant degradation. In these bioreporters, the bioluminescence operon (*lux*) is inserted as the sensor component into the biodegradation or resistance pathways of interest. When the pathway is expressed, the *lux* genes are expressed concurrently. A relatively simple measurement of light production can then be used to assess, for example, expression of a metal resistance gene or a hydrocarbon degradation pathway. Clearly, the potential advantage of bioreporters over chemical measurement lies in the possibility of determining bioavailable or bioactive concentrations of the contaminant of interest. Selifonova et al. (*124*) constructed the first Hg bioreporters, fusing the promotorless *lux* operon from *Vibrio fischeri* into the Tn21 mercury resistance operon (*mer*), and using *E. coli* as the host strain. The organisms showed semi-quantitative response to Hg in contaminated natural waters, at concentrations as low as a few nM. In constructing these first Hg bioreporters, the importance of understanding Hg transport pathways was recognized as being crucial if bioreporters were to be used to assess Hg bioavailability. The *mer* operon consists of a sequence of genes that encode active Hg transport and Hg reduction (*merA*), plus regulatory genes (*merR* and *merD*). Selifonova et al. constructed a set of three *mer-lux* fusions with and without the transport and reductase genes

(124). Interestingly, light production in response to Hg by strains with and without the transport genes was similar, suggesting Hg uptake was occurring by pathways other than *mer*-based active Hg transport. However, complicating factors, such as the potential energy and counter-ion requirements of Hg transport may cloud interpretation of these data.

Continued development of this bioreporter demonstrated that Hg-dependant light production in this strain was cell density-dependant (125) and dependant on the chemistry of the assay medium (126). The strain used in these studies contained a *mer-lux* fusion without Hg transport genes (pRB28). Reduction of cell density to about 10^5 cells per ml (at the high end of the range of cell densities found in natural waters) in Hg assays reduced the number of competitive binding sites for Hg, and therefore improved the sensitivity of the assay, into the pM range. Dissolved organic carbon also decreased the bioavailability of Hg to this strain. The assay buffer was manipulated to show that neutral Hg-Cl complexes induced more light production than negatively charged complexes, suggesting that uptake by this strain, under these conditions, was via diffusion.

Because light production is energy-dependent in these biosensors, it is necessary to separate factors that influence cellular activity in general from factors that influence Hg bioavailability specifically (127,128). Barkay et al. have used constitutive controls to achieve this goal, constructing an isogenic strain (pRB27) in which *lux* expression is constitutive, and therefore Hg-independent (126). Another approach is to construct biosensors with only a partial *lux* operon, so that the aldehyde precursors to *lux*-mediated light production are not produced by cells, but are supplied externally (129). This reduces the energy requirements of light production, but requires additional alteration of test media, and potentially affects Hg speciation. In order to examine Hg bioavailability under a wider range of conditions, Kelly et al. and Scott et al. transferred the pRB27 and pRB28 *mer-lux* fusions of Barkay et al. into *Vibrio anguillarum* (130,131). This host strain has wide salinity tolerance, and is a facultative anaerobe. Refinement of assay conditions also improved sensitivity to <0.5 pmol bioavailable Hg L^{-1}. This strain has been used to examine the bioavailability of trace level additions of Hg(II) to natural lake waters, and to examine the bioavailability of Hg in unamended natural waters. The percentage of ambient total Hg in lake and rain water available was found to be very low, as was the bioavailability of tracer additions of Hg to natural waters. Finally, as a first step in understanding Hg bioavailability in the conditions in which methylation occurs, Golding et al. (132) have worked with *E. coli* and *Vibrio anguillarum* bioreporters under anaerobic conditions. In some circumstances, Hg(II) uptake by both strains appears to occur via facilitated transport. This suggests that Hg uptake by these strains occurs by a different mechanism than Hg uptake by methylating SRB, which occurs via diffusion of neutral species (*30,86*). Differences in medium content could potentially account for these differences as, for example, facilitated transport of Hg bound to amino acids has been shown to occur across membranes of higher organisms (*133*).

Genetic engineering using the *mer* operon has also been applied to Hg bioremediation. The advantage of strains constructed with the *mer* operon lies in the specificity of the *mer*-based uptake and detoxification systems, which are often unaffected by the presence of other metals. For example, a mercury bioaccumulator has been engineered (*134*), as a potential aid in mercury bioremediation. To produce the bioaccumulator, an *E. coli* strain was constructed to express a *mer*-based Hg^{2+} transport system and to overexpress pea metallothionine (MT), which protects the cells from Hg toxicity and allows for continued accumulation of Hg-MT within cells. Accumulation of Hg by the strain was not affected by metal chelators such as EDTA and citrate. Organisms that overexpress organomercurial lyase have also been constructed as potential aids in clean-up of organomercury contaminated sites (e.g. *135*).

Mercury biosensors are a potentially valuable tool for assessing Hg bioavailability. To date they have been used to demonstrate that Hg complexation has a large influence on Hg bioavailability to these cells, and that only a small fraction of Hg dissolved in natural waters is generally available for uptake. Also, previously unidentified pathways for Hg transport may need to be considered. More studies should lead to a fuller understanding of Hg transport pathways in cells without the *mer*-based transport system, and allow comparison of those systems with the transport systems of methylating and demethylating microorganisms, and the broader spectrum of microorganisms at the bottom of the food web. An important issue in understanding the results of Hg biosensor studies is the role of Hg transport pathways coded by the *mer* operon and those of the host organism. The bioavailability of Hg to methylating organisms is perhaps the key to modeling Hg methylation rates. Bioreporters can potentially be used to define that fraction of the ambient Hg pool if it can be shown that Hg uptake by methylators and bioreporters are similar. This should be the focus of continued research. However, CH_3Hg production itself may be the best "bioreporter" of Hg bioavailability to methylating bacteria. Since CH_3Hg production by these microorganisms occurs intracellularly, CH_3Hg production depends on Hg transport and serves as a sensor for Hg bioavailability. We have used *D. propionicus* in this way to examine uptake of neutral HgS complexes (*30,86*).

References

1. Pacyna, J.M. In *Global and Regional Mercury Cycles: Sources, Fluxes and Mass Balances*; W. Baeyens; R. Ebinghaus; O. Vasiliev, Eds.; Kluwer Academic Publishers, Dordrecht, 1996, pp. 161-178.
2. Mason, R.P.; Fitzgerald, W.F.; Morel, F.M.M. *Geochim. Cosmochim. Acta* **1994**, *58*, 3191-3198.
3. Fitzgerald, W.F.; Engstrom, D.R.; Mason, R.P.; Nater, E.A. *Environ. Sci. Technol.* **1998**, *32*, 1-7.
4. Lindqvist, O.; Johannson, K.; Aastrup, M.; Andersson, A.; Bringmark, L; Hovsenius, G.; Hakanson, L.; Meili, M.; Timm, B. *Water Air Soil Pollut.*, **1991**, Special Issue, Vol. 55.
5. U.S. EPA. Mercury Study Report to Congress. EPA-452/R-97-004, US EPA Office of Air, 1997, Washington, DC.
6. Clarkson, T.W. *J. Trace Element Exp. Med.* **1998**, *11*, 303-317.
7. Mahaffey, K.R. *JAMA* **1998**, *280*, 737-738.
8. St. Louis, V.L.; Rudd, J.M.W.; Kelly, C.A.; Beaty, K.G.; Flett, R.J.; Roulet, N.T. *Environ. Sci. Technol.* **1996**, *30*, 2719-2729.
9. Krabbenhoft, D.P., Benoit, J.M., Babiarz, C.L., Hurley, J.P., Andren, A.W. *Wat. Air Soil Poll.* **1995**, *80*, 425-433.
10. St. Louis, V.L.; Rudd, J.M.,W.; Kelly, C.A.; Beaty, K.G.; Bloom, N.S.; Flett, R.J. *Can. J. Fish. Aquat. Sci.* **1994**, *51*,1065-1076.
11. Driscoll, C.T., Holsapple, J., Schofield, C.L., Munson, R. *Biogeochem.*, **1998** *40* , 137-146.
12. Waldron, M.C., Coleman, J.A., Breault., R.F. *Can. J. Fish Aquat. Sci.*, **2000**, *57*, 1080-1091.
13. Compeau, G.; Bartha, R. *Appl. Environ. Microbiol.* **1985**, *50*, 498-502.
14. Berman, M.; Bartha, R. *Bull. Environ. Contam. Toxicol.* **1986**, *36*, 401-404.
15. Ramlal, P.S.; Kelly, C.A.; Rudd, J.W.M.; Furutani, A. *Can. J. Fish. Aquat. Sci.* **1992**, *50*, 972-979.
16. Korthals, E.T.; Winfrey, M.R. *Appl. Environ. Microbiol.* **1987**, *53*, 2397-2404.
17. Jensen, S.; Jernelov, A. *Nature.* **1969**, *223*, 753-754.
18. Gilmour, C.C.; Riedel, G.S. *Water Air Soil Pollut.***1995**, *80*, 747-756.
19. Gilmour, C.C.; Henry, E.A.; Mitchell, R. *Environ. Sci. Technol.* **1992**, *26*, 2281-2287.
20. Branfireun, B.A.; Roulet, N.T.; Kelly, C.A.; Rudd, J.W.M. *Global Biogeochem. Cycles.* **1999**, *13*, 743-750.
21. Robinson, J.B.; Tuovinen, O.H. *Microbiol. Rev.* **1984**, *48*, 95-124.
22. Chen, Y.; Bonzongo, J.-C.J.; Lyons, W. B.; Miller, G.C. *Environ. Toxicol. Chem.* **1997**, *16*, 1568-1574.

23. King, J.K.; Saunders, F.M.; Lee, R.F.; Jahnke, R.A. *Environ. Toxicol. Chem.* **1999**, *18*, 1362-1369.
24. Gilmour, C.C.; Henry, E.A. *Environ. Poll.* **1991**, *71*, 131-169.
25. Watras, C.J. and 21 others. In *Mercury Pollution: Intergration and Synthesis.*, C.J. Watras and J.W. Huckabee, Eds., Lewis Publishers, Boca Raton, 1994, pp. 153-177.
26. Urban, N.R., Brezonik, P.L., Baker, L.A., Sherman, L.A. *Limnol. Oceanogr.*, **1994**, *39*, 797-815.
27. Steffan, R.J.; Korthals, E.T.; Winfrey, M.R. *Appl. Environ. Microbiol.* **1988**, *54*, 2003-2009.
28. Compeau, G.; Bartha, R. *Appl. Environ. Microbiol.* **1987**, *53*, 261-265.
29. Compeau, G.; Bartha, R. *Bull. Environ. Contam. Toxicol.*, **1983**, *31*, 486-493.
30. Benoit, J.M.; Mason, R.P.; Gilmour, C.C. *Appl. Environ. Microbiol.* **2001**, *67*, 51-58.
31. Devereaux, R.; Winfrey, M.R.; Winfrey, J.; Stahl, D.A. *FEMS Microbial. Ecol.* **1996**, *20*, 23-31.
32. King, J.K.; Kostka, J.E.; Frischer, M.E. *Appl. Environ. Microbiol.* **2000**, *66(6)*, 2430-2437.
33. Macalady, J.L. Mack, E.E., Nelson, D.C., Scow, K.M. *App. Environ. Microbiol.*, **2000**, *66*, 1479-1488.
34. Kotska, unpubl
35. Winfrey, M.R; Rudd, J.W.M. *Environ. Toxicol. Chem.* **1990**, *9*, 853-869.
36. Benoit, J.M.; Gilmour, C.C.; Mason, R.P.; Reidel, G.S.; Reidel, G.F. *Biogeochem.* **1998**, *40*, 249-265.
37. Gilmour, C.C.; Riedel, G.S.; Ederington, M.C.; Bell, J.T.; Benoit, J.M.; Gill, G.A.; Stordal, M.C. *Biogeochem.* **1998**, *40*, 327-345.
38. Henry, E.A. Ph.D. Thesis, Harvard University, **1992**.
39. Allan, C.J.; Heyes, A. *Water Air and Soil Pollut.* **1998**, *105*, 573-592.
40. Allan, C.J.; Heyes, A.; Roulet, N.T.; St. Louis, V.L.; Rudd, J.W.M. *Biogeochem.* **2000**, *52*, 13-40.
41. Heyes, A.; Gilmour, C.C.; Mason, R.P. Unpublished data.
42. Heyes, A.; Moore, T.R.; Rudd, J.W.M.; Dugoua, J.J. *Can J. Fish. Aquat. Sci.* **2000**, 57, 2211-2222.
43. Mason, R.P.; Lawson, N.M; Lawrence, A.L.; Leaner, J.J.; Lee, J.G.; Sheu, G-R. *Mar. Chem.* **1999**, 65, 77-96.
44. Mason, R.P.; Lawrence A.L. *Environ. Toxicol. Chem.* **1999**, *18*, 2438-2447.

45. Kannan, K.; Smith, R.J. Jr.; Lee, R.F.; Windom, H.L.; Heitmuller, P.T.; Macauley, J.M.; Summers, J.K. 1998. *Environ. Contam. Toxicol.*, **1998**, *34*, 109-118.

46. Bloom, N.S.; Gill, G.A.; Driscoll, C.; Rudd, J.; Mason, R.P. *Environ. Sci. Technol.* **1999**, *33*, 7-13.

47. Hines, M.E.; Horvat, M.; Faganeli, J.; Bonzongo, J.C.J.; Barkay,T.; Major, E.B.; Scott, K.J.; Baily, E.A.; Waewick, J.J.; Lyons, W.B. *Environ. Research*, **2000**, *83*,129-139.

48. Kannan, K; Falandysz. *Water Air Soil Pollut.* **1998**, *103*, 129-136.

49. Henry, E.A.; Dodge-Murphy, L.J.; Bighma, G.M.; Klein, S.M.; Gilmour, C.C. *Water Air Soil Poll.* **1995**, *80*, 489-498.

50. Watras, C.J.; Back, R.C.; Halvorsen, S.; Hudson, R.J.M.; Morrison, K.A.; Wente, S.P. *Sci. Tot. Environ.* **1998**, *219*, 183-208.

51. Suchanek, T.H.; Mullen, L.H.; Lamphere, B.A.; Richerson, P.J.; Woodmansee, C.E.; Slotten, D.G.; Harner E.J.; Woodward, L.A. *Water Air Soil Pollut.* **1998**, *104*, 77-102.

52. Verta, M.; Matilainen, T. *Water Air Soil Poll.* **1995**, *80*, 585-588.

53. Hurley, J.P.; Cowell, S.E.; Shafer, M.M.; Hughes, P.E. *Environ. Sci. Technol.* **1998**, *32*, 1424-1432.

54. Bonzongo, J.-C.; Heim, K.J.; Chen, Y.; Lyons, W.B.; Warwick, J.J.; Miller, G.C.; Lechler, P.J. *Environ. Toxicol. Chem.* **1996**, *15*, 677-683.

55. Hines, unpublished

56. Hintelman H.; Wilken, R.D. *Vom Wasser* **1994**, *82*, 163-173.

57. Miskimmin, B.M.; Rudd, J.W.M.; Kelly, C.A. *Can. J. Fish. Aquat. Sci.* **1992**, *49*, 17-22.

58. Miskimmin, B.M. *Bull. Environ. Contam. Toxicol.* **1991**, *47*, 743-750.

59. Guimaras, J..R.D.; Meili, M.; Hylander, L.D.; Silva, E.D.E.; Roulet, M.; Mauro, J.B.N.; de Lemos, R.A. *Sci. Total Environ.* **2000**, 261, 99-107.

60. Roulet, M.; Guimaraes, J.R.D.; Lucotte, M. *Water Air and Soil Pollution*, **2001**, *128*, 41-60.

61. Bodaly, R.A., Rudd, J.M.W., Fudge, R.J.P., Kelly, C.A. *Can. J. Fish. Aquat. Sci.* **1993**, *50*, 980-987.

62. Ramlal, P.S.; Kelly, C.A.; Rudd, J.W.M.; Furutani, A. *Can. J. Fish. Aquat. Sci.* **1993**, *50*, 972-979.

63. Barkay, T., Liebert, C., Gillman, M. *Appl. Environ. Microbiol.* **1989**, 196-1202.

64. Oremland, R. S.; Culbertson, C.W.; Winfrey, M.R. *Appl. Environ. Microbiol.* **1991**, *57*, 130-137.

65. Oremland, R.S.; Miller, L.G.; Dowdle, P.; Connel, T.; Barkay, T. *Appl. Environ. Microbiol.* **1995**, *61*, 2745-2753.

66. Pak, K.-R.; Bartha, R. *Bull. Environ. Contam. Toxicol.*1998, *61*, 690-694.
67. Pak, K.-R.; Bartha, R. *Appl. Environ. Microbiol.* 1998, *64*, 1013-1017.
68. Spangler, W.J.; Spegarelli, J.L.; Rose, J.M.; Miller, H.M. *Science.* 1973, *180*, 192-193.
69. Moore, M.J.; Distefano, M.D.; Zydowsky, L.D.; Cummings, R.T.; Walsh, C.T. *Acc. Chem. Res.* 1990, *23*, 301-308.
70. Marvin-Dipasquale, M.C.; Oremland, R.S. *Environ. Sci. Technol.* 1998, *32*, 2556-2563.
71. Marvin-DiPasquale, M.; Agee, J.; McGowan, C.; Oremland, R.S.; Thomas, M.; Krabbenhoft, D.; Gilmour, C.C. *Environ. Sci. Technol.*, 2000, *34*, 4908-4916.
72. Sellers, P.; Kelly, C.A.; Rudd, J.W.M.; MacHutchon, A.R. *Nature.* 1996, *380*, 694-697.
73. Choi, S.-C.; Chase, Jr., T; Bartha, R. *Appl. Environ. Microbiol.* 1994, *60*, 4072-4077.
74. Benoit, J.M.; Gilmour, C.C.; Mason, R.P. *Environ. Sci. Technol.* 1999, *33*, 951-957.
75. Gutknecht, J.J. *J. Membr. Biol.* 1981, *61*, 61-66.
76. Mason, R.P.; Reinfelder, J.R.; Morel, F.M.M. *Water Air Soil Pollut.* 1995, *80*, 915-921.
77. Mason, R.P.; Reinfelder, J.R.; Morel, F.M.M. *Environ. Sci. Technol.* 1996, *30*, 1835-1845.
78. Craig, P.J. Moreton, P.A. *Mar. Poll. Bull.* 1983, *14*, 408-411.
79. Choi, S-C.; Bartha, R. *Bull. Environ. Contam. Toxicol.* 1994, *53*, 805-812.
80. Schwarzenbach, G.; Widmer, M. *Chim. Acta.* 1963, *46*, 2613-2628.
81. Paquette, K.; Helz, G. *Water Air Soil Pollut.* 1995, *80*,1053-1056.
82. Paquette, K. 1994. Ph.D. Thesis. The University of Maryland, 1994.
83. Dyrssen, D. *Mar. Chem.* 1988, *24*,143-153.
84. Dyrssen, D; Wedborg, M. *Water Air Soil Pollut.* 1991, *56*, 507-520.
85. Benoit, J.M.; Mason, R.P.; Gilmour, C.C. *Environ. Toxicol. Chem.* 1999, *18*, 2138-2141.
86. Benoit, J.M.; Gilmour, C.C.; Mason, R.P. *Environ. Sci. Technol.* 2001, *35*, 127-132.
87. Jay, J.A.; Morel, F.M.M.; Hemond, H.F. *Environ. Sci. Technol.* 2000, *34*, 2196-2200.
88. Tossell, J.A. *J. Phys. Chem. A.* 2001, *105*, 935-941.
89. Stein, W.D.; Lieb, W.R. *Transport and Diffusion Across Cell Membranes*, Harcourt Brace Jovanovich, London, 1986.

90. Huerta-Diaz, M.A.; Morse, J.M. *Geochim. Cosmochim. Acta.* **1992**, *56*, 2681-2702.
91. Bono, A.B. M.S. Dissertation, McGill University, 1997. 177 pp.
92. Dmytriw, R., Mucci, A., Lucotte, M., Pichet, P. 1995. *Wat. Air Soil Pollut.* **1995**, *80*, 1099-1103.
93. Benoit, J.M.; Mason, R.P.; Gilmour, c.C.; Aiken, G.R. *Geochim. Cosmochim. Acta* **2001**, *65*, 4445-4451.
94. Reddy, M.M., Aiken, G.R *Wat. Air Soil Poll.* **2000**, *132*, 89-104.
95. Hudson, R.J.M.; Gherini, S.; Watras, C.; Porcella, D. In *Mercury as a Global Pollutant: Towards Integration and Synthesis.* C.J. Watras and J.W. Huckabee, Eds.; Lewis Publishers: Boca Raton, **1994**, pp. 473-526.
96. Miller, C.L.; Mason, R.P. ACS 222nd Meeting, Chicago, August, **2001**; ACS abstract, *41*(2), 514-518.
97. Mikac, N.; Kwokal, Z.; May, K..; Branica, M. *Mar. Chem.* **1989**, *28*, 109-126.
98. Muhaya, B.B.M.; Leemakers, M.; Baeyens, W. *Water Air Soil Poll.* **1997**, *94* , 109-123.
99. Xia, K.; Skyllberg, U.L.; Bleam, W.F.; Bloom, P.R.; Nater, E.A.; Helmke, P.A. **1999**, *33*, 257-261.
100. Kim, C.S.; Shaw, S.; Rytuba, J.J.; Gordon, E.B. Jr. ACS 222nd Meeting., Chicago, Il, August 26-30, **2001**, *4*, 497-503.
101. Spry, D.J.; Wiener, J.G. *Environ. Pollut.* **1991**, *71*, 243-304.
102. Ramlal, P.S.; Rudd, J.W.M.; Furutani, A.; Xun, L. *Can. J. Fish. Aquat. Sci.* **1985**, *42*, 685-692.
103. Xun, L.; Campbell, N.E.R.; Rudd, J.W.M. *Can. J. Fish. Aquat. Sci.* **1987**, *44*, 750-757.
104. Royals, H.E.; Lange, T.R. ACS Abstract, 1995 Meeting, Washington, DC, **1995.**
105. Hintelmann, H.; Evans, R.D.; Villeneuve, J.Y. *J. Anal. Atomic Spec.* **1995**, *10*, 619-624
106. Hintelmann, H.; Keppel-Jones, K.; Evans, R.D. *Environ. Toxicol. Chem.*, **2000**, *19*, 2204-2211.
107. Gilmour, C.C.; Riedel, G.S., Coates, J.S.; Lovley, D. Abstract, Amer. Soc. Microbiol., New Orleans, **1996.**
108. Choi, S.C.; Bartha, R. *App. Environ. Microbiol.* **1993**, *59*, 290-295.
109. Berman, M.; Chase, T.; Bartha, R. *Appl. Environ. Microbiol.* **1990**, *56*, 298-300.
110. Wood, J.M.; Kennedy, F.S.; Rosen, C.G. *Nature.* **1968**, *220*, 173-174.
111. Liebert. C.A. Hall, R.M., Summers, A.O. *Microbio. Mol. Biol. Rev.*, **1999**, *63*, 507-522.

112. Barkay, T. 2000. In *Encyclopedia of Microbiology*. 2nd edition. Academic Press, San Diego, pp. 171-181.
113. Summers, A.O. *Ann. Rev. Microbiol.* **1986,** *40*, 607-634.
114. Foster, T.J. *CRC Crit. Rev. Microbiol.* **1987,** *15*, 117-140.
115. Silver, S., Phung. L.T. *Ann. Rev. Microbiol.* **1996,** *50*, 753-789.
116. Osborn, A.M., Bruce, K.D., Strike, P., Ritchie, D.A.. *Appl. Environ. Microbiol.*, **1993,** *59*, 4024-4030.
117. Bruce, K.D. *Appl. Environ. Microbiol.* **1997,** *63* , 4914-4919.
118. Liebert, C.A. Wireman, J., Smith, T., Summers, A.O. *Appl. Environ. Microbiol.* **1997,** *63*, 1066-1076.
119. Ravel, J., DiRuggiero, J., Robb, F.T., Hill, R.T. *J. Bacteriol.* **2000,** *182*, 2345-2349.
120. Ravel, J., Schrempf, H., Hill, R.T. *Appl. Environ. Microbiol.* **1998,** *64*, 3383-3388.
121. Reyes, N.S., Frisher, M.E., Sobecky, P.A. *FEMS Microbial Ecol.* **1999,** *30*, 273-284.
122. Hines, M.E., Horvat, M., Faganeli, J., Bonzongo, J-C.J, Barkay, T., Major, E.B., Scott, K.J., Bailey, E.A., Warwick, J.J., Lyons, W.B. *Environ. Res.* **2000,** *83*, 129-139.
123. Jeffrey, W.H., Nazaret, S., Barkay, T. *Microbial Ecol.* **1996,** *32*, 293-303.
124. Selifonova, O.; Burlage, R.; Barkay, T. *Appl. Environ. Microbiol.* **1993,** *59*, 3083-3090.
125. Rasmussen, L.D.; Turner, R.R.; Barkay, T. *Appl. Environ. Microbiol.* **1997,** *63*, 3291-3293.
126. Barkay, T.; Gilman, M.; Turner, R.R. *Appl. Environ. Microbiol.* **1997,** *63*, 4267-4271.
127. de Weger, L. A.; Dunbar, P.; Mahafee, W.F.; Lugtenberg, B.J.J.; Sayler, G.S. *Appl. Environ. Microbiol.* **1991,** *57*, 3641-3644.
128. Hill, P. J.; Rees, C.E.D.; Winson, M.K.; Stewart, G.S.A.B. *Biotechnol. Appl. Biochem.* **1993,** *17*, 3-14.
129. Virta, M.; Lampinen, J.; Karp, M. *Anal. Chem.* **1995,** *67*, 667-669.
130. Kelly, C.A.; Scott, K.J.; Holoka, M.; Rudd, J.W.M. In preparation.
131. Scott, K.J.; Rudd, J.W.M.; Kelly, C.A. In preparation.
132. Golding, G.R.; Kelly, C.A.; Sparling, R.; Loewen, P.C.; Rudd, J.W.M.; Barkay, T. *Limnol. Oceanogr.* In press.
133. Leaner, J.J. PhD Dissertation, University of Maryland, College Park, 2001, 182 pp.
134. Chen, S.; Wilson, D.B. *Appl. Environ. Microbiol.* **1997,** *63*, 2442-2445.
135. Horn, J.M.; Brunke, M.; Deckwer, W.; Timmis, K.N. *Appl. Envir. Microbiol.* **1994,** *60*, 357-362.

Chapter 20

Mercury Contamination in the Seine Estuary, France: An Overview

D. Cossa[1], F. J. G. Laurier[1], and A. Ficht[2]

[1]Ifremer, Centre de Nantes, BP 21105, F.44311 Nantes cedex 3, France
[2]Service de Navigation de la Seine, Ile Lacroix, F.76000 Rouen, France

As mercury in the Seine estuary is mainly associated with suspended particulate matter (> 90 %), its dynamics is governed by sediment erosion-deposition cycles which are strongly developed in this macrotidal estuary. However, it can be mobilized by two reactions, methylation and reduction. The former favors bioaccumulation through the production of methylmercury, and the latter is conducive to atmospheric recycling through the formation of volatile elemental mercury. Freshwater inputs into the Seine estuary reflect anthropogenic influences; the average concentrations for the last five years are 2.9 ± 2.1 ng L^{-1} for the dissolved and 1.7 ± 0.7 μg g^{-1} for the particulate phase. Contamination of sediments is especially high in the upper estuary (> 1 μg g^{-1}), but decreases downstream (down to 0.2 μg g^{-1}). Yet the levels measured in animals are rarely higher than in the European guideline for human consumption. Temporal trends have shown a general decrease in concentrations.

The environmental dynamics of mercury is conditioned by physical, chemical and biological properties, which relate respectively to its volatility at ambient temperature, the stability of its bonds with carbon and sulfur, and its very high biological concentration and toxicity. The volatile elemental mercury ($Hg°$) constitutes the most mobile form, the insoluble mercury sulfide (HgS) and selenide (HgSe) the most stable forms, and monomethylmercury (CH_3Hg^+) the most toxic form. The last compound is found in the picomolar range in water but is concentrated up to ten million-fold in the top trophic chain of aquatic organisms, especially carnivorous fish. Thus, the consumer can be exposed to doses which in extreme cases eventually lead to severe neurotoxicity (1).

Mercury biogeochemical cycle involves conversions of chemical species that lead to phase changes and thus to very different fates in the environment (2). The primary sources are natural (mainly earth degassing and vegetation particles) and anthropogenic (mainly fossil fuel and waste incineration). Precipitation is the main source of mercury for the oceans. In rivers, in addition to the atmospheric inputs, soil and rocks provide particulate forms through the action of erosion and dissolved forms by weathering and leaching. At the interface between these two environments, estuaries constitute special entities where biogeochemical reactions, especially phase exchanges, are particularly active.

This paper provides a synthesis of current knowledge about mercury dynamics and the state of contamination of the Seine estuary. After a study of the sources and level of mercury contamination in the Seine basin and an evaluation of internal inputs into the estuary, this paper will consider the distribution of the metal between dissolved and particulate phases, the chemical reactions governing the speciation and bioavailability of mercury, the state of contamination in the biota, and temporal changes in contamination within the last decade.

Environmental Settings

The Seine River is 780 km long with a 75 000 km^2 drainage area. The mean river discharge is 490 m^3 s^{-1} and reaches 2000 and 60 m^3 s^{-1} during the flood and low flow conditions respectively. The hydrographic basin of the Seine includes one of the most heavily populated regions of Europe, where industrial and agricultural activities are largely developed, and as a consequence large amount of domestic and industrial wastes are carried by the Seine River. The upper limit of the Estuary is materialized by a dam (Poses) situated 163 km upstream of Le Havre and 202 km downstream of Paris (Figure 1). The salinity intrusion is up to

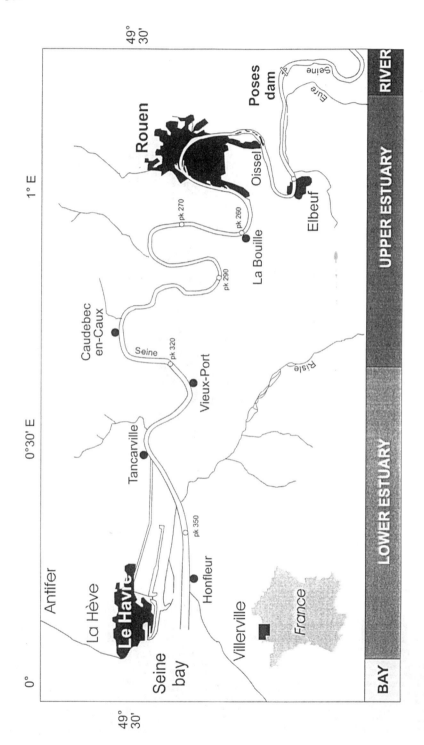

50 km upstream Le Havre. With a tidal range up to 5 m the Seine Estuary is a well mixed estuary with a very dynamic maximum turbidity zone (*3*).

Material and Methods

Mercury concentrations in waters, suspended particles, sediments and biota discussed here come from five cruises between 1991 and 1995 as contribution to the "Seine-Aval" programme and the monitoring programmes of the Seine Navigation Service (SNS) and the French Mussel Watch (Réseau National d'Observation de la Qualité du Milieu Marin, RNO). Additional information presented comes from various studies, particularly Coquery (*4*), Coquery et al. (*5*), Cossa et al. (*6*), Idlafkih (*7*), Mikac et al. (*8*), and Laurier et al. (*9*), as well as unpublished data from Ifremer and SNS. Methods for collections and analyses of waters and suspended particles are based on ultraclean techniques described by Coquery et al. (*5*), Idlafkih (*7*) and Laurier (*28*). Surface sediment samples and biota have been collected according to the "French Mussel Watch" procedures (*10,11*).

Results and discussion

The Seine basin: a contaminated system

Surface waters and sediments

Mean mercury concentrations in Seine freshwaters at Poses (the upstream limit of the estuary) during the last five years were 2.9 ± 2.1 ng L^{-1} for the dissolved fraction and 1.7 ± 0.7 µg g^{-1} for the particulate fraction. These levels of concentration place the Seine in the category of contaminated rivers including the Scheldt and the Elbe (Figure 2). The source of such a high concentration levels can not be accounted for by the influence of the natural substratum on the Seine watershed. The mercury content of suspended particulate matter in the upper basin has been estimated at 0.1 µg g^{-1} (*12*), and the preindustrial sediments at 0.02 µg g^{-1} (*7*), which is similar to the content in the earth's crust (0.02-0.05 µg g^{-1}). Thus, the distribution of mercury in the Seine watershed was not apparently abnormal before the industrial age, and the upper basin still shows little contamination today. However, the situation is tending to deteriorate rapidly downstream. Weighted mean mercury concentrations in the suspended particles for the hydrologic year 1994-1995 increased from 0.5 µg g^{-1} upstream of Paris (Morsang) to 0.8 µg g^{-1} downstream (Chatou) and to more than 1 µg g^{-1} at the entry to the estuary (Poses) (*7*).

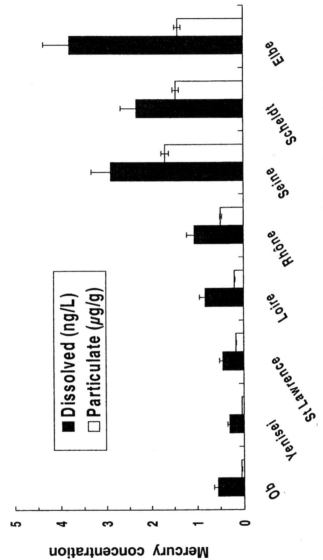

Figure 2. Comparison of mercury concentrations in different rivers worldwide

The results of monitoring dissolved mercury at Poses suggest, however, that the contamination of the Seine river by mercury has decreased in the last two years. Indeed, the average concentrations pass from 2.93 ± 2.50 ng L^{-1} (n=47) for 1995-1996, to 2.50 ± 1.15 ng L^{-1} (n=31) in 1997-1998 and 1.94 ± 0.75 ng L^{-1} (n=21) in 1999-2000 (Figure 3). Hydrologic and climatic conditions, as well as variations in human sources of waste, are likely to cause such differences.

On the basis of these data, a denudation rate of around 10 g of mercury per km^2 of basin and per year can be estimated. This figure is typical of urbanised watersheds in which human influence is dominant (13,14). A very slight fraction comes from the weathering of the rocks. In fact, rocks containing carbonate are mercury-poor, and the rate of alteration of aluminosilicates is 2.10^3 kg km^{-2} yr^{-1} (15), which means that only 0.1 g km^{-2} of mercury is mobilised per year (barely 8 kg per year for the entire Seine basin).

Atmospheric fallout

The mercury in precipitation of the Seine basin is not currently documented. Very recent measurements have been performed as a part of a pilot project supported by the French Agency for the Environment and Energy Resources (ADEME). Mercury concentrations during a period of 8 months in 2000-2001, ranged between 3.8 – 48.8 ng L^{-1} (n=21) (Dasilva, Colin and Cossa, unpublished results). These results confirm the observations made in other areas : the mercury concentrations in precipitation are one order of magnitude higher in rain water than in surface water or groundwater (16). Quémerais et al. (14) found a positive correlation between water discharge and dissolved mercury concentrations in the slightly contaminated St. Lawrence River, which illustrates this observation. Indeed, they interpreted this relationship as the result of the mixing of groundwater and hyporeic waters (with low mercury concentrations) which are dominant during the dry periods, with the surface runoff (with high mercury content) the contribution of which is increasing during the flood periods. Recent results from the Seine river suggest that this process occurs during the 1999-2000 period (Figure 4). Prior to 1999, there was a very large variability in mercury concentrations, including very high concentrations, especially for low water discharge (Figure 4). The overall tendency of decreasing concentrations with increasing water discharge may be interpreted as the result of the dilution of a roughly constant direct anthropogenic input by the runoff during flood periods (Figure 4).

On the basis of the present data and with an estimate of mercury concentrations in the troposphere, which range from 2 to 4 ng m^{-3}, and rainfall

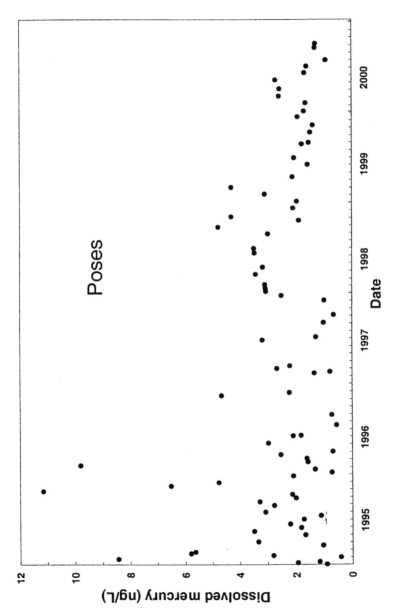

Figure 3. Time changes in mercury concentration in waters at the entry into the Seine estuary, the Poses dam

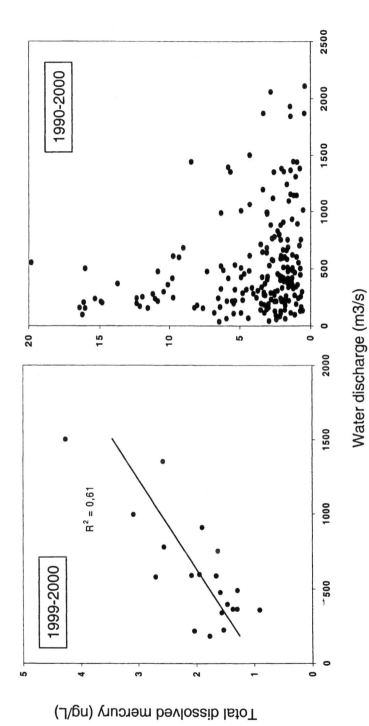

Figure 4. Total dissolved mercury as a function of the water discharge in the Seine river at the Poses dam

the total fallout for the Seine basin can be estimated at around 1000 kg yr^{-1}. Another evaluation based on the models of Mason et al. (*17*) and Galperin et al. (*18*) indicates that total fallout at this latitude would be 16 µg m^{-2} yr^{-1}, *i.e.* around 1300 kg yr^{-1} for the 78500 km^2 of the basin. These values are roughly the same as those for discharges into the atmosphere, estimated respectively at 1350 and 340 kg yr^{-1} for combustion of fossil fuels (coal and oil) and incineration of waste within the Seine basin (*19, 20*).

The Seine estuary and bay

Water column

Coquery et al. described the distribution of "dissolved" mercury (water filtered through a 0.8 µm glass fiber filter) within the Seine Estuary (*5*). Concentrations usually vary from 0.4 to 1.2 ng L^{-1} with the highest concentrations and variability at the freshwater end-member. In addition, unpublished results from a 1995 cruise pointed out higher concentrations (up to 1.8 ng L^{-1}) in the upper estuary, which suggest the present of anthropic influences from the Rouen industrialized area (Fig. 1) where an industrial discharges of more than 1.4 kg yr^{-1} were recorded until 1997. Coquery et al. (*5*) did not notice any systematic pattern between "dissolved" mercury and salinity. However, downstream the upper estuary, with the addition of results from a 1995 cruise, one can observed a slight tendency toward a concentration decrease in the very early mixing zone (Figure 5). This behavior can be assimilated to removal (aggregation and/or flocculation) from the "dissolved" phase of mercury-rich colloids. Flocculation is thought to occur in the Seine estuary when the increasing turbidity enhances the frequency of the contacts between particles (*21*) and when the ionic strength reduces the repulsion forces between particles. The distribution of the partition coefficient between particulate and "dissolved" phases (Kd = [particulate Hg]/[dissolved Hg] (mL g^{-1}), which exhibits increasing values supports this hypothesis (Figure 5). Additional results are needed to confirm this suggestion.

Particulate mercury concentrations are variable at Poses and in the upper estuary, sometimes exceeding 1 µg g^{-1}, which is in contrast with the stability of concentrations measured in the lower estuary where concentrations range between 0.50 and 0.75 µg g^{-1} (Figure 6). For salinities above 32 PSU in the bay, the distribution of mercury concentrations is indicative of mixture with less contaminated marine particles (Figure 6).

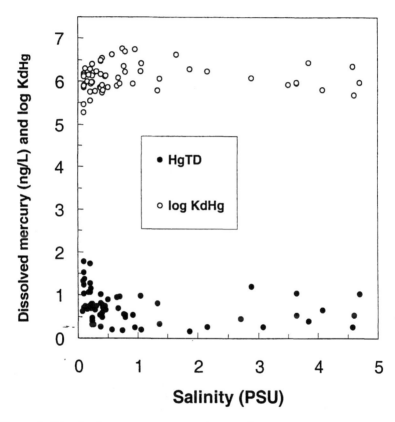

Figure 5. Dissolved mercury concentrations and partition coefficients (KdHg) as a function of salinity in the Seine estuary

308

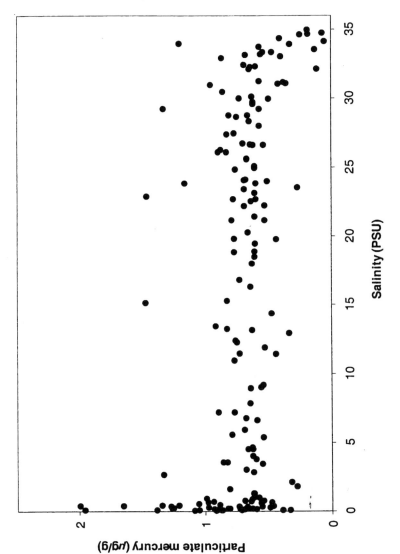

Figure 6. Particulate mercury distribution as a function of salinity in the Seine estuary

Sediments

The distribution of mercury in estuary deposited sediments is similar to that of suspended particles in the overlying water column (Figure 7). Surface sediments of the upper estuary (Poses to Vieux-Port) are characterized by elevated and quite variable concentrations ranging from 0.3 µg g^{-1} at Elbeuf to 11 µg g^{-1} in the Oissel sector (Figure 1). This latter measurement clearly confirms the anthropogenic influence already pointed out by the mercury measurements in the water. In the mid-part of the estuary, from upstream Tancarville (Courval) to Honfleur, the concentrations are relatively uniform (around 0.4 µg g^{-1}) and roughly the same as in the upper estuary downstream Rouen. In the lower estuary and Seine Bay, the concentrations are lower, sometimes at pre-industrial levels in the southwestern part of the bay. They are also more variable (range: 0.02-0.23 µg g^{-1}; average : 0.047 µg g^{-1}), attesting to variations in the nature of the sediments. In fact, the concentrations are directly proportional to the relative richness in organic carbon (Figure 8). The difference in the slope of the two curves indicates that sediment contamination is higher in the estuary and northeastern part of the bay than in the southwestern bay.

In a study based on sedimentary core samples collected in the lower estuary, Mikac et al. (*8*) showed that variations in mercury concentrations in sediments vary with depth and depend on their grain size and content in organic matter. The normalization of mercury concentrations with respect to organic matter shows the homogeneity of the metal in the sedimentary column. The same author also show that the sediment pore water is up to 100-fold mercury enriched compared to the overlying water column (*8*). This suggests that, during their mineralization by bacteria, particles rich in organic matter mobilize their associated mercury into pore water. Thus, molecular diffusion as well as pore water injection in the water column during erosion periods are mechanisms for mobilization of mercury from the sediment waters in the Seine estuary as in the bay. In any event, the surface sediments of the estuary and Seine Bay are unquestionably a mercury remobilization site favoring its bioavailability for organisms.

The bioconcentration of mercury in living organisms

Inorganic mercury and methylmercury are concentrated biologically by aquatic organisms. Relatively large variations have been observed for a given species, most often due to differences in age or diet. However, greater differences have been found between species. The bioconcentration factors (BF) generally range from 10^5 to 10^7 (*19*). The lowest BF are found in plankton, the highest in carnivorous fish, marine mammals and seabirds, and intermediary

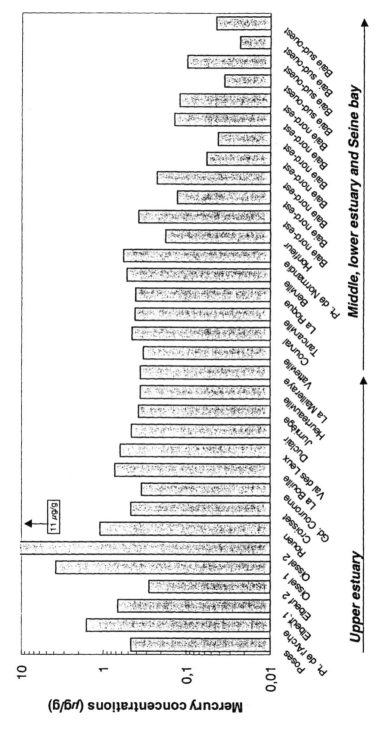

Figure 7. Mercury concentrations in surface sediments of the Seine estuary (data from reference 30)

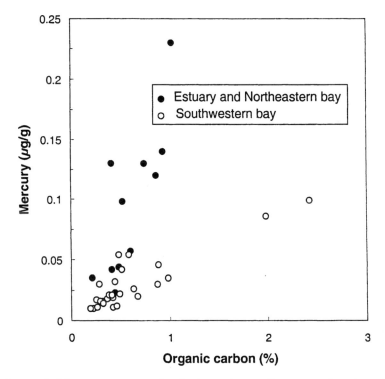

*Figure 8. Mercury concentration in the surface sediment as a function of the
particulate organic carbon content in the Seine estuary and bay*

values in mean trophic levels. The predator concentrates more than its prey. This is known as bioamplification, which is especially due to a particular chemical form, *i.e.* monomethylmercury. This chemical species, which constitutes less than 10 % of the total mercury in water (5), represents more than 95% in bass or congers (22). In intermediary levels, monomethylmercury accounts for 20 to 30 % in plankton and 40 to 80 % in mussels.

In the estuary and Seine Bay, total mercury and methylmercury concentrations have been determined for a limited number of species: the scallop (*Pecten maximus*), shrimp (*Crangon crangon*), flounder (*Platichthys flesus*), plaice (*Pleuronectes platessa*), sole (*Solea vulgaris*) and bib (pout) (*Trisopterus luscus*). The results reported in Table 1 do not indicate an acute contamination of the ecosystem capable of causing consumer risk. The maximal concentration set by the European Commission is 0.5 mg/kg (wet weight), *i.e.* 2.5 mg kg^{-1} in equivalent dry weight if the water content of the organisms is considered to be 80 %. In ecological terms, the concentrations are in accordance with the model described above: the highest are in fish and the lowest in filter-feeding molluscs of the secondary trophic level. Shrimp, whose diet can contain zooplankton, show intermediary concentrations.

There is an apparent discrepancy between these mercury concentration levels in biota compared to the high concentrations measured in water, suspended particles and sediment. Reasons for that situation are probably to be found in the hydrosedimentary and biological productivity regimes of the Seine bay. The Bay in widely open to the English Channel influences and the contaminants inputs from the Seine river are efficiently diluted. Indeed, the tidal amplitude and currents are strong to supply enough energy for an intense mixing. In addition, the nutrients inputs from the river are abundant allowing an intense primary production (200-400 grams of carbon per square meter per year according to Cugier) (24), which acts as a dilution agent (i.e. organic carbon) for the chemical contaminants taken up by the phytoplankton. Mercury speciation changes occurring in the estuary may also contribute to a relatively poor bioavailbility of mercury in this particular environment. Two major groups of reactions which can promote phase changes deserve to be discussed, reduction/oxidation and methylation/demethylation.

Reduction and methylation

In the Seine estuary, as in other coastal environments, mercury is involved in two main reactions, reduction and methylation, that use the same substrate (HgII). The first reaction leads to the recycling of mercury *via* the atmosphere, and the second favours its bioaccumulation in food webs.

Table I. Mercury concentration in some species sampled in Seine Bay in
1986 and 1987 (22).

Species	Total mercury (mg kg⁻¹, dry weight)		Mean proportion of methylmercury (%)
	Min – Max (number of determinations*)	Mean ± standard deviation	
Bib	0.28 – 1.50 (25)	0.76 ± 0.32	87
Mussel	0.06 – 0.22 (30)	0.13 ± 0.04	44**
Scallop	0.04 – 0.07 (14)	0.05 ± 0.02	71
Shrimp	0.09 – 0.35 (25)	0.18 ± 0.06	89
Flounder	0.15 – 1.36 (50)	0.54 ± 0.30	82
Plaice	0.15 – 0.77 (25)	0.35 ± 0.16	86
Sole	0.09 – 1.21 (25)	0.35 ± 0.29	77

*number of individuals, except for mussels (batches of 30 individuals).
**Percentage based on samples collected in 1996 (23).

The reduction of Hg° is due to photochemical reactions (25) and phytoplanctonic enzymatic reactions (26). The efficiency of photoreduction in this particular coastal environment needs to be explored. Coquery et al. (5) have shown that the quantity of dissolved gaseous mercury in the waters of the Seine estuary is proportional to the level of oxygen supersaturation, which is itself dependent on the intensity of the primary production. The same authors found that dissolved gaseous mercury was supersaturated in Seine Bay. In May 1991 the supersaturation reached 470 % and, in November 1992, up to 144 %. Based on these results, a rate of Hg° evasion into the atmosphere of 60 ng per square meter and per day has been calculated. In terms of the area of Seine Bay, this flux can be extrapolated to 80 kg of mercury per year.

It is generally accepted that the methylation of mercury results from the action of sulfatoreducing bacteria present in a suboxic environments, *i.e.* essentially in the top centimeters of sediment or in the anaerobic zone which develops occasionally in the water column of the estuary in summer. In the Seine estuary, monomethylmercury could not be quantified in dissolved form in the water column because the concentrations were lower than the detection limit (0.2 ng L⁻¹) of the method used (5). However, concentrations of 4 to 10 ng L⁻¹ were measured in interstitial waters of sediments (8). In suspended particulate matter in the Seine, monomethylmercury constitutes less than 10 % of total mercury and less than 1% of the mercury in sediments (5,8). By the action of diffusion and the resuspension of surface sediments, a part of this methylmercury is periodically transferred into the water column. As this process could conceivably extend to include all of Seine Bay, sediment source needs to be considered with respect to the bioaccumulation of methylmercury in the ecosystems. Bioconcentration in benthic organisms is probably the main source of mercury in trophic systems.

In order to explore the temporal trend in the mercury contamination within organisms, the mercury concentrations in the soft tissue of the blue mussel (*Mytilus* sp.), used in RNO monitoring, was monitored. Figure 10 illustrates temporal variations for the bivalve at four stations around the estuary since 1986: the north shore at Antifer, La Hève, Le Havre, and the south shore at Villerville (Fig. 1). The four series showed similar concentration patterns: large seasonal variations corresponding to changes in the condition index of molluscs and residual variations characterised by stability from 1986 to 1990, an increase from 1990 to 1993, and a slow decrease until 1998.

Fluxes

Riverine inputs

As a mass balance of suspended particle is not available for the estuary and Seine bay, such a budget for mercury is yet impossible. However, the presented research allow to propose some quantification of fluxes (Fig. 9). The gross (before any estuarine transformation) mercury flux at Poses at the head of the estuary was estimated varied between 500 and 1 500 kg of mercury in the last five years, with more than 90 % being associated with suspended particulate matter. The direct anthropic recorded inputs within the Seine estuary are around several ten or so kilograms per year according to data of the Seine Navigation Service. They adds to the upstream input, as well as an undetermined (probably small since mercury is efficiently trapped in soils) fraction contained in sludge from sewage treatment plants disposed on farmland (560 kg yr^{-1} in Seine-Maritime according to Merrant) (27). Conversely, dredging operations remove significant quantities, 50 to 100 kg of mercury per year, from the estuary.

Flux from the sediments

Based on pore water profile in sediment cores, Mikac et al. (8) calculated a diffusive flux of 40 to 60 ng cm^2 year^{-1}, which in relation to the surface area of the estuary (30 km^2) corresponds to an annual flux of 15 kg. Reported for the entire bay (4000 km^2) and taking into account the differences in mercury concentration in sediments (with a similar mercury partition between pore water and solid sediment), the annual flux would be at least 150 kg. Despite the imponderables relating to these estimations, they indicate the potential importance of mobilisation for the entire bay. Moreover, in a period of great

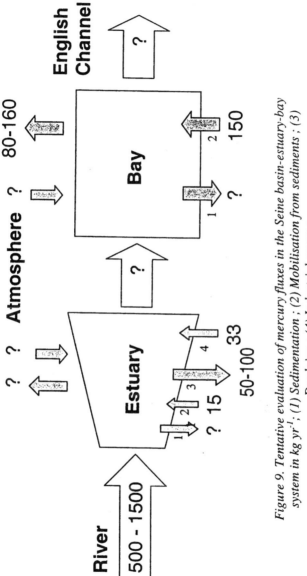

Figure 9. Tentative evaluation of mercury fluxes in the Seine basin-estuary-bay system in kg yr⁻¹; (1) Sedimentation ; (2) Mobilisation from sediments ; (3) Dredging ; (4) Industrial sewage

316

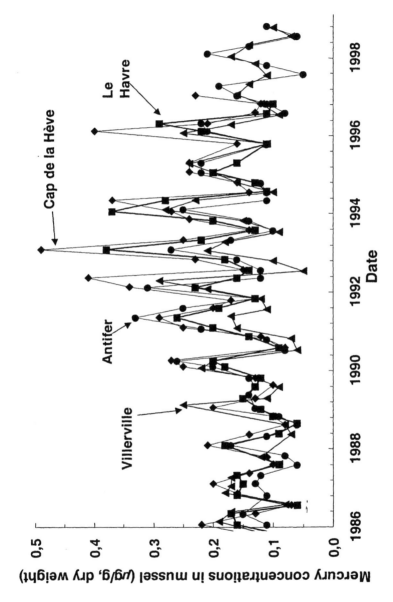

Figure 10. Time changes in mercury concentration in soft tissues of the mussel
(Mytilus edulis)

turbulence, erosion of the surface layer of sediments is another mechanism for an higher post depositional mercury mobilisation, which remains very difficult to quantify and which is not taken into consideration in the present rough estimate. More work is unquestionably needed on this aspect.

Flux to the atmosphere

Coquery et al. (*5*) proposed an evasion from the surface water of the Seine bay of 80 kg/year. Independent estimate can be obtained based on Mason et al. model (*17*), which proposes a production of 100 ng of Hg° per gram of planktonic carbon produced. Based on an estimation of annual primary production of 200-400 g of carbon per m^2 and per year (*24*), Hg° production would be 80-160 kg yr^{-1}. Thus, regardless of the calculation mode used, evasion flux accounts for more than 10 % of the total mercury carried by the Seine, *i.e.* more than the dissolved fraction. This implies that marine mercury and/or fluvial particulate mercury are mobilised in the estuary or the bay. Although the order of magnitude of this phenomenon seems certain, the estimation, based on a limited number of results, needs to be more precise, particularly after estimation of the efficiency of the photoreduction. However, this first estimate suggests that the mercury received in the Seine bay is intensively recycled *via* the atmosphere and in part return the Seine basin with precipitations ; thus, the mercury exported to the marine environment still remains a continental water problem.

Conclusions

The waters, suspended particulate matter and sediments of the Seine show mercury concentrations indicative of the marked contamination typical of an urbanized and industrialized river. Diffuse atmospheric fallout from human sources is the main cause. Discharges from various sources are added in the estuary. However, in recent years the contamination has decreased.

Particulate mercury from the River is mixed with marine and estuarine particles (*i.e.* those mineralized from continental or marine sources and/or originating from erosion of the adjacent coastal environment). One part of the contamination is exported into the English Channel, a second part is recycled into the atmosphere, and a third part is stored in the sediments of the estuary and the bay. Although precise evaluation of these fluxes is difficult in our present state of knowledge, a first approximation indicates that the mercury dynamics of the Seine estuary-bay system is dominated by water-atmosphere and sediment-particle exchanges.

Most of the mercury in the estuary is associated with suspended particulate matter. A large portion is mobilized and made available for methylation or reduction. The latter reaction, through the generation of volatile elemental mercury, leads to the atmospheric recycling of an amount equivalent to more than 10 % of river input. Quantification of the methylation rate is not available. In any event, the contamination of the Seine estuary is more characterized by inorganic mercury compounds than by methylated species. Finally, intraestuarial dredging activities remove 10% of river input each year.

Despite heavy contamination of the abiotic environment, concentrations in the living organisms studied rarely reach levels likely to involve risk for consumers of marine products. This is probably related to the high primary production of the area which dilutes the contaminants and refrains the biological amplification in the food webs. In addition, time changes in recent years for mercury concentrations in river waters and mussels tend to indicate that contamination is decreasing.

References

1. Fitzgerald, W. F.; Carlsson, T. W. *Environ. Health Persp.* **1997**, *96*, 159-166.
2. Fitzgerald, W. F.; Mason, R. P. In *Metal ions in biological systems. Vol. 34.*; Sigel, A. and Sigel, H. Eds.; Marcel Dekker, Inc: New York, 1997, pp53-111.
3. Salomon, J. C. In *Hydrodynamics of Estuaries, Vol. II*, Kjerfve, B. Ed. CRC Press, 1988, pp79-89.
4. Coquery, M. Ph.D. Thesis, University of Paris 6. Speciality: Oceanology, Meteorology and Environment, 1994.
5. Coquery, M., Cossa, D.; Sanjuan, J. *Mar. Chem.* **1997**, *58*, 213-227.
6. Cossa, D.; Meybeck, M.; Idlafkih, Z.; Bombled, B. *Rapport IFREMER.* R.INT.DEL/94.13/Nantes, **1994**, 151 pp.
7. Idlafkih, Z. 1998. Ph.D. Thesis, University of Paris 6. Speciality: Earth Sciences. 1998.
8. Mikac, N.; Niessen, S.; Oudane, B.; Wartel, M. *Appl. Organometal. Chem.*, **1999**, *13*, 1-11.
9. Laurier, F.; Cossa, D.; Coquery, M.; Sarazin, G.; Chiffoleau, J. F. 1999.. Colloque Seine-Aval, Rouen, **1999**, November 17-19.
10. Aminot, A.; Chaussepied, M. Manuel d'analyses en milieu marin. Masson, 1982.
11. Claisse, D. *Mar. Pollut. Bull.* **1989**, *20*, 523-528.
12. Perreira-Ramos, L. *Etude et exploitation critique de résultats analytiques de métaux sur sédiments.* Campagnes sur les " grandes rivières " du bassin

Seine-Normandie de 1981 à 1986. Rapport AFBSN, IHC (UPMC). 1988, pp55.

13. Hurley, J. P.; Benoit, J. M.; Babiarz, C. L.; Schafer, M. M.; Andrens, A. W.; Sullivan, J. R.; Hammond, R.; Webb, D. A. *Environ. Sci. Technol.* **1995**, *29*, 1867-1875.

14. Quémerais, B.; Cossa, D.; Rondeau, B.; Pham, T. T.; Gagnon, P.; Fortin, B. *Environ. Sci. Technol.* **1999**, *33*, 840-849.

15. Roy, S.; Gaillardet, J.; Allègre, C. J. *Geochim. Cosmochim. Acta* **1999**, *63*, 1277-1292.

16. Guentzel, J. L.; Landing, W. M.; Gill, G. A.; Pollman, C. D. *Environ. Sci. Technol.* **2001**, *35*, 863-873.

17. Mason R. P.; Fitzgerald, W. F.; Morel, F. F. M. *Geochim. Cosmochim. Acta*, **1994**, *58*, 3191-3198.

18. Galperin, M., Gusev, A.; Davidova, S.; Koropalov, V.; Nesterova, E.; Sofiev, M. *EMEP/MSC-E Tech. Rep. 7/94. Co-operative program for monitoring and evaluation of long-range transmission and air pollutants in Europe.* Meteorological Synthesizing Center, Kedrova St. 8-1, Moscow, Russia, 1994.

19. Cossa, D.; Thibaud, Y.; Romeo, M.; Gnassi-Barelli, M. *Le mercure en milieu marin : biogéochimie et écotoxicologie.* Rapport Scientifique et Technique de l'Ifremer No. 19. 1990, pp130.

20. MATE, Principaux rejets industriels en France, bilan de l'année 1997. Direction de la prévention des pollutions et des risques, Service de l'environnement industriel. Ministère de l'Aménagement du Territoire et de l'Environnement. 1998, pp233.

21. Dupont, J.P.; Guézennec, L.; Lafite, R.; Le Hir, P.; Lesueur, P. 2001. Matériaux fins: le cheminement des particules en suspension. Rapport N° 4. Programme Seine–Aval. Edition Ifremer, Brest, France, 2001, pp.39.

22. Cossa, D.; Auger, D.; Averty, B.; Luçon, M.; Masselin, P.; Sanjuan, J. Niveaux de concentration en métaux, métalloïdes et composés organochlorés des produits de la pêche côtière française. Publication Ifremer, Brest. 1990, pp57.

23. Claisse, D.; Cossa, D.; Bretaudeau-Sanjuan, J.; Touchard, G. Bombled, B. *Mar. Pollut. Bull.* **2001**, *42*, 329-332.

24. Cugier, P. Ph.D. Thesis, University of Caen, France. Speciality: Earth-Fluid Envelope. 1999, pp249.

25. Amyot, M.; Gill, G. A.; Morel, F. F. M. *Environ. Sci. Technol.***1997**, *37*. 3606-3611.

26. Mason, R. P.; Fitzgerald, W. F. 1996. In: *Global and regional mercury cycles.* Baeyens, W.; Ebinghaus, R.; Vasiliev, O. Eds., NATO ASI Series. 2. Environment – Vol. 21. Kluwer Acad. Publish, 1996, 249-272.

27. Merrant, F. Les sources de contamination par le mercure de trois gisements de moules de Seine Maritime. Training report. Service Santé-Environnement, DDASS Seine Maritime, 1996.
28. Laurier, F. J. G. Ph.D. Thesis, University Denis Diderot (Paris 7), France, Speciality: Chemical oceanography. 2001. pp173.
29. Guézennec, L. Ph. D. Thesis, University of Rouen. Rouen, France, 1999.
30. RNO, *Surveillance du milieu marin.* Travaux RNO-édition 1999. Ministère de l'environnement and Ifremer. BP 21105, F.44311 Nantes cedex, 1999.

Chapter 21

Determination of Stable Mercury Isotopes by ICP/MS and Their Application in Environmental Studies

Holger Hintelmann[1] and Nives Ogrinc[1,2]

[1]Department of Chemistry, Trent University, Peterborough, Ontario
K9J 7B8, Canada
[2]Jozef Stefan Institute, 1000 Ljubljana, Slovenia

Inductively coupled plasma mass spectrometry (ICP/MS) is a powerful tool for mercury determinations. Introducing mercury in form of gaseous species into a dry plasma greatly reduces memory effects, achieving absolute detection limits of less than 100 pg of Hg. The capability of ICP/MS to take advantage of speciated isotope dilution methods makes this technique suitable for very precise and accurate measurements. Equally important, multiple stable tracer experiments to study the fate of Hg species in the environment become possible. This novel concept allows the investigation of multiple transformation processes simultaneously. However, since isotope enrichments are often well below 100 %, a system of linear equations must be solved to exactly calculate individual isotope concentrations. This paper describes the necessary calculations and demonstrates how stable isotopes improve methylmercury measurements.

Mercury is an ubiquitous pollutant in the environment and of great concern as a large number of aquatic organisms in many ecosystems from all over the world show elevated levels of Hg. Several freshwater fish are known to be contaminated with high levels of toxic and bioavailable methylmercury (MeHg) (*1*). Consequently, many countries have issued advisories to manage the consumption of fish, representing the main entry of MeHg into the human diet and thus, being the main concern from a human health point of view.

Numerous studies have been conducted to elucidate the fate of Hg species in the environment and to understand the factors controlling MeHg formation and bioaccumulation. Similarly, many different methods have been developed to accurately determine the concentrations of Hg species in various environmental matrices (*2*). Only recently has inductively coupled plasma mass spectrometry (ICP/MS) emerged as a competitive technique for Hg species analysis (*3-6*). A couple of unfounded myths regarding the applicability of ICP/MS for Hg (species) analysis had to be overcome, namely its perceived lack of sensitivity (compared to the de-facto standard of highly sensitive atomic fluorescence measurements) and suspected potential for exhibiting severe memory effects.

However, the appeal of employing highly precise isotope dilution techniques and the intriguing possibility of conducting stable isotope tracer experiments to study the fate of Hg have prompted us early on to develop ICP/MS methods (*7*). This paper gives a comprehensive description of the various methods in use in our laboratory for Hg species determinations in a variety of matrices, focusing especially on water column measurements. We will furthermore describe the concept of speciated isotope dilution and stable tracer techniques introducing a novel approach to accurately calculate individual Hg isotope concentrations in multiple isotope addition experiments.

Materials and Methods

Accurate and precise Hg species measurements at ultra trace levels, typically encountered in pristine water samples, presents the analytical chemist with a number of challenges. Most notably is the need for vigorously cleaned laboratory-ware and the challenge to preserve the integrity of the individual Hg species during all sample preparation and measurement steps (no degradation or de-novo formation of Hg species).

Cleaning of Teflon and glassware for ultra trace level measurements

Cleaning methods for equipment used in Total Hg and MeHg measurements are identical, except that Teflon bottles used for MeHg procedures are not subjected to the BrCl soaking step. Teflon bottles designated for MeHg analysis must never been in contact with such solutions, since BrCl might diffuse into Teflon, leach back out into subsequently prepared samples and effectively decomposes MeHg in the new sample.

Teflon bottles, vials and tubing are rinsed first with tap water, then with Milli-Q water and are immersed in (or filled with) a 50% (v/v) HNO_3 solution and heated to 65°C for 2 days. This is followed by at least 4 rinses with Milli-Q water and soaking in 10% (v/v) conc. HCl at room temperature for another 3 days. The Teflon is rinsed 4x with Milli-Q water and again soaked in a solution of 250 ml of Milli-Q water, 1 ml of concentrated HCl and 2 ml 0.2 N BrCl. Prior to the next rinsing step, 100 µl of $NH_2OH \cdot HCl$ per 250 mL of solution is added to neutralize BrCl. After repeated rinsing with Milli-Q water (at least 4x), all Teflon bottles are filled to 25% of their volume with 1% low Hg-grade HCl, capped and double bagged in new polyethylene bags for storage. For new bottles and equipment, the initial nitric acid soaking may be omitted.

Glass-ware is first rinsed 4x with Milli-Q water and then transferred into a container filled with 10% (v/v) HCl. After overnight soaking at room temperature, the equipment is rinsed 4x with Milli-Q water and soaked for another night at room temperature in a BrCl solution (16 l water, 160 ml HCl conc. and 80 ml 0.2 N BrCl). Finally, the equipment is rinsed 4x with Milli-Q water. Distillation bridges are stored in plastic containers, while vessels are filled to 25% with 1% low Hg-grade HCl, capped and stored in the plastic container.

Quartz fiber filters (QF/F, 47 mm diameter) used to collect particles from water samples are cleaned prior to use by heating overnight in a muffler furnace at 500 °C. Individual filters are stored in precleaned (procedure as for new teflon equipment) polyethylene petri-dishes.

Sample digestion for total mercury determinations

Water samples and particles collected on QF/F quartz fiber filters are treated with a strong oxidant (0.2 N BrCl) to oxidize all Hg species to Hg^{2+}. All sample manipulation are performed in a clean room under a class 100 laminar flow hood. Water samples are digested by adding 0.5 mL of BrCl per 100 mL of sample. At the same time, 20 µL of an internal [201]Hg(II) standard solution (2.3 ng/mL) are added. Bottles are sealed again in plastic bags for a minimum digestion period of 12 h (room temperature). Oxidation is considered to be complete if excess of BrCl is present after 12 h as determined by a yellow tint in the sample. If BrCl has been consumed, additional BrCl must be added and allow to react for another 12 h.

A slightly modified (8) method for digesting particles collected on QF/F filters has been employed. The filter paper is fitted into 30 mL Teflon vials and 20 µL of the internal [201]Hg(II) standard solution (2.3 ng/mL) are added directly onto the filter, which is then covered with 10 mL of Milli-Q water. The particles are digested by adding 0.1 mL HCl (conc.) and 0.1 mL BrCl (0.2N). Vials are tightly closed and heated overnight to 65°C on a hot plate.

Vegetation, sediment and biota samples are digested by weighing the sample (typically 100 mg of freeze dried plants, 1 g of wet sediment or 0.5 g of wet fish) into 25 mL glass or Teflon vials. After spiking with an appropriate amount of [201]Hg(II) standard solution (approximately equaling the amount of Hg expected to be in the sample), samples are digested by adding 5 to 10 mL of a HNO_3/H_2SO_4 (7:3) mixture. Samples are heated to 80°C until the formation of brown NO_x gases has ceased. Samples are diluted to volume and analyzed.

Distillation of methylmercury from water and particles

Methylmercury can be isolated from complex sample matrices by atmospheric pressure water vapor distillation. Our procedure has been slightly modified from procedures described in detail elsewhere (9, 10). This distillation step is required for water and particle samples to remove matrix components that may interfere with the subsequent analytical procedures.

Water samples are distilled in an all glass distillation apparatus consisting of two 50 mL glass tubes connected via an air cooled glass distillation bridge (i.d. 8 mm). 50 g of water are weighed into the distillation tube and 20µl of 1 ng/ml $CH_3^{201}Hg^+$ solution (internal standard), 500 µl of 9M H_2SO_4 and 200µl of 20% KCl solution are added to each sample. 5 mL of Milli-Q water are placed into the receiver tube. Both tubes are connected to the distillation bridge and the set-up is placed into an aluminum heating block set to 140 °C. Hg-free N_2 gas (60 – 80 mL/min, controlled by individual flow meters) is connected to each unit to hasten the distillation process. Approximately 85 % (42.5 mL) of the sample are distilled, which should take between 5 and 7 h. Samples can be stored in the dark at 4 °C in the receiver tubes without additional preserving agents for up to one week.

Sediments and particles collected on filters are distilled in a similar fashion, except that 30 mL Teflon bottles are used instead of the glass distillation system. Filters (or sediments) are placed in the vial, 10 mL of Milli-Q water, 20 µl of the $CH_3^{201}Hg^+$ standard solution (1 ng/ml), 0.5 ml of H_2SO_4 (9 M) and 0.2 ml of KCl (20%, w/v) are added. 9 mL of sample are distilled within 60 to 90 min.

Measurement of total mercury in water and particles

Total Hg in water and particles is determined using an off-line cold vapor method involving preconcentration of Hg on gold traps. Cleaned gas-wash bottles (glass, 125 mL) are filled with 100 mL of Milli-Q water and 1 mL of $SnCl_2$ reducing solution (20% w/v in 20% HCl, v/v) and are purged with Hg-free N_2 for 1 hour to remove Hg from the system. Another 0.5 mL $SnCl_2$ is added, pre-cleaned gold traps are attached to the bubbler outlet and the solution is purged at 100 mL/min for 20 minutes to give the system (or bubbler) blank.

An additional 0.5 ml of 0.2N BrCl, together with 0.1 ml of NH$_2$OH·HCl, 0.5 mL of SnCl$_2$ are measured to obtain the reagent blank. Typically, 100 mL of the BrCl treated sample is weighed into the bubbler and excess BrCl is pre-reduced by addition of 0.1 mL of NH$_2$OH·HCl. 0.5 mL of SnCl$_2$ are added and the sample is purged for 20 minutes collecting liberated Hg^0 on the gold traps. Particle digests can be measured by transferring the 10 mL digest to the bubbler without removing previous samples. Sample analysis will only proceed if the following two criteria are met: a) no bubbler blank must contain more than 20 pg of Hg and b) the standard deviation of the mean response for three replicate standards must be <10%. Gold traps are analyzed using a "single amalgamation" procedure. The traps are connected to Hg-free carrier Ar (200 ml/min) and heated for 3 minutes at 500 °C to release Hg. An additional ~800 ml/min Ar make up gas is added prior to connection to the ICP/MS torch.

Measurement of total mercury in fish and sediments

Such samples usually contain enough Hg to be measured without preconcentration of Hg on gold. Reduction and cold vapor generation of dissolved Hg in digests is accomplished in an on-line flow injection system as shown in Figure 1 and described in more detail in (7).

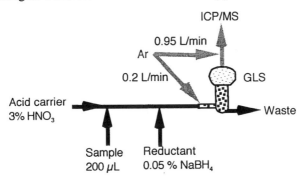

Figure 1: Schematic of the flow injection system for total Hg determinations

500µL of sample are transported in a 3 % (v/v) nitric acid carrier and mixed with a sodium borohydride reducing solution (0.05 % NaBH$_4$, w/v, in 0.05 % NaOH, w/v) at flow rates of approximately 1 ml/min each. The Hg^0 formed is stripped from the solution by introduction of Hg-free Ar (200 mL/min). Separation of gas from liquids takes place in a nylon block by means of a hydrophobic Teflon membrane. At the same time, excess solution is pumped to waste. After addition of approximately 800 ml/min of makeup Ar, the outlet of the gas-liquid-separator is directly connected to the ICP-MS torch.

Measurement of methylmercury

MeHg in sample distillates and alkaline leaching solutions is ethylated at room temperature using tetraethyl borate buffered to pH 4.5. This reagent derivatizes MeHg to volatile MeHgEt, which can be purged and preconcentrated on Tenax traps. Derivatization is carried out in 125 mL gas-wash bottles (glass). Initially, the system (bubbler) blank is determined by adding 100 mL of Milli-Q water, 0.2 mL of acetate buffer (2M) and 0.1 mL of NaBEt$_4$ solution (1%, w/v in 1% KOH, w/v) into the bubbler. Tenax traps are connected to the exit ports and the gas inlets are capped. After a reaction time of 20 minutes, gas purging lines are reconnected to the bubbler inlets and derivatized Hg species are purged with Hg-free N$_2$ onto Tenax for 20 min at a rate of 100 mL/min. Traps are removed and solutions in bubblers are discarded. After 3x rinsing with Milli-Q water, standards and samples are processed in a similar fashion. Typically, 50 mL of distilled water, 10 mL of particle distillates are processed. Tenax traps are reconnected to a packed GC column (15% OV-3 on chromosorb W-AW (DMCS) 80/100) and Hg species are released by heating the traps to 200 °C for 30 seconds. Mercury species are isothermally separated at 100 °C with an Ar flow of 30 mL/min. After addition of approximately 1000 mL/min of Ar make-up gas, the gas line is directly connected to the ICP/MS torch. Sample analysis will only proceed if the following two criteria are met: a) no bubbler blank must contain more than 5 pg of MeHg and b) the standard deviation of the mean response for three replicate standards must be <10%.

Quality control and quality assurance

A minimum of one procedural blanks and where available, one certified reference material (CRM) are analyzed for every 10 samples. Unfortunately, CRM's and RM's certified for MeHg are limited to fish, oyster and sediment samples. No suitable water, zooplankton, benthos or vegetation reference materials are commercially available. Additionally, the only two available MeHg certified sediment RM have severely elevated concentrations of total and MeHg. No low level materials, being more representative for uncontaminated pristine sediments exist. The situation is slightly better with RM's certified for total Hg. However, low level water RM's are non-existing as well.

Preparation of isotopically enriched methylmercury

While isotopically enriched inorganic Hg is commercially available (typically in form of HgO or Hg^0), enriched methylmercury needs to be synthesized. Several methods are described in the literature (5). We have successfully applied the following two methods. The concentrations of CH$_3{}^{201}$Hg$^+$ in the final working standards was determined by reverse isotope dilution analysis. At the same time, the isotopic composition of the chromatographic peak representing inorganic Hg(II) (derivatized to diethylmercury) is evaluated for deviations from the natural Hg isotope pattern, which would indicate the presence of ^{201}Hg(II) impurities in the enriched $CH_3^{201}Hg^+$ product. We never found such contamination, ensuring a pure isotope enriched $CH_3^{201}Hg^+$ solution for isotope addition experiments.

Tetramethyltin method (11)

$CH_3^{201}Hg^+Cl$ was synthesized by dissolving 2.5 mg (23 μmol) of ^{201}HgO in 4 mL HCl (2 M) and addition of a 1 mL methanolic solution of $(CH_3)4Sn$ (1.8 M). The mixture was shaken for 2 h and the generated CH_3HgCl was extracted twice with 5 mL of toluene. The combined organic phases were washed several times with Milli-Q water. To remove trace quantities of inorganic Hg(II) and organotin compounds, CH_3Hg^+ was extracted into 1 ml of a 1 mM sodium thiosulfate solution and washed three times with toluene. After addition of 0.2 mL copper sulfate (1 M) and 0.5 mL sodium chloride (0.5 M) CH_3HgCl was finally extracted into toluene (3 x 3 mL) to give a CH_3HgCl solution with a purity of > 99%.

Cobalamin method (12)

$CH_3^{201}HgBr$ was synthesized by dissolving 100 μg of ^{201}HgO in 10 μL of HCl (conc.). After dilution with 500 μL sodium acetate buffer (0.1 M, pH 5), 500 μg methylcobalamine dissolved in another 500 μL of buffer were added. The reaction mixture was incubated for 3 h at room temperature, 200 μL KBr solution were added (0.3 M KBr in 2 M H_2SO_4) and the formed $CH_3^{201}HgBr$ was extracted 3x with 400 μL of toluene. The combined toluene extracts were dried over sodium sulfate and 100 μL of this primary stock solution were diluted with 10 mL of isopropanol. Working solutions were prepared fresh daily by diluting the secondary isopropanol stock with deionized water as needed.

Tuning and optimisation of the ICP/MS for mercury measurements

The ratios of Hg isotopes in different environmental samples were determined using a Platform-ICP/MS (Micromass Ltd., Manchester, UK). The instrument consists of a standard ICP source and quadrupole mass analyser with a unique ion transfer system based on an RF-only hexapole collision cell. The instrument provides computer control of the xyz-torch positions. In addition to an RF-voltage, a DC-voltage may be applied to the hexapole as a bias to reject low energy ions originating from non-plasma sources. Typically, He and H_2 are introduced simultaneously into the collision cell at flow rates of 0-10 mL min^{-1}.

The platform is uniquely equipped with a scintillator-photomultiplier type of detector, which has a large dynamic linear range extending over 8 orders of magnitude. In contrast to commonly used electron multiplier detectors, it does not switch between pulse and analogue counting modes and hence, no cross-calibration or dead time corrections are needed. Both parameters must be diligently controlled on instruments sporting electron multipliers to obtain the most precise isotope ratios over a wide range of concentrations.

Although modern ICP-MS instruments provide relatively good short- and medium-term signal stability, there is a strong dependence of the signal response on plasma operating conditions. For optimum performance the instrument requires daily optimisation and monthly tuning. For this purpose a continuous stream of elemental mercury is introduced into a dry plasma to determine optimal operation conditions. The use of a dry plasma avoids many problems, in particular extended wash-out times associated with introduction of ionic mercury species into a wet plasma. A steady mercury signal was generated by gently heating a quartz tube containing a small HgS crystal. Parameters most critical for obtaining maximum sensitivity are nebuliser gas flow rate, torch position, hexapole gas flow rates, cone voltage and hexapole bias.

The dependence of signal intensity as a function of the nebuliser flow rate is illustrated in Figure 2. Optimal flow rates ranged between 0.8 and 1.05 L min^{-1} with the analyte intensities between 5×10^5 and 1.3×10^6 counts per second (cps). Beyond these limits a reduction in sensitivity is observed and measured Hg isotope ratios deviate randomly.

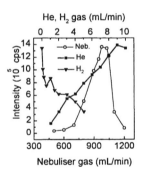

Figure 2: Influence of nebuliser and collision cell gas flows

The measurement of sensitivity as a function of the collision cell gas flow rates was performed changing both gas flow rates independently. Figure 2 shows that an increase in He leads generally to an improvement of ion sensitivity in this instrument. This observation was expected as intensities of heavy ions are generally improved under such conditions. This effect is commonly referred to as the thermalization effect in collision cell instruments (13). On the other hand the presence of H₂ in the collision cell had an unexpected effect. Even the slightest amount of H_2 (the minimum allowed flow rate is 0.1 ml min^{-1}) significantly quenched the Hg signals and changed measured Hg isotope ratios. This peculiar effect could not be explained satisfactorily and is still under investigation. Hence, the ICP/MS instrument is operated with zero H_2 in the collision cell, when used for Hg measurements.

The influence of the cone lens voltage and the retarding hexapole DC-bias potential (HBP) on the intensity of Hg ions are illustrated in Figure 3. The HBP is employed as a barrier, which low energy ions are not able to surmount. Changing the HBP to only –0.1 V will already reduce the Hg signal by 50%. Since none of the mercury ions is interfered by any residual molecular gas ions, the HBP is set to 0 V to achieve the highest sensitivity possible (13). Table I summarizes ICP/MS operating conditions providing the highest signal response for Hg measurements.

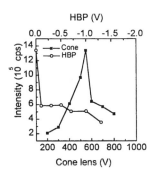

Figure3: Influence of cone lens and hexapole bias potential

Table I. Optimized parameters for Hg isotopes measurements ICP/MS

	ICP		*Analyser/Collision cell*
Incident power	1350 W	Cone lens	550V
Plasma gas	13.98 L min^{-1}	Hexapole exit lens	400 V
Nebulizer gas	1.03 L min^{-1}	H_2 gas flow	0 mL min^{-1}
Carrier gas	0.03 L min^{-1}	He gas flow	10 mL min^{-1}

All methods will generate transient signals, where quantification is based on integrating peak areas of Hg isotopes. Generally, a minimum of 10 data points is required to adequately describe a peak for integration purposes. However, we feel that this minimum requirement is insufficient to determine isotope ratios with high precision. We typically adjust scan times to obtain one data point per second per isotope. Average peak width are in the order of 30 sec, resulting in a minimum of 30 points to fully describe the peak. Faster scan rates introduce a higher baseline noise level, which makes reproducible peak-start and peak-end detection and isotope ratio measurements less precise.

Figures of merit

The described methods and techniques proved to be extremely sensitive, overcoming the perceived shortcomings of ICP/MS for Hg species analysis and allowed ultra trace determinations of Hg species in pristine water samples. Using a dry plasma and introducing only gaseous Hg species (Hg^0 or peralkylated organomercury species) resulted in relatively short washout times with negligible memory. Even after introducing high concentrations of Hg for continuously for more than one hour, the baseline went back to acceptable low values within minutes after removing the Hg source.

Absolute instrumental detection limits for Hg were determined to be 100 fg (30 pg of ^{202}Hg) for both methylmercury and total Hg using the gold trap preconcentration technique. These values compare favorable with instrumental LOD's reported for AFS (300 fg, *14*), the previous state-of-the-science technique for Hg measurements. Typical procedural blanks routinely achieved in water analysis are 1.9 ± 0.6 pg of MeHg and 6.9 ± 3.5 pg for total Hg. These blanks translate into detection limits of 35 pg/L for MeHg measurements (50 mL sample volume) and 100 pg/L for total Hg determinations (100 mL sample volume). Table II compares results obtained with methods described in this paper with certified values of reference materials. Generally, there is no statistical difference among measured and certified concentrations.

Table II: Comparison of measured with certified total Hg values in CRM

Reference material	Method	Certified (ng/g)	Measured (ng/g)
NIST 1515 (apple leaves)	FIA	44 ± 5	48 ± 5
NIST 1515 (apple leaves)	Au-trap		43 ± 2
MESS-3 (marine sediment)	FIA	91 ± 9	87 ± 9
MESS-3 (marine sediment)	Au-trap		96 ± 8
DORM-2 (dogfish muscle)	FIA	4640 ± 260	4350 ± 350

Most of our measurements are based on isotope ratio determinations. Hence the techniques and ICP/MS methods are all optimised with the premise to give most precise ratio determinations. This approach also introduces a new concept for limits of detection. Particularly in tracer experiments, we are faced with the challenge to detect a variation in the isotope ratio from the natural ratio, which would then be an indication for the presence of the tracer isotope. Since we will always measure a background of ambient Hg species, the detection limit is affected with our precision to measure those isotope ratios. Hence, our predisposition with isotope ratio measurements. Generally LOD's for isotope concentrations are limited by the precision of the isotope ratio determination and the background concentration of Hg (the more Hg already in the sample, the more tracer isotope is needed to change the natural ratio by 1%). LOD's for isotopes can be expressed as:

$$\text{LOD (isotope)} = \text{ambient Hg} \times A \times 3\text{RSD (of ratio measurement)}$$

with A = natural abundance of tracer isotope. Using optimum RSD's (0.5 %) and ^{200}Hg as the tracer isotope, at least 0.35 % of the total Hg must be in form of the ^{200}Hg tracer isotope to change the isotope ratio enough allowing to detect its presence in the sample (*11*).

Stable tracer studies

In contrast to a traditional isotope dilution experiment, where the amount of tracer isotope is exactly known, a stable tracer experiment designed to study the fate of Hg species follows a different concept and requires a different approach for calculating individual isotope concentrations. In an stable isotope experiment, known amounts of one or more stable Hg isotopes are added to the system under investigation. This will change the isotopic composition of the added species in the sample or in the source compartment. At the beginning of the experiment (at time zero) the isotopic composition of the species to be generated or of Hg in the target compartment is identical to the natural Hg pattern. Only if the tracer isotope reacts to a new species (by means of methylation, demethylation or redox processes) or is transferred from the source to the target compartment (e.g. uptake into organisms or adsorption from water onto particles) it will alter the isotopic Hg pattern of the new species or in the target compartment. The magnitude of the change in isotope ratios is then related to the amount of new Hg species generated or transferred.

Figure 4 illustrates this concept, showing the isotope pattern of ambient Hg according to IUPAC (*17*) (4a) and the composition of a typical isotope enriched solution of Hg (4b). Here, the solution is enriched with ^{199}Hg(II) (97.36 %). As an example, adding this solution to a sediment sample will change the isotope pattern of the inorganic Hg in the sediment.

The concentration of $CH_3^{199}Hg^+$ is now enriched in the sample relative to the fraction of $CH_3^{199}Hg^+$ in the original sample and the magnitude of the relative change relates to the concentration of ^{201}Hg(II) methylated. At the same time, any other isotope can be used to determine the concentration of ambient MeHg in the same sample (e.g. ambient MeHg = $^{202}A \times CH_3^{202}Hg^+$, ^{202}A = natural abundance of ^{202}Hg).

However, at the beginning of the experiment (time zero), the isotope pattern of MeHg in the sediment remains unchanged, since no isotope enriched MeHg was added. Only if a fraction of the added ^{199}Hg(II) is converted to $CH_3^{199}Hg^+$ will the isotope pattern of MeHg change. Figure 5 represents a GC-ICP/MS chromatogram of a sediment extract from a methylation assay.

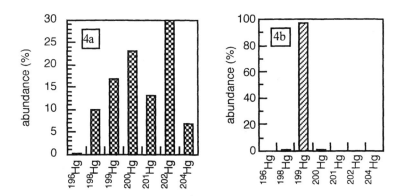

Figure 4: Hg isotope abundance in nature (4a) and ^{199}Hg tracer solution (4b)

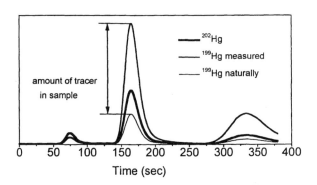

Figure 5: GC-ICP/MS chromatogram of a sediment extract after incubation with ^{199}Hg(II) showing the relative increase in the $CH_3^{199}Hg^+$ signal

Calculation of stable isotope concentrations

Concentrations of isotopes in enriched solutions can be determined using a simple reversed isotope dilution calculation approach (*15, 16*), as long as either the concentration of the enriched isotope or the concentration of the natural standard are exactly known. However, reversed isotope dilution is limited to the calculation of the concentration of one unknown isotope concentration. Certain studies, though, employ multiple tracers, e.g. to determine gross rates of opposing processed such as methylation and demethylation Typically, in such studies, the isotope concentration to be determined is unknown, i.e. the amount of transformed or accumulated isotope. In situations, where unknown concentrations of enriched isotopes in the presence of unknown concentrations of ambient Hg need to be determined, reverse isotope dilution will fail and only the linear equation approach can yield exact results. Hence, the solution described here is more general and applicable to indefinite numbers of stable tracer isotopes.

The total concentration of each isotope is the sum of individual contributions from different sources, such as the ambient sample, isotope enriched standards and/or tracer solutions:

$$\sum{}^1I = {}^1a + {}^1b + ... + {}^1n \tag{1.1}$$
$$\sum{}^2I = {}^2a + {}^2b + ... + {}^2n \tag{1.2}$$
$$\sum{}^iI = {}^ia + {}^ib + ... + {}^in \tag{1.i}$$

where the superscripts 1 to i refer to the individual isotopes of the element I, while parameters a to n refer to the different sources of isotopes (isotopes in tracer solutions or the ambient sample). The relative contribution of each isotope is defined by its relative abundance in those sources. Thus, individual isotope concentrations can be expressed as the ratio of this isotope and one common isotope (usually the most abundant one), e.g. isotope ratios in sources a to n:

$$
\begin{array}{lll}
{}^1a/{}^1a = R_{11} = 1 & {}^1b/{}^2b = R_{12} & {}^1n/{}^in = R_{1i} \\
{}^2a/{}^1a = R_{21} & {}^2b/{}^2b = R_{22} = 1 & {}^2n/{}^in = R_{2i} \\
{}^ia/{}^1a = R_{i1} & {}^ib/{}^2b = R_{i2} & {}^in/{}^in = R_{ii} = 1
\end{array}
$$

Substituting this set of ratios into the Eq. (1.1)- (1.i) yields

$$\sum{}^1I = {}^1a + R_{12}{}^2b + ... + R_{1i}{}^in \tag{2.1}$$
$$\sum{}^2I = R_{21}{}^1a + {}^2b + ... + R_{2i}{}^in \tag{2.2}$$
$$\sum{}^iI = R_{i1}{}^1a + R_{i2}{}^2b + ... + {}^in \tag{2.i}$$

leading to a system of i linear nonhomogeneous equations with m + 1 = i unknowns, with m being the number of stable isotopes applied in a multiple tracer experiment. Isotopic signals ($\sum{}^1I - \sum{}^iI$) and original source ratios R of isotopes are determined. The coefficients ${}^1a$ to ${}^in$ are the unknowns to be calculated. They represent the isotope of interest (1, ..., i) in each source. Employing a matrix inversion approach is the most convenient technique to solve this system of equations. Rewriting the problem in matrix form yields:

$$AX = B$$

with

$$A = \begin{bmatrix} 1 & R_{12} & \cdots & R_{1i} \\ R_{21} & 1 & \cdots & R_{2i} \\ \vdots & \vdots & \ddots & \vdots \\ R_{i1} & R_{i2} & \cdots & 1 \end{bmatrix}, \quad X = \begin{bmatrix} {}^{1}a \\ {}^{2}b \\ \vdots \\ {}^{i}n \end{bmatrix}, \quad \text{and} \quad B = \begin{bmatrix} \sum {}^{1}I \\ \sum {}^{2}I \\ \vdots \\ \sum {}^{i}I \end{bmatrix}$$

where the unknowns can be written as an n x 1 column vector called X. Thus, the solution vector X is obtained using the inverse of A:

$$X = A^{-1} B$$

Using this formula, we can solve for all variables in a system of linear equation at the same time. The final solution, giving the individual isotope contributions from the different sources, can be programmed as a spreadsheet using matrix operations. The applicability of our calculation is demonstrated with the following example, where the problem is reduced to the most simple situation involving only one tracer solution. To determine the exact concentration of $CH_3{}^{201}Hg^+$ in the isotope enriched $CH_3{}^{201}Hg^+$ standard solution, a 50 µL aliquot was spiked with exactly 250 pg of CH_3Hg^+, changing the natural isotope ratio and the total concentrations of ^{201}Hg and ^{202}Hg isotopes depend on their concentrations in both, solution a (ambient CH_3Hg^+ standard) and b ($CH_3{}^{201}Hg^+$ standard). Thus, by using equations (2.1) – (2.2) we can write

$$\sum [^{202}Hg] = {}^{202}a + R_{12}{}^{201}b$$
$$\sum [^{201}Hg] = R_{21}{}^{202}a + {}^{201}b.$$

Further, the $^{201}Hg/^{202}Hg$ ratios in both solutions are required. They can either be obtained through direct measurement of isotope ratios in the original standards or by calculation using IUPAC values (17) and abundances given with the isotope certificates. Since ratios in the resulting mixed sample have to be measured anyway, it is recommended to employ measured isotope ratios at least for ambient mercury as well. This is of particular importance, when concentrations of the tracer isotope approach the detection limit. Theoretically, relative abundances of the isotope enriched solutions need to be measured as well. However, this is often unpractical and less precise, owing to the extreme ratios involved. The use of certified isotope abundances in this case is acceptable, as long as the mass bias introduced by the ICP/MS measurement is only moderate. Our determinations typically exhibited an overall deviation from IUPAC ratios of less than 2% for all ratios involved as shown in Table III.

Table III. Precision of isotope ratio measurements for cinnabar under optimal instrument conditions and comparison wit the IUPAC values

	$^{201}Hg/^{202}Hg$	$^{200}Hg/^{202}Hg$	$^{199}Hg/^{202}Hg$
Measured ratio	0.442	0.774	0.571
IUPAC	0.444	0.776	0.565
F (true/measured)	1.003	1.002	0.990

Considering the significant uncertainty that the peak integration process introduces into the MeHg isotope ratio determination, particular in >95% enriched isotope solutions, our experience has shown that greater accuracy is obtained by using certificate based isotope ratios for all calculations. To continue with our example, the measured ratios in the standard solution a and the isotope enriched solution b (1.18% ^{202}Hg and 98.11% ^{201}Hg) are

$$[^{201}Hg]/[^{202}Hg] = R_{21} = 0.444 \pm 0.0056$$
$$[^{202}Hg]/[^{201}Hg] = R_{12} = 0.012 \pm 0.0002$$

Peak areas (arbitrary units) in a mixed solution were determined to be 997238 for $CH_3^{202}Hg^+$ and 1034148 for $CH_3^{201}Hg^+$. Thus our equations are as follows:

$$997238 = {}^{202}a + 0.012 \, {}^{201}b$$
$$1034148 = 0.444 \, {}^{202}a + {}^{201}b$$

Rewriting the system of equations in the matrix form:

$$A = \begin{bmatrix} 1 & 0.012 \\ 0.444 & 1 \end{bmatrix}, \quad X = \begin{bmatrix} {}^{202}a \\ {}^{201}b \end{bmatrix}, \quad \text{and} \quad B = \begin{bmatrix} 997238 \\ 1034148 \end{bmatrix}$$

and applying the inverse of A, A^{-1}

$$A^{-1} = \begin{bmatrix} 1.005362 & -0.01209 \\ -0.4458 & 1.005362 \end{bmatrix}$$

gives the solution

$$X = \begin{bmatrix} 990080 \\ 595124 \end{bmatrix}$$

The areas of 990080 and 595124 represent the concentration of $CH_3{}^{202}Hg^+$ in the ambient CH_3Hg^+ standard and the concentration of $CH_3{}^{201}Hg^+$ in the isotope enriched $CH_3{}^{201}Hg^+$ standard, respectively. The concentration of the spiked solution b can now be determined by:

$$\left[CH_3{}^{201}Hg^+\right] = \frac{{}^{201}b \times \left[CH_3Hg^+\right] \times {}^{202}A}{{}^{202}a \times V}$$

where ${}^{202}a$ and ${}^{201}b$ are the blank corrected peak areas, $[CH_3Hg^+]$ is the concentration of methylmercury in the non-enriched standard, ${}^{202}A$ is the natural abundance of $CH_3{}^{202}Hg^+$ and V the volume of the $CH_3{}^{201}Hg^+$ solution added. Typically, this method allows the determination of isotope concentrations with a relative standard deviation of 0.87 % (total Hg) to 1.16 % (MeHg). This high precision is of great importance, since concentrations of synthesized isotope enriched CH_3Hg^+ can usually not be determined gravimetrically. The stated precision is maintained, even if up to five individual isotope concentrations have to be calculated. Only for concentrations close to the detection limit does the RSD increase up to 8%, when 5 linear equations are involved.

Application of enriched methylmercury as internal standard

The determination of methylmercury in environmental samples is a challenge. Levels in water, soils and sediments are typically low. Numerous methods for isolation of MeHg from the sample matrix have been developed. Alas, those techniques usually do not provide quantitative recoveries. Consequently, to obtain accurate results isolation yields must be determined for each individual sample, which is possible using standard addition techniques. Since this is very time consuming, particularly in combination with tedious MeHg determinations, often recoveries were only determined once for a single samples and the value obtained was applied to all samples of the same batch.

Stable isotopes now allow a convenient assessment of individual recoveries without any additional sample preparation. A known amount of $CH_3{}^{201}Hg^+$ is added prior to distillation as an internal standard. Using the concepts described earlier, we can determine the amount of $CH_3{}^{201}Hg^+$ recovered and correct the concentration of all other isotopes in the sample accordingly (ambient MeHg as well as potential enriched tracer isotopes). Of course, this approach assumes that the added spike is in equilibrium with the ambient MeHg at time of distillation and as such its behavior is indistinguishable from MeHg originally present in the sample. However, conventional standard addition methods are based on the same assumption. Figure 6 shows results of individual recoveries for a series of water and particle distillations. Although average recoveries from water (88.3 ± 12.3 %) and particles (92.8 ± 22.9 %) are similar to recoveries typically reported by other researchers, it is clear that individual yields can vary greatly.

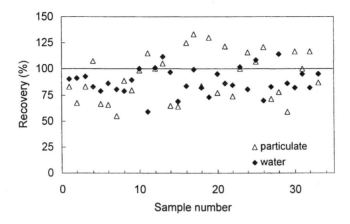

Figure 6: Distillation recoveries of CH$_3$^{201}Hg$^+$ (20 pg) from water and particles

Applying an average (or "historically" obtained) recovery factor would potentially introduce a fairly large uncertainty. The use of stable CH$_3$^{201}Hg$^+$ as an internal standard greatly minimizes the uncertainty. Total mercury levels in replicate (n = 3) environmental samples are routinely measured with a relative standard deviation (RSD) of 2.9 % (tracer isotopes) to 3.7 % (ambient Hg), while RSD's for MeHg range from 2.0% (tracer isotopes) to 2.2 % (ambient Hg). In fact, the correction not only corrects for variable recoveries, but also for random errors associated with subsequent derivatization and detection steps, e.g. the internal standard will compensate for drifts in instrumental response.

References

1. *Mercury Report to Congress* (EPA 452/R-97-0003) **1997**.
2. Clevenger, W. L.; Smith, B. W.; Winefordner, J. D, *Crit. Rev. Anal. Chem.* 1997, *27*, 1.
3. Hintelmann, H.; Keppel-Jones, K; Evans R. D. *Environ. Toxicol. Chem.* **2000**, *19*, 2204.
4. Lauretta, D. S.; Klaue, B.; Blum, J. D.; Buseck, P. R. *Geochim. Cosmochim. Acta* **2001**, *65*, 2807.
5. Lambertson, L.; Lundberg, E.; Nilsson, M.; Frech, W. *J. Anal. At. Spectrom.* **2001**,*16*, 1296.
6. Christopher, S. J.; Long, S.E.; Rearick, M. S.; Fassett, J. D. *Anal. Chem.* **2001**, *73*, 2190.

7. Hintelmann, H.; Evans, R. D; Villeneuve, J. Y. *J. Anal. At. Spectrom.* **1995**, *10*, 619.
8. Morrison, K.A.; Watras, C.J. *Can. J. Fish. Aquat. Sci.* **1999**, *56*, 760.
9. Horvat,M.; Liang, L.; Bloom, N. S. *Anal. Chim. Acta* **1993**, *281*, 135.
10. Bloom, N. S.*Can. J. Fish. Aquat. Sci.***1989**, *46*, 1131.
11. Hintelmann, H.;Evans, R. D. *Fresenius J. Anal. Chem.* **1997**, *358*, 378.
12. Hintelmann, H. *Chemosphere* **1999**, *39*, 1093.
13. Feldmann, I.; Jakubowski, N.; Stuewer, D. *Fresenius J. Anal. Chem.* **1999**, *365*, 415.
14. Bloom, N. S.; Fitzgerald, W. F. *Anal. Chim. Acta* **1988**, *208*, 151.
15. Snell, J. P.; Stewart, I. I.; Sturgeon, R. E.; Frech, W. *J. Anal. At. Spectrom.* **2000**, *15*, 1540.
16. Longerich, H. P. *Atom. Spectrosc.* **1989**, *10*, 112.
17. Rosman, K. J. R.; Taylor, P. D. P. *International Union of Pure and Applied Chemistry Isotopic Composition of the Elements* **1997**.

Chapter 22

Selenium Speciation in Soils and Plants

Patricia M. Fox, Danika L. LeDuc, Hussein Hussein,
Zhi-qing Lin, and Norman Terry[*]

Plant and Microbial Biology, University of California, Berkeley, CA 94720

With respect to living organisms, selenium has a dual character. At low concentrations, it may be an essential nutrient. When present at higher concentrations in the environment, however, it may be toxic. Whether selenium is a nutrient or a toxicant depends not only on its concentration but also on its chemical form. This fact has spurred great interest in its biogeochemistry and environmental cycling. Selenium may exist as both organic and inorganic species with different bioavailabilities and toxicities. Selenium speciation is strongly affected by the oxidation state and pH of the soil (or water) environment. It is also dramatically affected by transformations mediated by living organisms, especially plants and microbes. Recent technological advances such as the use of X-ray absorption spectroscopy and HPLC-ICP-MS have been particularly useful in speeding the study of selenium speciation. Here we review the research contributions from many fields including soil science, ecology, plant physiology and biochemistry, and microbiology, to provide a more complete picture of the speciation and transformation of selenium in soils and plants.

In recent years, much attention has been paid to the biogeochemistry and environmental cycling of the trace element selenium (Se). There is a narrow range between Se's action as a toxicant and as a nutrient to humans and animals. Selenium is also a toxicant and possibly a nutrient to plants. Selenium levels exceeding 2 mg kg^{-1} are found in many seleniferous soils throughout the western U.S., Ireland, Australia, Israel, and other countries (*1*). These soils are derived from marine parent materials containing high levels of Se. Selenium contamination may also result from human activities such as oil refining, mining, and fossil fuel combustion. The toxicity of Se is dependent not only on overall levels of Se, but also on the chemical form, or speciation. Thus, much research has been performed in the past few years focusing on the speciation of Se in the environment. The purpose of this review is to summarize the recent findings on Se speciation, with a particular emphasis on reactions and processes important in the field. In nature, selenium can be found in oxidation states of –II, 0, IV, and VI. A variety of chemical forms exist at these oxidation states; for instance, Se(VI) is usually present as the oxyanion selenate (SeO_4^{2-}), Se(IV) is present as selenite ($HSeO_3^-$, SeO_3^{2-}), and Se(0) is present as a solid, including both red monoclinic and gray hexagonal forms (*2,3*). Se(-II) is found in a variety of organic compounds, including selenoamino acids and volatile Se forms such as dimethylselenide (DMSe) and dimethyldiselenide (DMDSe). Inorganic selenides, such as hydrogen selenide and metal selenides, are also possible.

Selenium Speciation in Soil

Techniques for Se Speciation in Soil

In soils, the primary Se forms include selenate, selenite, elemental Se, and organic Se species. Traditionally, the speciation of Se in soils has been determined using a sequence of chemical extractants, which target specific pools or phases of an element in soil (*4,5*). Such pools include soluble, specifically adsorbed, organic matter associated, and oxide bound, among others. More recently, researchers have combined sequential extraction schemes with Se speciation by hydride-generation atomic absorption spectrometry (HGAAS) to achieve a greater understanding of Se speciation rather than just association (*6,7*). Because HGAAS only detects selenite in solution, selective oxidation and reduction of water samples or soil extracts allows one to distinguish between selenate, selenite, and Se(-II) compounds at levels down to 1 µg L^{-1}. These procedures are relatively easy to perform using commonly available equipment.

However, due to the indirect nature of the methods, pools of Se identified may be operationally defined, particularly because chemical extraction may alter the speciation of Se in the soil.

A recently exploited alternative to conventional extraction procedures is x-ray absorption spectroscopy (XAS) (2,3,8,9). X-ray absorption near-edge structure (XANES), with edge-fitting from model compounds, can accurately and quantitatively determine the oxidation states of Se present in soils (2). With this technique, the sample of interest is examined directly, without any extractions or treatments that may alter its composition or speciation. However, XAS facilities are not widely available, require specialized knowledge, and require higher concentrations (10-100 mg kg^{-1}) for accurate analysis, so traditional extraction techniques are still quite useful.

Comparative Solubility and Sorption of Different Se Species

Selenium may be incorporated into the solid phase of soil through either precipitation, co-precipitation, adsorption onto soil surfaces, or absorption into minerals or organic matter. Elrashidi et al. (10) reviewed the solubilities of a wide variety of Se mineral and precipitate forms. Because of their high solubility, metal-selenates and metal-selenites are unlikely to persist in soils, particularly at alkaline pH (10). By contrast, metal-selenides and elemental Se are quite insoluble in soils under reducing conditions (10).

Due to the high solubility of selenate and selenite precipitates and minerals, the solubility of Se in aerobic environments is more likely to be controlled by adsorption and complexation processes. While selenite is strongly adsorbed to soil surfaces, selenate is much more weakly adsorbed (11-13). Studies on Se adsorption onto soils have indicated that the most important soil components are Fe- and Al-oxides and oxyhydroxides, Mn-oxides, clays, organic matter, and carbonates (11,13-15). The levels of Se adsorption by Fe- and Mn-oxides are similar, with Al-oxides adsorbing slightly lower levels (16-18). Kaolinite adsorbs greater levels of selenite than montmorillonite below pH 6, even when the minerals have similar surface areas (13,19). Selenite adsorption is sometimes correlated with organic carbon in soils (14,15). However, negatively charged organic substances may also block adsorption sites, preventing Se adsorption (20,21).

The adsorption of both selenite and selenate depends on a number of soil characteristics. Adsorption is highly pH dependent, with greater sorption occurring at lower pH on soils (11-13) and soil minerals such as Fe-oxides (17,22-24), Al-oxides (25-27), Mn-oxides (17,18,28), and clays (13,19). Unlike these soil minerals, selenite sorption to calcite and hydroxyapatite is low in the acidic region and peaks at pH 8 (13,29). Studies on Mn- and Fe-oxides have demonstrated that the magnitude of selenite adsorption increases with increasing mineral surface area (*18,22,28*). In addition to pH and mineral characteristics, Se adsorption is affected by the presence of other anions in soil solution which may compete for adsorption sites. Competition of many anions with selenate is a function of ionic strength (*23,25,27*). Phosphate, molybdate, silicate, and citrate most effectively compete with selenite for adsorption sites, whereas other anions decrease adsorption to a much lesser extent (*17,24,30,31*). By contrast, certain anions such as carbonate, formate, and acetate promote selenate adsorption onto aluminum oxide, and $CaCl_2$ increases selenite adsorption to soil (*25,31*).

Distribution of Se Species in Soils and Sediments

In both aerobic and anaerobic soils, Se is present mostly as Se(0) (26-66 %) and Se(-II) (27-60 %), with only low amounts present as selenate (1-11 %) and selenite (1-16 %) (*7,32,33*). However, in some soils selenite may be more important; Martens and Suarez (*6*) reported that selenite comprised 68% of the total Se in a soil from Sumner Peck Ranch in Fresno, CA. In anaerobic environments such as sediments, Se may be present in much larger quantities (e.g., 83 mg kg^{-1}) than in aerobic soils (e.g., 1.0 mg kg^{-1}) (*32*).

The most important factors affecting the speciation of Se in soil are oxidation-reduction status, pH, and microbial communities. Redox potential has long been known to control the speciation of Se in soils and solution. Under oxidizing conditions (>300 mV) selenate is the most dominant species, whereas under moderately reducing conditions (0-200 mV) selenite becomes the dominant species and under strongly reducing conditions (<0 mV), selenides dominate (*10,34,35*). The oxidation and reduction of Se is also controlled by pH, with reduction favored at lower pH (*10*). Several researchers have shown that accumulation of Se occurs under reducing conditions, such as in ponded sediments, through the reduction of selenate to selenite and further reduction to Se(0) and Se(-II) (*3,9,34,35*). Reduced Se species in solution will be rapidly oxidized to selenite and selenate in aerobic environments, but oxidation of solid phase adsorbed or precipitated Se occurs more slowly (*10,35*).

As discussed above, Se(-II) compounds may represent either inorganic or organic selenides. The formation of numerous metal selenides under strongly reducing conditions is thermodynamically possible (*10*) but has not been directly observed. The organic Se(-II) compounds selenomethionine (SeMet) and DMSe have been identified in soil organic matter and soil solutions, respectively (*34,36,37*). However, these low molecular weight organic Se compounds do not persist in soils (*38,39*), and their importance is unclear when compared to the bulk of the organic selenides, which may be tied up in complex organic structures.

Microbially Mediated Transformation of Se Species

Microbial populations can transform soil Se via oxidation, reduction, and volatilization, through both assimilatory and dissimilatory mechanisms (*40-44*). Microbially mediated Se transformations are affected by a variety of factors influencing the growth of microorganisms, including the presence of nutrients, temperature, moisture, O_2 availability, and pH (*45*).

Several bacterial strains have been isolated from Se-contaminated environments that use selenate as an electron acceptor for growth, including *Thauerea selenatis, Sulfurospirillum barnesii,* and *Enterobacter cloacae* (*46-49*). *Thauerea selenatis* reduces SeO_4^{2-} to SeO_3^{2-} via selenate reductase and SeO_3^{2-} to Se(0) via nitrate reductase (*50*). Blum et al. (*51*) isolated the anaerobic bacteria, *Bacillus selenitireducens* and *Bacillus arsenicoselenatis* from the anoxic mud of Mono Lake, CA. *Bacillus arsenicoselenatis* grew by dissimilatory reduction of Se(VI) to Se(IV), while *Bacillus selenitireducens* grew by dissimilatory reduction of Se(IV) to Se(0) (*51*). When the two strains were cocultured, a complete reduction of Se(VI) to Se(0) was achieved.

Various groups of bacteria and fungi can transform Se species, including SeO_4^{2-}, SeO_3^{2-}, S(0) and various organoselenium compounds, into volatile forms by assimilatory reduction of extracellular Se to organoselenide compounds (*52*). The production of volatile Se compounds, such as DMSe and DMDSe is thought to be a protective mechanism used by microorganisms to avoid Se toxicity in contaminated ecosystems. Microorganisms are also involved in the oxidation of reduced Se (Se(-II,0)) to selenite and selenate (*53,54*). However, information on these processes is limited.

Possible Abiotic Se Transformations

While most environmentally important Se transformations occurring in soil are biologically mediated, Se(VI) reduction to Se(IV) and Se(0) can also occur abiotically (*55,56*). Green rust, an Fe(II,III) oxide common in hydromorphic soils, reduces Se(VI) to elemental Se at rates comparable to those observed for microbially mediated reduction (*56*). Although limited information is available, abiotic Se oxidation may also be possible. Wright (*57*) demonstrated theoretically that Se(-II,0) may be oxidized to selenite and selenate by nitrate and provided evidence for this process using batch experiments on a seleniferous shale.

Selenium Speciation in Plants

Selenium Speciation Techniques in Plants

Detection and identification of Se-containing compounds in plants, following either acid or enzymatic digestion, requires the following three steps: 1) separation of compounds, 2) element specification, and 3) structure confirmation. The separation of compounds is routinely accomplished with high performance liquid chromatography (HPLC) methods. Ion pair, ion exchange, and reverse phase chromatography methods have all been used to separate Se containing species (58-65). Separation methods must provide high resolution, quantification of low (pg) concentrations, and reproducibility of retention times. Differentiating between Se and S is the next crucial step in the identification of Se compounds because the S analogues of most Se compounds are present in the plant, often in much higher concentrations. Inductively coupled plasma emission spectroscopy (ICP) and electrothermal atomic absorbtion spectrometry (ETAAS) are two straightforward methods used to confirm the presence of Se in the HPLC fractions (58-61). Often, though, these methods detect unidentifiable Se containing compounds. Structural information is then required to make a definitive identification. Electrospray ionization mass spectrometry (ESI-MS) provides this necessary information (62). ESI-MS data give the mass of the compound, the masses of fragments dissociated through collision, and the protonation state of the Se atom. Further, Se's characteristic isotope pattern is readily observed. With this information a structural model can be proposed and confirmed with the S analog. This method suffers from a loss of sensitivity caused by matrix elements, having a detection limit on the order of 100-5000 ng/mL. The technique remains a viable option, however, as it requires only 5 μL, which can be achieved through purification and preconcentration of HPLC fractions. The lower detection limit (0.1-50 ng/mL) and linearity of ICP-MS makes it an attractive alternative to ESI-MS, especially when quantitating known compounds present in low amounts (58-61). Gas chromatography techniques are particularly suited to the separation of selenoxides and volatile Se compounds produced by plants, such as DMSe, DMDSe, and methaneselenol. Again, this method is more effective when used in combination either with mass spectrometry (GC-MS) to yield the mass of the resolved compounds or with an atomic emission detector (GC-AED) to distinguish the Se and S compounds (66,67).

Uptake, Transport, and Assimilation of Se Species in Plants

Although there is no conclusive evidence that plants require Se, they are nevertheless able to take up, transport, and metabolize various Se species to organic forms (68,69). Selenium is chemically similar to sulfur, and the mechanisms responsible for uptake, transport, and metabolism closely parallel those for sulfur (70-72).

Selenium uptake, accumulation, and metabolism vary among different plant species. Plants can be classified by their ability to tolerate and accumulate Se. Primary hyperaccumulators, including *Oonopsis condensata*, *Stanleya pinnata*, and some species of the genus *Astragalus*, are found only on highly seleniferous soil where they accumulate 1000's of mg kg^{-1} of Se, mostly in organic forms (73). These plants have evolved intriguing biochemical mechanisms not only to tolerate Se, but to accumulate it to concentrations higher than in soil. Secondary accumulators have Se concentrations of 100's of mg kg^{-1} and can be found on soils with a wide range of Se concentrations. Members of this group include the large family *Brassicaceae* and *Allium* species, such as garlic and onions (74). These species are able to volatilize Se and often accumulate the same organic forms found in the hyperaccumulators, only to a lesser extent. The growth of non-accumulators, however, is inhibited by the presence of Se in the soil. Most common crop and forage plants fall under this category.

Uptake and Transport of Se Species

The rate of Se uptake is dependent on the external Se concentration, as well as the chemical species present. In aerated soils, SeO_4^{2-} is the predominant form of soluble Se available to plants. Selenate is actively transported from the soil into plant roots by sulfate permeases (75). Accumulation of SeO_4^{2-} has been shown to increase linearly with external SeO_4^{2-} concentration up to 200 µM (76). Sulfate/selenate permeases responsible for the transport of these species over the cellular membrane and between subcellular compartments have now been identified and cloned from a number of plant species (77-79). The presence of a SeO_4^{2-} specific transporter in Se hyperaccumulators has been suggested based on the high Se:S ratio found in their tissues, but as of yet, none has been definitively identified (74). Uptake of selenite in plants is believed to be passive (80,81). Unlike SeO_4^{2-}, which can be accumulated against a concentration gradient, the rate of selenite uptake and maximum accumulation are bounded by the external concentration. For example, tomato roots supplied with SeO_4^{2-} accumulated Se to concentrations 6 to 13 times the external concentration, whereas those treated with selenite accumulated Se to concentrations lower than the external supply (80). Organic forms of Se, such as SeMet have been shown to be actively taken up at higher rates than either SeO_3^{2-} or SeO_4^{2-}, most likely by a metabolically driven process (82,83), but normally constitute very little of the total soluble Se in the soil.

Transport and localization of Se also depend on the chemical form supplied. *Brassica juncea* plants supplied with SeO_4^{2-} accumulate mostly SeO_4^{2-} in their shoots (76), while plants supplied with SeO_3^{2-} accumulate organic Se in their roots (84). Plants supplied with SeO_4^{2-} accumulate 10 times as much Se in their shoots as those supplied with the same concentration of selenite, even though the concentration in the roots is similar (84). This indicates both that SeO_4^{2-} uptake is much greater than that of selenite and that SeO_4^{2-} is translocated to the shoot whereas SeO_3^{2-} remains in the roots where it is converted to organic forms. This same phenomenon had been previously noted in bean and tomato plants (80,81).

Selenium can also be assimilated from the atmosphere, from volatile compounds or Se adsorbed to airborne particulates (85). This method of absorption can be a major source of Se uptake by the plant, especially when grown on soils with low Se concentrations (86). The translocation of leaf-adsorbed Se was recently studied in a greenhouse experiment with soybean leaves exposed to ^{75}Se from aerated cellulose particles (87). Selenium from the atmosphere was accumulated mainly in the leaves and then transported to the seeds; only small amounts were found in the roots and even less in the soil.

Environmental Influences on Uptake and Transport

High soil sulfate concentration is perhaps the most direct inhibitor of Se uptake by plants, presumably because it competes with SeO_4^{2-} for the same transporter (88). In broccoli, there was a 90% decrease in SeO_4^{2-} uptake when the sulfur concentration was increased from 0.25 to 10 mM, whereas uptake of SeO_3^{2-} and SeMet were decreased only 33% and 15.3% respectively (76). Fly-ash and gypsum, SO_4^{2-} containing compounds used to decrease soil acidity, also decrease the amount of Se accumulated by plants (89,90). The extent of this effect is dependent on plant species as well as timing of applications and harvests. Hyperaccumulators are able to preferentially accumulate Se even in the presence of high SO_4^{2-} concentrations, providing further evidence for a Se specific transporter (91). Sulfur starvation conditions trigger increased expression of the SO_4/SeO_4 transporter and enzymes involved in S/Se metabolism, leading to increased SeO_4^{2-} uptake in a number of plants (92,93).

Other environmental conditions such as pH and water content of soils can affect Se uptake by changing the solubility and/or availability of Se in the soil. For instance, lowering soil pH has been shown to reduce Se accumulation in soybean and tomato plants (94). Chemicals other than sulfur can also decrease Se uptake, although an exhaustive list of these compounds is not available.

Their effects are most likely due to effects on plant physiology as well as modifying soil Se availability. For example, the presence of orthophosphate and nitrate resulted in decreased levels of Se accumulated in tissues of forage crops, rape plants, and excised barley roots, although the mechanism of inhibition remains unknown (95-99). Other common soil ingredients, such as NaCl, however, have no competitive effect on Se uptake but could decrease biomass depending on species-specific salinity tolerance (100). Water stress increases the rate of Se uptake in tall fescue (101). However, plant growth is decreased under water stress, resulting in a decrease in total Se accumulated. Plants grown on low pH soil or under conditions of nutrient starvation, other than S, will also generally produce less biomass and, therefore, take up less total Se.

Role of Plant-Microbe Interactions on Uptake and Transport

Rhizosphere bacteria have been shown to promote the accumulation of Se in plant tissues. Rhizosphere bacteria enhanced Se accumulation and volatilization by *Brassica juncea*, whereas the application of ampicillin led to the inhibition of tissue Se accumulation by about 70% and Se volatilization by about 35% (102). In a similar manner, axenic saltmarsh bulrush (*Scirpus robustus* Pursh) plants supplied with different rhizosphere bacteria accumulated 70-80% higher Se concentrations in their roots and 40-60% higher Se concentrations in shoots than plants grown under axenic conditions (103). There are several possible mechanisms by which bacteria could enhance Se accumulation in plant tissues. Bacteria may stimulate the plants to produce compounds which enhance Se accumulation (102). For example, a root exudate isolated from plants inoculated with bacteria enhanced Se accumulation in *Brassica juncea* (102). Rhizosphere bacteria could also increase root surface area, thereby increasing element uptake (102,104). Another possibility is that bacteria may also transform Se to forms more readily taken up by roots such as SeMet (76,102).

Plant-Based Transformations of Se Species

Selenium is most likely metabolized via the S assimilation pathway (71,72). The S assimilation pathway reduces inorganic forms of S to the sulfur amino acids, cysteine and methionine. The detection of the Se analogs, selenocysteine (SeCys) and SeMet, was an early hint that Se and S may be metabolized via the same enzymatic pathway (105,106). Later experiments demonstrated that sulfur and selenium analogs compete for the same enzymatic steps (107-109).

A fairly complete model of Se transport and metabolism in plants has been developed. As mentioned earlier, SeO_4^{2-} is first removed from the soil by sulfate permeases present in the roots. It is then quickly transported via the xylem to the shoot without chemical modification (*80*). Most enzymes along the S/Se assimilation pathway are localized in the chloroplasts. Indeed, 77-100% of the enzymatic activity leading to the biosynthesis of SeCys is contained in the chloroplasts (*107*). Here, SeO_4^{2-} is slowly reduced to SeO_3^{2-} by the enzyme ATP sulfurylase (*110,111*). The reduction of SeO_4^{2-} to SeO_3^{2-} is the primary barrier in converting inorganic forms of Se to organic species in many plants (*84*).

Selenite is sequentially reduced to SeCys, the branching point for further Se metabolism in the plant. The conversion of SeO_3^{2-} to organic Se occurs quickly and SeO_3^{2-} is not accumulated in plants (*84*). The enzymes involved in these reduction steps have not yet been clearly identified, although candidates have been proposed based on analogy with other organisms (*107*). Although SeCys is extremely unstable and difficult to detect free in plants, many of its derivative products have been identified. SeCys can be incorporated into protein, metabolized to form Se analogs of organic S compounds such as glutathione, or further transformed along the sulfur assimilation pathway to SeMet. The production of SeMet from SeCys is catalyzed by cystathionine-γ-synthase, which produces Se-cystathionine from SeCys, cystathionine-β-lyase, which then cleaves Se-cystathionine into Se-homocysteine, and finally methionine synthase (*112*).

Selenium taken up by plants can be further assimilated into volatile forms, such as DMSe or DMDSe (*113*). Rates of volatilization are dependent upon the Se species supplied. It is energetically easier for plants to produce volatile Se from SeMet than from Se oxyanions because it is further downstream along the S assimilation pathway (*72*). For example, Se volatilization by *Brassica juncea* was 20 times faster from SeMet than from selenite and was 2 times faster from selenite than selenate (*92*). Hyperaccumulators have an additional volatilization pathway which proceeds from SeCys to DMDSe (*113,114*). The rate of production of volatile Se depends upon a number of factors. Different plant species have different abilities to metabolize and volatilize Se (*115,116*). A recent field study showed that pickleweed (*Salicornia bigelovii*) plants produced volatile Se at rates 10-fold higher than other plant species, including saltgrass (*Distichlis spicata* L.), saltbush (*Atriplex nummularia* L.), and cordgrass (*Spartina gracilis* Trin.) (*115*). When pickleweed plants were supplied with selenate, a large fraction of the supplied Se was converted into organic forms, indicating that the formation of volatile Se from selenate in pickleweed plants was not limited by the reduction of selenate as it is in many other plants (*117*).

Selenium Toxicity

Selenium bioavailability and toxicity are critically dependent on chemical form. Although the mechanisms of Se toxicity remain unclear, plausible hypotheses fall in two general categories: replacement of sulfur compounds by their Se analogues or generation of free radical species. The replacement of the sulfur amino acids, Cys and Met, with their selenium analogues, SeCys and SeMet, in enzymes could detrimentally alter their structure and function. SeCys residues form longer and weaker bonds than Cys residues (107). Also, whereas Cys is mainly protonated at pH 7, the SeH group of SeCys is almost completely ionized (118), a difference that could be critical in enzymes catalyzing redox reactions and acid/base chemistry. Since a protein with all Met residues replaced with SeMet retained full activity, methionine's other roles, as an initiator of protein translation and as a methyl donor may be more crucial. Selenomethionyl tRNA has been shown to be a less effective substrate than methionyl tRNA (119). Discrimination at this level could decrease the general rate of protein production. This hypothesis is consistent with the observation that Se/S ratios of free SeMet/Met are higher than SeMet incorporated into proteins. Differences in strength between a $Se-CH_3$ bond and a $S-CH_3$ bond could also contribute to SeMet's toxic effects by changing methylation pathways in exposed cells (106). Alternatively, the ability of some selenium compounds, such as selenite and selenocystine, to generate free radicals and other reactive oxygen species could explain their toxicity (120).

Conclusion

Recent advances in techniques for Se speciation in plants and soils are leading to a new body of literature which provides both a better mechanistic understanding of the processes involved in Se cycling in the environment and the pools of Se available for uptake and assimilation by plants and animals. Due to the variety of oxidation states and chemical forms possible, the speciation of Se in the environment is fairly complex, and is dependent on a variety of factors. In soils, factors such as redox status, pH, and microbial transformations are important considerations, while in plants, metabolic processes, forms of Se supplied, and plant species differences largely control Se speciation.

References

1. Mayland, H.F.; James, L.F.; Panter, K.E.; Sonderegger, J.L. In *Selenium in Agriculture and the Environment*; Jacobs, L.W. Ed; SSSA: Madison, WI, 1989; pp. 15-51.,

2. Pickering, I.J.; Brown Jr., G.E.; Tokunaga, T.K. *Environ. Sci. Technol.* **1995**, *29*, 2456-2459.
3. Tokunaga, T.K.; Pickering, I.J.; Brown Jr., G.E. *Soil Sci. Soc. Am. J.* **1996**, *60*, 781-790.
4. Chao, T.T.; Sanzolone, R.F. *Soil Sci. Soc. Am. J.* **1989**, *53*, 385-392.
5. Lipton, D.L. Ph.D. dissertation, Univ. of Calif., Berkeley, CA, 1991.
6. Martens, D.A.; Suarez, D.L. *Environ. Sci. Technol.* **1997**, *31*, 133-139.
7. Zhang, Y.Q.; Moore, J.N. *Environ. Sci. Technol.* **1996**, *30*, 2613-2619.
8. Tokunaga, T.K.; Brown Jr., G.E.; Pickering, I.J.; Sutton, S. R.; Bajt, S. *Environ. Sci. Technol.* **1997**, *31*, 1419-1425.
9. Tokunaga, T.K.; Sutton, S.R.; Bajt, S.; Nuessle, P.; Shea-McCarthy, G. *Environ. Sci. Technol.* **1998**, *32*, 1092-1098.
10. Elrashidi, M.A.; Adriano, D.C.; Workman, S.M.; Lindsay, W.L. *Soil Science* **1987**, *144*, 141-152.
11. Neal, R.H.; Sposito, G.; Holtzclaw, K.M.; Traina, S.J. *Soil Sci. Soc. Am. J.* **1987**, *51*, 1161-1165.
12. Neal, R.H; Sposito, G. *Soil Sci. Soc. Am. J.* **1989**, *53*, 70-74.
13. Goldberg, S.; Glaubig, R.A. *Soil Sci. Soc. Am. J.* **1988**, *52*, 954-958.
14. Dhillon, K.S.; Dhillon, S.K. *Geoderma* **1999**, *93*, 19-31.
15. Christensen, B.T.; Bertelsen, F.; Gissel-Nielsen, G.; *J. Soil Sci.* **1989**, *40*, 641-647.
16. Wijnja, H.; Schulthess, C.P. *J. Colloid Interface Sci.* **2000**, *229*, 286-297.
17. Balistrieri, L.S.; Chao, T.T. *Geochim. Cosmochim. Acta* **1990**, *54*, 739-751.
18. Saeki, K.; Matsumoto, S.; Tatsukawa, R. *Soil Science* **1995**, *160*, 265-272.
19. Bar-Yosef, B.; Meek, D. *Soil Science* **1987**, *144*, 11-19.
20. Saeki, K.; Matsumoto, S. *Commun. Soil Sci. Plant Anal.* **1994**, *25*, 3379-91.
21. Masset, S.; Monteil-Rivera, F.; Dupont, L.; Dumonceau, J.; Aplincourt, M. *Agronomie* **2000**, *20*, 525-535.
22. Parida, K.M.; Gorai, B.; Das, N.N.; Rao, S.B. *J. Colloid Interface Sci.* **1997**, *185*, 355-362.
23. Su, C.; Suarez, D.L. *Soil Sci. Soc. Am. J.* **2000**, *64*, 101-111.
24. Balistrieri, L.S.; Chao, T.T. *Soil Sci. Soc. Am J.* **1987**, *51*, 1145-1151.
25. Wijnja, H.; Schulthess, C.P. *Soil Sci. Soc. Am J.* **2000**, *64*, 898-908.
26. Schulthess, C.P.; Hu, Z. *Soil Sci. Soc. Am. J.* **2001**, *65*, 710-718.
27. Wu, C.H.; Lo, S.L.; Lin, C.F. *Colloids Surf. A.* **2000**, *166*, 251-259.
28. Parida, K.M.; Gorai, B.; Das, N.N. *J. Colloid Interface Sci.* **1997**, *187*, 375-380.
29. Monteil-Rivera, F.; Fedoroff, M.; Jeanjean, J.; Minel, L.; Barthes, M.G.; Dumonceau, J. *J. Colloid Interface Sci.* **2000**, *221*, 291-300.

30. Dhillon, S.K.; Dhillon, K.S. *J. Plant Nutr. Soil Sci.* **2000**, *163*, 577-582.
31. Neal, R.H.; Sposito, G.; Holtzclaw, K.M.; Traina, S.J. *Soil Sci. Soc. Am. J.* **1987**, *51*, 1165-1169.
32. Martens, D.A.; Suarez, D.L. *J. Environ. Qual.* **1997**, 26, 424-432.
33. Gao S.; Tanji, K.K.; Peters, D.W.; Herbel, M.J. *J. Environ. Qual.* **2000**, *29*, 1275-1283.
34. Masscheleyn, P.H.; Delaune, R.D.; Patrick Jr., W.H. *J. Environ. Qual.* **1991**, *20*, 522-527.
35. Jayaweera, G.R.; Biggar, J.W. *Soil Sci. Soc. Am. J.* **1996**, *60*, 1056-1063.
36. Abrams, M.M.; Burau, R.G.; and Zasoski, R.J. *Soil Sci. Soc. Am. J.* **1990**, *54*, 979-982.
37. Abrams, M.M.; Burau, R.G. *Commun. Soil Sci. Plant Anal.* **1989**, *20*, 221-237.
38. Martens, D.A.; Suarez, D.L. *Soil Sci. Soc. Am. J.* **1997**, *61*, 1685-1694.
39. Martens, D.A.; Suarez, D.L. *Soil Biol. Biochem.* **1999**, *31*, 1355-1361.
40. Ehrlich, H.L. *Geomicrobiology;* Marcel Dekker, Inc.: New York, NY, 1996.
41. Lucas, F.S.; Hollibaugh, J.T. *Environ. Sci. Technol.* **2001**, *35*, 528-534.
42. Losi, M.E.; Frankenberger, Jr., W.T. *J. Environ. Qual.* **1998**, *27*, 836-843.
43. Oremland, R.S.; Hollibaugh, J.T.; Maest, A.S.; Presser, T.S.; Miller, L.G.; Culbertson, C.W. *Appl. Environ. Microbiol.* **1989**, *55*, 2333-2343.
44. White, C.; Wilkinson, S.C.; Gadd, G.M. *Intl. Biodeterioration Biodegradation* **1995**, *35*, 17-40.
45. Frankenberger Jr., W.T.; Karlson, U. In *Selenium in the Environment;* Frankenberger Jr., W.T.; Benson, S., Eds.; Marcel Dekker, Inc.: New York, NY, 1994; pp 369-387.
46. Macy, J.M.; Michel, T.A.; Kirsch, D.A. *FEMS Microbiol.* **1989**, *61*, 3556-3561.
47. Macy, J.M.; Rech, S.; Auling, G.; Dorsch, M.; Stackenbrandt, E.; Sly, L.I. *Int. J. Syst. Bacteriol.* **1993**, *43*, 135-142.
48. Oremland, R.S.; Switzer-Blum, J.; Culbertson, C.W.; Visscher, P.T.; Miller, L.G.; Dowdle, P.; Strohmaier, F.E. *Appl. Environ. Microbiol.* **1994**, *60*, 3011-3019.
49. Losi, M.E.; Frankenberger, Jr., W.T. *Appl. Environ. Microbiol.* **1997**, *63*, 3079-3084.
50. Macy, J.M. In *Selenium in the Environment*; Frankenberger Jr., W.T.; Benson, S., Eds.; Marcel Dekker, Inc.: New York, NY, 1994; pp. 421-444.
51. Blum, J.S.; Bindi, A.B.; Buzzelli, J.; Stolz, J.F.; Oremland, R.S. *Arch. Microbiol.* **1998**, *171*, 19-30.
52. Karlson, U.; Frankenberger Jr., W.T. *Soil Sci. Soc. Am. J.* **1988**,*52*,678-81.
53. Dowdle, P.R.; Oremland, R.S. *Environ. Sci. Technol.* **1998**, *32*, 3749-3755.

352

54. Losi, M.E.; Frankenberger Jr., W.T. *J. Environ. Qual.* **1998,** *27,* 836-843.
55. Refait, P.; Simon, L.; Génin, J.M.R. *Environ. Sci. Technol.* **2000,** *34,* 819-825.
56. Myneni, S.C.B.; Tokunaga, T.K.; Brown Jr., G.E. *Science* **1997,** *278,* 1106-1109.
57. Wright, W.G. *J. Environ. Qual.* **1999,** *28,* 1182-1187.
58. Thomas, C.; Jakubowski, N.; Stuewer, D.; Klockow, D.; Emons, H. J. *Anal. Atom. Spectrom.* **1998,** *13,* 1221-1226.
59. Bird, S.M.; Uden, P.C.; Tyson, J.F.; Block, E.; Denoyer, E. J. *Anal. Atom. Spectrom.* **1997,** *12,* 785-788.
60. Ponce de León, C.A.; Sutton, K.L.; Caruso, J.A.; Uden, P.C. J. *Anal. Atom. Spectrom.* **2000,** *14,*1103-1107.
61. Montes Bayón, M.; B'Hymer, C.; Ponce de León, C.A.; Caruso, J.A. *J. Anal. Atom. Spectrom.* **2001,** *16,* 492-497.
62. Casiot, C.; Vacchina, V.; Chassaigne, H.; Szpunar, J.; Potin-Gautier, M.; Lobiński, R. *Anal. Commun.* **1999,** *36,* 77-80.
63. Janák, J.; Billiet, H.A.H.; Frank, J.; Luyben, K.; Hušek, P. *J. Chromatography A* **1994,** *677,* 192-196.
64. Kotrebai, M.; Birringer, M.; Tyson, J.F.; Block, E.; Uden, P.C. *Anal. Commun.* **1999,** *36,* 249-252.
65. Kotrebai, M.; Birringer, M.; Tyson, J.F.; Block, E.; Uden, P.C. *Analyst* **2000,** *125,* 71-78.
66. Uden, P.C.; Bird, S.M.; Kotrebai, M.; Nolibos, P.; Tyson, J.F.; Block, E.; Denoyer, E. *Fresenius J. Anal. Chem.* **1998,** *362,* 447-456.
67. Vásquez Peláez, M.; Montes Bayón, M.; Garcia Alonso, J.I.; Sanz-Medel, A. *J. Anal. Atom. Spectrom.* **2000,** *15,* 1217-1222.
68. Schwarz, K.; Foltz, C.M. *J. Am. Chem. Soc.* **1957,**_79_,3292-3293.
69. Rotruck, J.T.; Pope, A.L.; Ganther, H.E.; Swanson, A.B.; Hafeman, D.G.; Hoehstra, W.G. *Science* **1973,**_179_,588-90.
70. Anderson, J.W.; Scarf, A.R. In *Metals and Micronutrients: Uptake and Utilization by Plants*; Robb, D.A.; Pierpoint, W.S., Eds.; Academic Press: London, 1983; pp 241-275.
71. Lauchli, A. *Bot. Acta* **1993,** *106,*455-468.
72. Terry, N.; Zayed, A.M.; de Souza, M.P.; Tarun, A.S. In *Annu. Rev. Plant Physiol. Plant Mol. Biol.* Jones, R.L.; Bohnert, H.J.; Delmer, D.P., Eds.; Palo Alto, CA, 2000; pp 401-432.
73. Trelease, S.F.; Beath, O.A. *Selenium;* New York, NY, 1949.
74. Rosenfield, I.; Beath, O.A. *Selenium, Geobotany, Biochemistry, Toxicity, and Nutrition;* Academic Press: New York, NY, 1964.
75. Ulrich, J.M.; Shrift, A. *Plant Physiol.* **1968,**_43_,14-19.
76. Zayed, A.; Lytle, C.M.; Terry, N. *Planta* **1998,**_206_,284-292.
77. Hawkesford, M.J.; Davidian, J.-C.; Grignon; C. *Planta* **1993,**_190_,297-304.

78. Smith, F.W.; Ealing, P.M.; Hawkesford, M.J.; Clarkson, D.T. *Proc. Natl. Acad. Sci. USA* **1995**, *92*, 9373-9377.
79. Saito, K. *Current Opinion in Plant Biology* **2000**, 188-195.
80. Asher, C.J.; Evans, C.S.; Johnson, C.M. *Aust. J. Biol. Sci.* **1967**,*20*,737-48.
81. Arvy, M.P. *J. Exp. Bot.* **1993**, *44*, 1083-1987.
82. Abrams, M.M.; Shennan, C.; Zasoski, R.; Burau, R.G. *Agron. J.* **1990**, 1127-1130.
83. Williams, M.C.; Mayland, H.F. *J. Range Manage.* **1992**, *45*, 374-378.
84. deSouza, M.P.; Pilon-Smits, E.A.H.; Lytle, C.M.; Hwang, S.; Tai, J.C.; Honma, T.S.U.; Yeh, L.; Terry, N. *Plant Physiol.* **1998**, *117*, 1487-1494.
85. Zieve, R.; Peterson, P.J. *Planta* **1984a.**,*160*, 180-184.
86. Zieve, R; Peterson, P.J. *Trace Sub. in Environ. Health* **1984b**, *18*, 262-267.
87. Shinonaga, T.; Ambe, S. *Water, Air, and Soil Pollution* **1998**, *101*, 93-103.
88. Barak, P.; Goldman, I.L. *J. Agric. Food Chem.* **1997**, *45*,1290-4.
89. Matzi, T.; Keramidas, V.Z. *Environ. Pollution* **1999**, *104*,107-112.
90. Woodbury, P.B.;Arthur, M.A.; Rubin, G.; Weinstein, L.H.; McCune, D.C. *Water, Air, and Soil Pollution* **1999**, *110*, 421-432.
91. Bell, P.F.; Parker, D.R.; Page, A.L. *Soil Sci. Soc. Am. J.* **1992**,*56*,1818-24.
92. Zayed, A.M.; Terry, N.J. *Plant Physiol.* **1992**,*140*,646-652.
93. Bolchi, A.; Petrucco, S.; Tenca, P.L.; Foroni, C.; Ottonello, S. *Plant Mol. Biol.* **1999,** *39*,527-537.
94. Wang, H.F.; Takematsu, N.; Ambe, S. *Appl. Radiation and Isotopes* **2000**, *52*, 803-811.
95. Raymond, W.F. *Adv. Agron.* **1967**, *21*, 1.
96. Ali, M.B. *Nucleus* **1970**, *7*,126.
97. Gissel-Nielsen, G. *Z. Pflanzenernaehr Bodenkd.* **1975**, *138*, 97.
98. El Kobia, T. *Agronchimica* **1975**, *19*, 5.
99. Mikkelsen, R.L.; Page, A.L.; Bingham, F.T. In *Selenium in Agriculture and the Environment;* Jacobs, L.W. Ed.; SSSA Spec. Pub. 23, Madison, WI 1989; p. 65.
100. Bañuelos, G.S.; Zayed, A.; Terry, N.; Wu, L.; Akohoue, S.; Zambrzuski, S. *Plant and Soil* **1996**, *202*, 138-146.
101. Tenant, T.; Wu, L. *Arch. Environ. Contam. Toxicol.* **2000**, *38*, 32-29.
102. de Souza, M.P.; Chu, D.; Zhau, M.; Zayed, A.; Ruzin, S.E.; Schicnes, D.; Terry, N. *Plant Physiol.* **1999,** *119*, 565-573.
103. de Souza, M.P.; Huang, C.P.A.; Chee, N.; Terry, N. *Planta* **1999**, *209*, 259-263.
104. Kapulink, Y. In *Plant Roots: The Hidden Half;* Waisel, Y.; Eshel, A.; Kafkazi, U., Eds. Marcel Dekker, Inc.: New York, NY, 1996; pp. 769-781.
105. Butler, G.W.; Peterson, P.J. *Aust. J. Biol. Sci.* **1967**, *20*, 77.
106. Brown, T.A.; Shrift, A. *Plant Physiol.* **1981**, *67*, 1051-3.
107. Ng, B.H.; Anderson, J.W. *Phytochemistry* **1979**, *18*,573-580.

108.Burnell, J.N. *Plant Physiol.* **1981**, *67*, 316-324.

109.Dawson, J.C.; Anderson, J.W. *Phytochemistry* **1989**, *28*, 51-55.

110.Shaw, W.H.; Anderson, J.W. *Biochem. J.* **1972**, *127*, 237-247.

111.Dilworth, G.L.; Bandurski, R.S. *Biochem. J.* **1977**, *163*, 521-529.

112.Hell, R. *Planta* **1997**, *202*, 138-146.

113.Lewis, B.G.; Johnson, C.M.; Broyer, T.C. *Plant and Soil* **1974**,*40*,107-118.

114.Evans, C.S.; Asher, C.J.; Johnson, C.M. *Aust. J. Biol. Sci.* **1968,** *21*, 13-20.

115.Terry, N.; Lin, Z.Q. *Managing high selenium in agricultural drainage water by agroforestry system: Role of selenium volatilization*, DWR B-80665, CA DWR: Sacramento, CA, 1999, 67 pp.

116.Pilon-Smits, E.A.H.; de Souza, M.P.; Hong, G.; Amini, A.; Bravo, R.C. *J. Environ. Qual.* **1999,** *28*, 1011-1018.

117.Lee, A.; Lin., Z.Q.; Pickering, I.J.; Terry, N. *Planta* **2001**, *213*, 977-980.

118.Huber, R.E.; Criddle, R.S.*Arch. Biochem. Biophys.* **1967**, *122*, 164-173.

119.Eustice, D.C.; Kull, F.J.; Shrift, A. *Plant Physiol.* **1981**, *67*, 1054-1058.

120.Spallholz, J.E. *Biomed Environ Sci.* **1997**, *10*, 260-270.

Chapter 23

Soil Methylation–Demethylation Pathways for Metabolism of Plant-Derived Selenoamino Acids

Dean A. Martens[1] and Donald L. Suarez[2]

[1]Southwest Watershed Research Center, Agricultural Research Service, U.S. Department of Agriculture, Tucson, AZ 85719
[2]U.S. Salinity Laboratory, Agricultural Research Service, U.S. Department of Agriculture, 450 Big Springs Road, Riverside, CA 92507

There is conflicting field information about Se toxicity in waterfowl and fish, based on criteria of total Se concentration. At least part of this uncertainty is due to the difference in toxicity associated with various Se species. There is toxicity data on the selenoamino acid, selenomethionine (SeMet) to avian species, but little is known on the environmental transformations of SeMet and the possible intermediates of organic Se decomposition. To determine the potential decomposition of Se amino acids, methylation and demethylation pathway intermediates for the transformations of sulfur (S) amino acids, identified from aerobic marine sediments were compared to potential analog Se intermediates synthesized for this study. Two terrestrial soils with apparently different pathways for metabolizing SeMet were treated with 25 μg S intermediate-S g^{-1} soil and the soil headspace analyzed for the methylation pathway gas dimethylsulfide (DMS) or the demethylation pathway gas dimethyldisulfide (DMDS). Addition of S-methyl-methionine (MMet), and dimethylsulfoniopropionic acid (DMSP) to the Panhill and Panoche soils resulted in only DMS evolution; addition of 3-methiopropionic acid (MTP) resulted in DMDS in the soils and 3-mercaptopropionic acid (MCP) addition was not volatized confirming that terrestrial soil S pathways are similar to documented marine pathways. The Panhill soil evolved only DMDS as a result of the methionine (Met) demethylation pathway and the Panoche soil evolved only DMS from the methylation of Met. The evolution of Se gases dimethylselenide (DMSe) and dimethyldiselenide (DMDSe) from addition of SeMet, methyl-selenomethionine (MSeMet), dimethylselenopropionic acid (DMSeP) followed the same pattern as noted with the S products. DMSe evolved from a methylation pathway and DMDSe evolved from a demethylation metabolism. Selenocystine (SeCys) and a methylated selenocysteine (MSeCys) added to the two soils showed limited volatilization as DMSe. A large portion of the Se not volatilized from soil was found as a non amino acid organic selenide compound(s) and these unidentified intermediate compounds may be present in significant concentrations in some environments. The different metabolic pathways of Se in soils may explain why in certain waterfowl areas Se-induced problems have not been found where predicted based on total Se concentrations.

The geologic setting and climate of the west-central San Joaquin Valley of California has resulted in soil salinization of land used for agriculture. Agricultural engineering solutions to decrease naturally occurring salinity have in turn created problems with respect to disposal of the highly saline irrigation drainage waters. A major problem with disposal of return waters from the west-central San Joaquin Valley of California is the inadvertent cycling and concentration of Se in evaporation pond sediments. As plants and organisms in the evaporation ponds assimilated the dominant inorganic Se forms, selenate (Se+6) and selenite (Se+4) present in the waters into organic Se forms, the death and decomposition of the biomass released organic forms of Se back into the environment. Although the Se uptake by plants is competitively inhibited by the presence of sulfate this appears not to be the case for aquatic organisms (1).

Selenium toxicity results from the alteration of the three-dimensional structure of proteins and the impairment of enzymatic function by substitution of Se for sulfur (S) in S-amino acids (2). Of the naturally occurring Se compounds tested for toxicity, the S amino acid analog SeMet was the most toxic to waterfowl (3). The formation of the toxic Se-amino acids in nature can be from plant or microbial sources, but the synthesis of SeMet by soil microorganisms has been reported to be low (4) and many plant species have been found to assimilate large amounts of inorganic Se into Se-amino acid analogs Met and cysteine (Cys)(5).

There is a similarity between the biogeochemistry of organic S compounds such as Met and cysteine and the Se analogs SeMet and SeCys (6). Both Met and SeMet decomposition in soil result in volatilization of the methylated compounds dimethylsulfide (DMS) and dimethylselenide (DMSe), respectively. The S cycle has received much attention because DMS accounts for nearly 50% of the global biogenic S entering the atmosphere and influences climate by promoting cloud formation (7). In the S cycle, Met is oxidized by several pathways resulting in DMS (methylation) or dimethyldisulfide (demethylation), promoted by both anaerobic and aerobic organisms (8). The vast majority of the S work has involved marine sediments with a very limited research effort occurring in terrestrial soils. Martens and Suarez (6) reported the occurrence of Se methylation and demethylation pathways in two terrestrial soils treated with Met and SeMet. The methylation pathway resulted in the majority of Se being evolved as DMSe and the demethylation pathway evolved DMDSe and resulted in the accumulation of an unidentified nonamino acid organic Se compound. They also detected MSeMet and DMSeP by hydroelimination analysis (9) as possible intermediates in the soil exhibiting the methylation pathway. Reamer and Zoller (10) speculated that different microorganisms are responsible for the formation of each volatile Se species and that shifts in volatile species composition may be due to the differing tolerances of microorganisms to environmental stress. Despite the intensive research efforts on Se cycling, little is known about the resulting speciation of Se with mineralization of organic Se compounds and no information is available for determining the importance of the different decomposition pathways for organic S and Se present in terrestrial soils.

Predicting Se-induced problems in waterfowl based only on Se concentration in the receiving waters and waterfowl have not proved to be a consistent for identifying areas likely to have Se problems (11). This study was conducted to determine the decomposition of S and Se pathway intermediates in soil and evolution of volatile S and Se species indicative of the metabolizing pathways. Speciation of nonvolatilized Se, following soil incubation with the Se pathways intermediates was also determined because the activity of methylation or demethylation pathways in soil may influence the accumulation of organic Se compounds.

Materials and Methods

Reagents and Standards

A description of the properties of the Panhill and Panoche soils used in this study was given by Martens and Suarez (6). The Met, SeMet, S-methyl-methionine (MMet), 3-mercaptopropionic acid (MPA), 3-bromopropionic acid and selenocystine (SeCys) were obtained from Sigma Chemical Company (St. Louis, MO); 3-methiolpropionic acid (MTP) was obtained by alkaline hydrolysis (6) of methyl-3-(methiol)propionic acid (Aldrich Chemical Company, Milwaukee, WI); dimethylsulfoniopropionic acid (DMSP) was obtained from Research Plus, Inc. (Bayonne, NJ) and dimethyl selenide (DMSe) was obtained from Strem Chemicals, Inc. (Newburyport, MA).

Synthesis of DMSeP, MSeMet, and Methylselenocysteine (MSeCys)

Dimethylselenopropionate was synthesized as proposed for synthesis of DMSP by Challenger and Simpson (12). Equimolar quantities (neat) of 3-bromopropionic acid and dimethyl selenide were refluxed together at 55°C for 6 h. The resulting gelatinous mass was thoroughly washed with successive 5 mL aliquots of ethyl ether to remove the unreacted substrates and then solubilized over night in absolute alcohol. A white precipitate remaining was separated from the ethanol, rinsed with ether, then alcohol and dried over sulfuric acid. This material was found to be the trimethylselenonium ion (TMSe$^+$) by C and Se content analysis. The remaining ethanol material was concentrated and washed with ethyl ether and separated from the formed white precipitate until no further white precipitation (TMSe$^+$) was noted with ether addition. The DMSeP was a viscous liquid, not a granular solid as reported for DMSP synthesis, dried over sulfuric acid and the stored at -25°C in a desiccator.

Methylselenomethionine was synthesized by the method proposed by Toennies and Kolb (13) for synthesis of MMet. Selenomethionine (0.1961 g) was mixed with 1.7 mL formic acid, 0.5 mL acetic acid, 0.3 mL deionized water, and 0.56 methyl iodide (Sigma Chemical Co.), and incubated in a Teflon stoppered round bottom flask for 5 days in the dark at ambient temperatures (22°C). The mixture was then reduced to one-fourth volume under reduced pressure and 10 mL of methanol was added. The resulting white precipitate was washed with methanol, dissolved in a minimum of 50% ethanol, and crystallized with addition of 50 mL ethanol. The MSeMet was dried over sulfuric acid, weighed and stored at -25°C in a desiccator.

The method of Foster and Ganther (*14*) was used to synthesize MSeCys. Selenocystine (20.3 mg) was dissolved in 2 mL of 0.1 $\underline{M}$ NaOH in a round bottom flask under a nitrogen purge and treated with 2.5 mg sodium borohydride and stirred 0.5 h to form selenocysteine. Three additions of 0.2 mL methyl iodide were added over one h with stirring and subsequent acidification to pH 1.5 with 6 $\underline{M}$ HCl, and the volume reduced under reduced pressure (37°C). The MSeCys was purified by ion-exchange chromatography on SP-Sephadex (Sigma Chemical Co.) by applying the sample in water to a 1.5 x 17.5 cm column equilibrated with 50 mL of 0.05$\underline{M}$ formic acid, pH 2.5. The column was then washed with 75 mL 0.05$\underline{M}$ ammonium formate, pH 4.0 and the MSeCys was eluted with 150 mL 0.1$\underline{M}$ ammonium formate. The sample was lyophilized at ambient temperatures to remove water and then at 30°C to remove the remaining buffer and the MSeCys was dried over sulfuric acid, weighed and stored in a desiccator at -25°C.

Compound identity and purity were confirmed by C and Se content analysis, electron impact and thermospray mass spectrometry, NMR and hydroelimination as described by Fan et al. (*15*).

Aerobic Soil S and Se Pathway

Organic S and Se mineralization experiments were conducted with 5 g of air-dry soil added to duplicate 125 mL screwtop Erlenmeyers equipped with a Mininert[TM] gas sampling valves and incubated for up to 7 d at −34 kPa moisture tension after addition of specified amount of organic S-S or Se-Se compound flask[-1]. Volatile Se and S evolution as DMSe, DMS, DMDS or DMDSe was determined by gas chromatography as described by Martens and Suarez (*16*) and measured daily. Selenium speciation following decomposition was determined by sequential extraction as outlined by Martens and Suarez (*16*). Briefly, DI water was used to wash the soil quantitatively into 40 mL Teflon[TM] centrifuge tubes, shaken and centrifuged to remove water-soluble Se compounds. The samples were then treated with a 0.1 $\underline{M}$ phosphate buffer (pH 7.0) and centrifuged to determine adsorbed Se, followed by a 0.1 $\underline{M}$ NaOH extraction for organic Se and tightly-held Se+4. The samples were then extracted with 17 $\underline{M}$ nitric acid for determination of elemental Se (Se°). Hydride generation atomic absorption spectrometry was conducted with a Perkin Elmer 3030B instrument under conditions given by Martens and Suarez (*16*).

Results and Discussion

Synthesis of DMSeP, MSeMet, and MSeCys

Challenger and Simpson (12) reported that the bromide salt of DMSP was a granular white solid, which was confirmed by synthesis in our laboratory. However using the same method for DMSeP resulted in formation of a white TMSe$^+$ precipitate and a very viscous DMSeP liquid. Elemental analysis of the final DMSeP product yielded values in good agreement with theoretical: C content 21.3 measured vs. 22.9% expected, Se content 29.1% measured vs. 30.1% expected. Both TMSe$^+$ and DMSeP were highly resistant to hydrolysis by concentrated HNO_3 (16 h; 130°C). Extended (3 h; 130°C) oxidization by H_2O_2 (30%) and potassium persulfate (0.1 $\underline{M}$) were necessary for Se analysis.

A theoretical yield of 75% S-methyl methionine was reported by Toennies and Kolb (13) for the outlined synthesis pathway. However we obtained a yield of only 10% for the synthesis of MSeMet using the published S method. The resulting white granular precipitate was resistant to HNO_3 oxidation, but was oxidized by the peroxide and persulfate treatment, which yielded 23.0% Se and 21.7% C, again in good agreement with the theoretical values (23.1% Se and 21.3% C).

A 90% recovery of MSeCys was determined using the method described by Foster and Ganther (14). The MSeCys was found to be very hydroscopic and sensitive to elevated temperatures, and following peroxide and persulfate oxidation resulted in 42.2 vs. 43.4% Se; C analysis found 23.9 vs. 26.4% C. The hydroscopic nature of this product may account for the slight deviation from theoretical values.

Decomposition Met and Pathway Intermediates

Table I shows the amount of volatile S and the species evolved from aerobic soils treated with Met, MMet, DMSP, MTP and MCP. A low percentage of the Met added to the Panhill (24% as DMDS) or Panoche (34% as DMS) soil was volatilized during the 4 d incubation suggesting that soil microorganisms initially conserved Met and was not utilized as an energy source. Hadas et al. (17) found that ^{14}C-alanine was rapidly assimilated during short-term incubations with increased ^{14}C mineralization only after alanine was no longer present in solution. Martens and Suarez (6) reported more extensive volatilization of DMS from Met additions to the Panhill and Panoche and ion chromatography analysis detected very low levels of SO_4 with incubations longer than 10 d. The Panhill soil evolved exclusively DMDS indicating that the organisms with a demethylation metabolism pathway dominated (Figure 1). Taylor and Gilchrist (*7*) reported that DMDS evolution in aerobic marine sediments was via demethylation of MTP and resulted in methane thiol, which was in turn oxidized to DMDS. Met addition to the Panoche soil resulted in only DMS. Taylor and Gilchrist (*7*) reported that MMet is formed from the methylation of Met before volatilization of DMS occurred. Addition of MMet to the Panhill and Panoche soils resulted in near quantitative DMS evolution suggesting that methylation of Met is the rate limiting step in the volatilization of S from aerobic soils tested (Table I). It is apparent from the results that both soils have the enzyme(s) required to convert MMet to DMS. An isolated soil bacteria using MMet as a C source was found to express the enzyme Met sulfonium lyase, which when purified, converted MMet to DMS (*18*). Dimethylsulfoniopropionic acid also resulted in nearly quantitative evolution of DMS from the Panoche soil, but a limited volatilization of DMS from DMSP additions was noted with the Panhill soil, suggesting organisms using the demethylation pathway do not express enzyme systems for synthesis of MMet or DMSP.

Table I. Cumulative evolution of volatile S species after addition of 25 μg organic compound-S g^{-1} soil to a Panhill and Panoche soil in a sealed screw-top Erlenmeyer for various times.

		Sulfur Species									
Soil	*Time*	*Met*		*MMet*		*DMSP*		*MTP*		*MCP*	
	(h)	*DMS*[a]	*DMD*	*DMS*	*DMD*	*DMS*	*DMD*	*DMS*	*DMD*	*DMS*	*DMD*
		---------- μg S evolved g^{-1} soil ----------									
Panhill	12	0.0	0.0	5.0	0.0	2.0	0.0	0.0	0.0	0.0	0.0
	24	0.0	0.0	10.5	0.0	2.8	0.0	0.0	0.0	0.0	0.0
	48	0.0	2.6	18.6	0.0	4.6	0.0	0.0	3.2	0.0	0.0
	108	0.5	6.0	23.0	0.0	5.0	0.0	0.0	4.8	0.0	0.0
Panoche	12	0.8	0.2	4.3	0.0	4.1	0.0	0.3	0.7	0.0	0.0
	24	1.6	0.3	7.6	0.0	7.8	0.0	0.4	1.2	0.0	0.0
	48	2.6	0.1	15.3	0.0	16.4	0.0	0.4	2.3	0.0	0.0
	108	8.6	0.0	22.0	0.0	23.5	0.0	0.3	4.6	0.0	0.0

[a]DMS, dimethylsulfide; DMD, dimethyldisulfide.

362

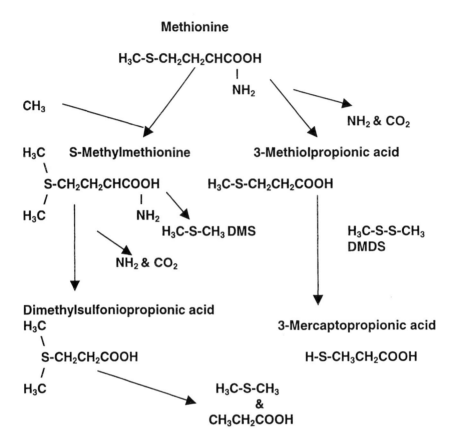

Figure 1. Methylation (MMet) and demethylation (MTP) pathways for Met as suggested by Visscher and Taylor (8).

Decomposition SeMet and Pathway Intermediates

When samples of the Panoche and Panhill soil were steam sterilized (2 h, 120°C, 104 kPa) on two consecutive days and sterilized solutions of S and Se intermediates were added, no volatilization was noted for the tested intermediates for up to four d (data not presented), confirming the biological nature of the two pathways. The speciation results for nonsterile Panoche and Panhill soils incubated with SeMet and SeCys for 2 and 7 d are presented in Table II. Approximately 80% of the SeMet-Se added to the Panoche soil was evolved as DMSe confirming the findings of Martens and Suarez (6). Only 4.6%

of the added Se remained in a Se(-2) state at the end of 7 d, suggesting that the methylation pathway limits accumulation of Se organic compounds in the Panoche soil. In contrast, the Panhill soil evolved about half of the added Se (44%) as a DMDSe volatile form compared to the Panoche soil and the majority of the remaining Se (41.5%) was found as a nonamino acid Se(-2) form (Table II). Martens and Suarez (6) first identified the compounds DMSeP and MSeMet by hydroelimination analysis (9) in the Panoche soil treated with SeMet indicating that SeMet followed pathways similar to the S methylation pathway identified in marine sediments, but the same Se intermediates were not noted in the Panhill soil. In contrast to the Se volatilization pattern noted for SeMet, after 2 d SeCys-Se incubation, no or low levels of volatile Se were noted and 44% (Panoche) and 28% (Panhill) of the remaining Se was present as nonamino acid Se(-2) as shown by Martens and Suarez (6), suggesting that the majority of the Se volatilization from the environment results from the methylation pathways of SeMet. Figure 2 suggests that the failure to identify

Table II. Selenium recovery and speciation after incubation of 50 μg SeMet-Se g^{-1} soil and 25 μg SeCys-Se g^{-1} soil in a Panhill and Panoche soil for various times (average ± standard deviation).

Soil	Se Form	Day	Se(+6)	Se(+4)	Se(-2)	Se^o	Volatile Se^a	Total Se
			---------------------- μg Se g^{-1} soil --------------------					
Panoche	SeMet	2	2.3	6.5	15.3	0.8	27.3	52.2±1.3
Panoche	SeMet	7	2.2	5.3	2.3	0.9	40.3	51.0±0.4
Panhill	SeMet	2	3.2	1.3	30.1	0.5	14.5	49.6±1.2
Panhill	SeMet	7	2.7	4.2	21.3	0.6	22.5	51.3±0.5
Panoche	SeCys	0.25	0.4	6.1	15.1	1.8	1.9	25.3±0.1
Panoche	SeCys	2	1.1	7.5	11.3	3.1	2.8	25.8±0.4
Panhill	SeCys	0.25	0.4	3.1	19.2	2.1	0.0	24.8±0.6
Panhill	SeCys	2	0.9	13.5	7.1	3.7	0.0	25.2±0.2

[a]Panoche soil evolved dimethylselenide; Panhill soil evolved dimethyldiselenide.

the Se compounds DMSeP and MSeMet in the Panhill soil was due to the presence of a demethylation pathway. The evolution of DMDSe (from the oxidation of methaneselenol) from the Panhill soil reported here confirms the work of Martens and Suarez (6) providing further evidence that different Se metabolizing pathways can be present in the different soils. Studies to determine the amount of organic Se assimilated in alfalfa grown in the presence of selenate (a model non-Se accumulating plant) found that the majority (75%) of the soluble organic Se was present as SeCys and MSeCys with lower levels of SeMet (19). Following residue microbial decomposition, Se speciation of the seleniferous alfalfa (19) reflected the speciation pattern noted for SeCys additions to soil (Table II). The presence of SeCys and SeMet in the plant biomass is no doubt the dominant source of the organic Se compounds to the environment, but virtually no information is available on the decomposition pathways of the organic forms of Se. This information is essential to understanding the toxicity potentials in evaporation ponds as up to 60% of the total dissolved Se in aquatic systems may be present as these organic forms (20).

To determine the importance of the methylation-demethylation pathways shown in Figure 3 for Se speciation, the Panoche and Panhill soils were incubated for up to 7 d after addition of 10 μg g^{-1} soil of each intermediate (Table III). Since no commercial sources of the Se pathway compounds were available, the Se compounds had to be synthesized for these mineralization studies. Results show that both soils tested were as efficient for removal of the MSeMet additions as SeMet additions (Table II), but were less efficient for volatilization of DMSeP. The Panoche soil evolved 95% of the sulfur analog DMSP addition as DMS confirming DMSP as the next proposed step in the S volatilization pathway (Table I), but the Se data suggested that volatilization of Se occurred following the methylation of SeMet (Table II, Figure 2). Lewis et al. (21) reported that an enzymatic fraction isolated from cabbage (*Brassica oleracea*) was active for catalyzing the release of DMSe from MSeMet and DMS from MMet suggesting the importance of the formation of MSeMet in Se volatilization. This is divergent from the evidence presented for the S pathway (Table I) (7) and suggests that conversion of MSeMet to DMSeP limits loss of Se from the soil. Even the Panhill soil, which did not volatilize DMSe from SeMet additions (Table III) evolved nearly all of the MSeMet-Se added as DMSe. This level of volatilization was not found when Se was added as DMSeP, suggesting that the pathways for methylation of SeMet was limited in the Panhill soil, but the enzyme(s) are still present for volatilizing MSeMet. The additional data supports the findings of Martens and Suarez (6) that the demethylation pathway results in accumulation of nonamino acid Se(-2) species and the volatile Se species will be DMDSe. Table III also shows the Panoche and Panhill soil Se speciation data when exposed to the TMSe$^+$ ion. That data shows that extensive volatilization of TMSe$^+$ (45%) can occur under soils exhibiting the methylation pathway as in the Panoche, but the Panhill soil treated with TMSe$^+$ accumulated organic Se (74%) as measured after 7d incubation. Cooke and Bruland (20) reported the identification of organic Se compounds MSeMet and TMSe$^+$ in ground water from beneath the Kesterson Wildlife Refuge and the Salton Sea in California. The identification of the organic Se compounds suggests that production and persistence of organic Se metabolites occurs in soil systems contaminated by inorganic Se.

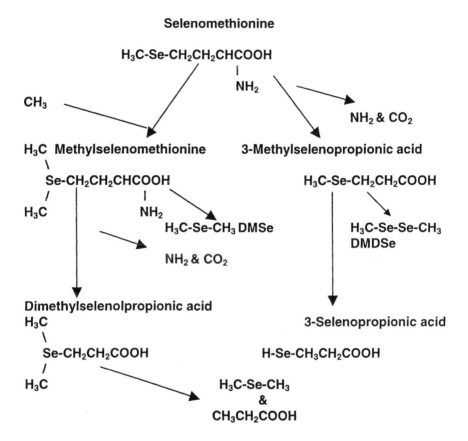

Figure 2. Methylation (MSeMet) and demethylation (3-methylselenopropionic acid) pathways for SeMet as suggested for Met (8).

Table III. Selenium recovery and speciation (average ± standard deviation) after incubation of 10 µg Se g^{-1} soil as MSeMet, DMSeP, MSeCys and TMSe$^+$ (50 µg Se total) in the Panhill and Panoche soils for up to 7 d.

Soil	Se Form	Day	Se(+6)	Se(+4)	Se(-2)	Se^o	Volatile Se^a	Total Se
				----------------------	µg Se Recovered	-------------------		
Panoche	MSeMet	2	1.1	9.2	13.0	0.6	27.2	51.1± 1.2
Panoche	MSeMet	7	0.0	7.5	6.2	1.8	37.5	52.9± 0.9
Panhill	MSeMet	2	1.5	3.3	27.4	0.6	17.3	50.1± 0.7
Panhill	MSeMet	7	1.9	5.6	17.1	0.6	26.4	51.5± 1.3
Panoche	DMSeP	2	2.1	8.5	32.9	1.1	7.6	52.2± 1.4
Panoche	DMSeP	7	0.0	7.0	19.4	4.2	21.0	51.6± 0.8
Panhill	DMSeP	2	3.8	3.6	40.3	0.7	1.8	50.2± 1.2
Panhill	DMSeP	7	1.4	6.2	35.2	1.1	5.5	49.4± 0.9
Panoche	MSeCys	2	0.7	15.6	27.1	2.0	5.0	48.4± 1.1
Panoche	MSeCys	7	0.3	15.8	20.9	5.6	8.4	51.0± 1.0
Panhill	MSeCys	2	6.4	6.4	37.2	1.6	0.0	51.6± 1.5
Panhill	MSeCys	7	2.7	18.8	20.3	2.0	5.9	49.7± 1.6
Panoche	TMSe$^+$	2	0.3	8.5	35.6	1.3	2.6	48.4± 0.6
Panoche	TMSe$^+$	7	0.6	8.7	14.5	2.4	23.1	51.0± 1.2
Panhill	TMSe$^+$	2	0.5	12.8	34.8	0.6	0.0	51.6±.09
Panhill	TMSe$^+$	7	0.0	6.8	36.7	1.3	2.8	49.7± 1.5

[a]The volatile Se was recovered as dimethylselenide from both soils.

The pathway for metabolism of SeCys is less complicated for SeMet (Figure 3). Addition of MSeCys determined that the soils were not efficient for Se volatilization from MSeCys additions and as noted with SeCys additions (Table II), the majority of the Se remaining was present in the Se(-2) species. The soil Se speciation pattern noted for SeCys and MSeCys (Tables II and III) is nearly identical to the soil Se speciation in the Panhill and Panoche following addition of seleniferous alfalfa (*19*), which contained the majority of extractable Se amino acids as SeCys and MSeCys. The formation and persistence of nonamino acid organic Se with SeMet additions to soils exhibiting a demethylation population and from SeCys or MSeCys suggest that certain soils may have a potential to accumulate organic Se compounds that have not been tested for toxicity to avian species. Future research needs to address the possible formation in soils and sediments of nonamino acid Se with special attention to the toxicity of the organic Se compounds 3-methylselenopropionic or 3-selenopropionic acid to avian species.

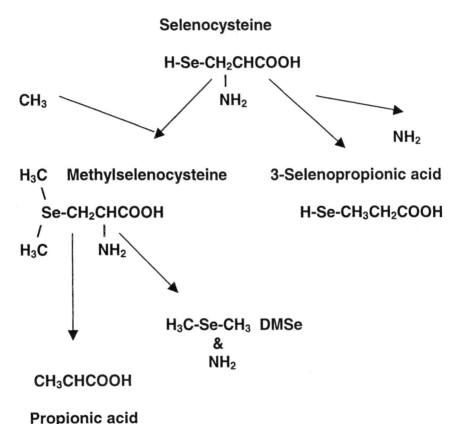

Figure 3. Methylation (MSeCys) and demethylation (3-selenopropionic acid) pathways for SeCys as suggested for Met (8).

Soils metabolizing SeMet via the methylation pathway will result in loss of Se from the system with lower Se(-2) accumulation, but the demethylation pathway will result in potential SeMet accumulation as 3-methylselenopropionic or 3-selenopropionic acid as noted by Martens and Suarez (6). A simple deamination of SeCys forming 3-selenopropionic acid may also result in an accumulation of organic Se compounds (Figure 3) and may explain why Martens and Suarez (6) found a rapid loss of SeCys from the soil system when analyzed

368

for the amino acid, and a comcomitant accumulation of a nonamino acid organic Se.

The toxicological significance of these proposed intermediates is not known as the only organic Se compound tested for toxicity is SeMet. While low levels of the organic Se forms accumulate with SeMet, elevated levels of persistent organic Se accumulate with MSeCys or SeCys additions. The organic Se compounds are important because under field conditions, MSeCys and SeCys are the major organic Se compounds present in plant residues and available for decomposition. The different pathways for metabolism of organic Se compounds in some instances result in formation of significant quantities of intermediates. The differences may explain why Se-induced problems with waterfowl have not been detected in wildlife refuges that had been predicted to be problem spots based on total Se concentrations (11).

References

1. Skorupa, J. P. In *Environmental Chemistry of Selenium*, Frankenberger, Jr. W. T.; Engberg, R. A., Eds. Marcel Dekker, Inc: NY, 1998; p 315-354.
2. Frost, D. V.; Lish, P. M. *Annu. Rev. Pharmacol.* **1975**, *15*, 259-284.
3. Heinz, G. H.; Hoffman, D. J; Gold, L. G. *Arch. Environ. Contam. Toxicol.* **1988**, *17*, 561-568.
4. Frankenberger, Jr. W. T.; Karlson, U. *Soil Sci. Soc. Am. J.* **1989**, *53*, 1435-1442.
5. Shrift, A. In *Organic Selenium Compounds: Their Chemistry and Biology.* Klayman, D. L.; Gunther, W. H. H., Eds, Wiley: NY, 1973; p 763-814.
6. Martens, D. A.; Suarez, D. L. *Soil Sci. Soc. Am. J.* **1997**, *61*, 1685-1694.
7. Taylor, B. F.; Gilchrist, D. C. *Appl. Environ. Microbiol.* **1991**, 57, 3581-3584.
8. Visscher, P. T.; Taylor, B. F. *App. Environ. Microbiol.* **1994**, *60,* 4617-4619.
9. White, R. H. *Mar. Res.* **1982**, *40*, 529-536.
10. Reamer, D. C.; Zoller, W. H. *Science,* **1986**, *208,* 500-502.
11. O'Toole, D.; Raisbeck, M. F. In *Environmental Chemistry of Selenium,* Frankenberger, Jr. W. T.; Engberg, R. A., Eds. Marcel Dekker, Inc: NY. 1998; p 355-395.
12. Callenger, F.; Simpson, M. I. *J. Chem. Soc.* **1948**, *3*, 1591-1597.
13. Toennies, G.; Kolb, J. J. *J. Am. Chem. Soc.* **1945**, *67*, 849-851.
14. Foster, S. J.; Ganther, H. E. *Anal. Biochem.* **1984**, *137*, 205-209.
15. Fan, T. W.-M.; Lane, A. N.; Martens, D. A.; Higashi, R. M. *Analyst,* **1998,** *123*, 875-884.
16. Martens, D. A.; Suarez, D. L. *Environ. Sci. Technol.* **1997**, *31,* 133-139.

17. Hadas, A.; Sofer, M.; Molina, J. A. E.; Barak, P. Clapp, C. E. *Soil Biol. Biochem.* **1992**, *24*, 137-143.
18. Mazelis, M.; Levin, B.; Mallinson, N. *Biochem. Biophy. Acta* **1965**, *105*, 106-109.
19. Martens, D. A.; Suarez, D. L. *HortSci.* **1999**, *34*, 34-39.
20. Cooke, T. D.; Bruland, K. W. *Environ. Sci. Technol.* **1987**, *21*, 1214-1219.
21. Lewis, B. G.; Johnson, C. M.; Broyer, T. C. *Plant Soil.* **1974**, *40*, 107-118.

Chapter 24

Sorption Behavior of Butyltin Compounds in Estuarine Environments of the Haihe River, China

S. G. Dai, H. W. Sun, Y. Q. Wang, W. P. Chen, and N. Li

College of Environmental Science and Engineering, Nankai University, Tianjin, 300071, Peoples Republic of China

Experiments were performed to disclose sorption behavior of tributyltin (TBT) and its breakdown products, dibutyltin (DBT) and monobutyltin (MBT) using sediments of the Haihe River Estuary, Tianjin, China. Sorption constants varied with butyltin species, changing in an order of MBT (2.79×10^3 L/Kg) >DBT (1.52×10^3 L/Kg) >TBT (5.76×10^2 L/Kg). These sorption constants were pH-dependent, with highest sorption occurring in an acidic pH range, indicating that butyltins bind to sediment mainly through the corresponding cations. Distribution percentage of butyltins decreased in residue sediment and increased in humic matter with reduction in the number of butyl groups. Sorption constant of TBT varied with the type of sediment. Sediment -3, which contained the highest content of organic matter, clay, and permanent negative charge gave the highest sorption. Sorption of TBT decreased as salinity increased. It is concluded that at acidic condition (pH<4), TBT^+ ion binds mainly to the permanent negative mineral charge , while at medium pH (4-10), ion exchange with =XOH(M) groups in the mineral and complexion with =ROH(M) groups in humic matter with TBT^+ ion appear to be the main mechanisms. When pH was higher than 10, hydrophobic partitioning of TBTX (X=OH, Cl etc.) neutral molecules into humic matter seemed to control the sorption and gave the second peak sorption. For DBT and MBT, sorption is most likely controlled by their cationic character.

Introduction

Triorganotin compounds, especially tributyltin (TBT) and triphenyltin (TPT), have been used as the active component in antifouling paints since the 1960s; meanwhile they also have found a broad spectrum of other applications including fungicides, bactericides, other pesticides and wood preservatives (*1*). Production of organotins as pesticides in the world is reported to be 5×10^7 kg/year, and it is estimated that 3×10^6 kg of TBT and TPT are introduced into the aquatic environment every year (*2*). TBT is much more important than TPT from an environmental point of view because it has been produced on a larger scale. TBT compounds in aquatic environments display high toxicity to aquatic organisms, being toxic to oysters and other non-target molluscan species at ng/L concentrations or less (*3*). Recent reports on sublethal effects of TBT, including genotoxicity, embryotoxicity, immunotoxicity and endocrine- disrupting effects (*4-7*) have drawn a renewed attention to contamination from TBT.

Because of adverse effects of TBT on the aquatic eco-system, France first adopted controlled use of TBT in 1982. To date, many countries around the world have also enacted related regulations. Usually, application of TBT is prohibited on vessels shorter than 25m, and its release rate should be less than $4\mu g/cm^2/d$ (*8*). These measurements have led to a significant decrease in TBT concentrations in aquatic environments. However, the use of TBT as the biocide in antifouling paints has not been prohibited completely. There is no related standard or regulatory control in China and some other Asian countries. TBT contamination in aquatic environments is still present worldwide and at levels of concern (*9-12*).

Another reason for the long-term, relatively high level of TBT contamination in the water column after restrictions were established has been suspected to be that the highly contaminated sediment may act as a source for long-term water column contamination of TBT (*13,14*). TBT binds strongly with sediments and is highly persistent under anoxic sediment conditions (*15,16*). Therefore, profound knowledge of its sorption behavior on sediments is necessary for predicting thelong-term contamination pattern in aquatic environments, and also for evaluating its bioavailability. There have been several reports on the sorption behavior of TBT (*17-21*). However, results are conflicting, and general conclusions cannot be attained. For example, sorption constants of TBT are cited to fall in a range of 6.0 to 2.92×10^5 kg/L, covering several orders of magnitude (*22*). This indicates that sorption behavior of TBT may vary greatly according to sediment type and water-phase conditions. It has been reported that TBT sorption varied with sediment type, increasing in the order of sandy-silt < silty-sand < silty-clay (23). TBT is an ionizable organic compound. In the water phase, it ionizes into TBT^+ cation, and meanwhile it may associate with predominant anions (such as CI^-, OH^- and HCO_3^- etc.) in surrounding water to give neutral molecules. Hence, TBT may show both hydrophobic and cationic characteristics according to conditions. Factors that influence the chemical speciation of TBT are considered to influence the environmental behavior and fate of TBT. Hence, results of the studies on TBT sorption are quite disparate because conditions for experiments are different. Concerning the mechanisms of sorption, some studies have shown that TBT behaves like classic hydrophobic organic pollutants, and sorption of TBT is mainly associated with its undissociated organic phase (17,18), while other studies show that sorption behavior of TBT is similar to that of cationic metal ions (19,20). Besides, TBT sorption exhibits a strong salinity (24,25) and pH (18,24) dependence; the effect of salinity appearing either negative (19, 24) or positive (25). Furthermore, sorption constants of TBT and its metabolites, dibutyltin (DBT) and monobutyltin (MBT) are also reported in a different order. Berg et al. (17) reported that sorption constants changed in the sequence of TBT>DBT>MBT. However, the reverse sequence was reported by other authors (20,26).

In the present study, the influence of several factors on sorption of TBT, DBT and MBT was measured. Distribution patterns of the three butyltins in different fractions of natural sediment were investigated. Furthermore, mechanisms of the association between TBT and minerals and humic matter in sediment were discussed.

Materials and Methods

Chemicals

Tributyltin chloride, dibutyltin chloride and monobutyltin chloride were purchased from Aldrich. Stock solutions of 1000 μg Sn/mL were prepared in ethanol, and stored in refrigerators at 4°C under darkness before use. Tropolone (used for extraction) was purchased from Sigma. Grignard reagent, pentyl magnesium bromide, was synthesized in our laboratory.

Sediments and Water Samples

Sediments were collected from the vicinity (sea, estuary and river environments, respectively) of Haihe River Estuary, Tianjin, China. Sediment samples were collected by a sediment collector and stored in glass containers. Immediately after collection, samples were taken to the laboratory. The samples were air-dried then sieved through 100-mesh sieve. Water samples above the sediments were collected 0.5 m beneath water surface. They were transported to the laboratory, filtered through a 0.45-μm filter, and then refrigerated at 4°C before use. pH and salinity values of water phases were determined.

Properties of the sediment were determined. Organic matter content of the sediment was determined by $K_2Cr_2O_7$ volumetric method as described by Alzieu (27); Clay content was determined by the Central Laboratory of Nankai University using X-ray diffraction. Potential of zero charge (PZC) and permanent charge at zero charge (σ_0) of the mineral fraction of the sediment were determined by potentiometric acid-base titration under different ion strengths (28).

Distribution of Butyltins in Natural Sediment

In order to find the binding sites of the three butyltins, butyltin content was determined in total sediment and in different fractions of the sediment. Ten grams of dried and sieved sediment-1 (see Table I) was shaken with 100 mL of 0.1N NaOH for 24h, then centrifuged (3000 rpm) for 10 min. The residue acquired was called residue sediment (after extraction of humic matter); and the supernatant was an alkaline solution of humic matter. HBr (40%) was added to the above supernatant solution until no more precipitate occurred (pH 1 to 2). Then the dissolved phase and precipitate were separated by centrifugation (3000 rpm) for 10 min. The precipitate was the fulvic acid fraction and supernatant was the humic acid fraction. Butyltins determined in each fraction were assumed to be associated with the corresponding fraction.

Sorption Constant

Batch systems of 50ml water phase and 1g dried sediment in Erlenmeyer flasks were adopted for determining sorption constants. Under pre-determined conditions, the sorption isotherm was measured by shaking for 24 hrs at controlled temperature. For each isotherm experiment, five TBT concentration levels were used, and each concentration was duplicated. Then sorption constants were calculated by regression analysis of isotherm equations.

To determine the influence of pH on sorption, the above batch system was first acidified or made basic by 1N HCl or 1N NaOH to obtain pH value of about 3.0, 4.0, 5.0, 6.0, 6.5, 7.0, 8.0, 9.0 and 10.0, respectively. Because of the buffering capacity of sediments, samples were well shaken in order for the pH value to become stable. Then 1µg Sn/mL of butyltin was added into the system, and shaken for 24h to allow TBT to equilibrate between sediment and water phases.

Analysis

After separation from sediment by centrifugation, butyltins in water samples were extracted by 10 mL of 0.1% tropolone/benzene solution after acidification to pH of 1.0 by HBr. Then butyltins in the benzene phase were akylated by addition of pentyl Grignard reagent. Extraction of butyltins in natural and residue sediment samples was carried out by shaking with 30 mL of glacial acetic acid solution for 4h, and then extracted with 15 mL of 0.2% tropolone/benzene solution by shaking for 1h. The alkylation procedure was the same as above. Analysis of butyltins was carried out using a slightly modified procedure as reported by Tolosa et al. (29). Gas chromatography (M600D, Yong Lin Company, Korea) equipped with HP-5 capillary column (Hewlett-Packard Company, USA) and flame photometric detector (FPD) with a 610nm cut-off interference filter was used. Injector temperature was set at 250°C; column temperature was kept at 175°C during analysis. Nitrogen was used as carrier gas at a flow rate of 20 mL/min. Flow rate of air and hydrogen for the detector at 225°C were 100 and 120 mL/min. Splitless injection (2µL) was employed. Concentrations of butyltins in sediments for sorption constant experiments were calculated by subtraction of the water phase concentration from the corresponding initial total concentration.

Extraction of Humic Matter

The dried and sieved (1-mm sieve) sediment sample was stirred in deionized water at a ratio of 1:10 for 2hrs at pH of 2.0. Then the supernatant was decanted after standing for 30 min. The residue was dissolved by adjusting pH to 10 with 0.5 N NaOH, and the alkaline solution was shaken for 24 hrs to extract humic matter from the solid into solution. The solution was centrifuged, and the supernatant was adjusted to pH 2.0 to let humic matter precipitate. The precipitate was washed with deionized water for several times to pH neutral and dried at 40°C under vacuum.

Determination of Titration Curve of Humic Matter

There are several acidic ligand functional groups in humic matter. In order to characterize the functional groups in humic matter of the studied sediment, a titration curve of the above humic matter was measured according to the method described by Wang (30). The experiment was performed under nitrogen protection. Humic matter (100.0 g) was weighed and dissolved in deionized water at pH 10. After dissolution, 20 mL of 0.1N HCl was added to get humic acid to gel and the pH value was determined after equilibrium was achieved. Then titration was started by addition of 0.10 mL of 0.2N NaOH successively until pH reached 12.5. The pH values were measured at equilibrium after every addition. A titration curve was plotted by the first differential graphic method. Then pKa and the amount of ligand functional groups can be calculated from the titration curve.

IR Spectrum of Humic Matter

Pure humic matter was well ground in an agate grinder, and then dried at 55°C for 2 hrs. An IR spectrum was produced using KBr pellets . The IR spectrum of pure TBT was observed by liquid film. Humic matter (60mg) was added into 100 mL of 200 μgSn/mL TBT solution and shaken for 24 hrs. The solids were separated and treated for two more times in 100 mL 200 μgSn/mL TBT solution. The solid was dried under vacuum. An IR spectrum was observed using KBr .

Results and Discussion

Properties of Sediments

Three kinds of sediment were collected from the vicinity of the Haihe River Estuary, Tianjin, China. They came from three different water environments: sediment-1 (salt water), sediment-2 (estuarine water), and sediment-3 (fresh water). Their properties are given in Table I. Sediment-3 has the highest organic matter content and clay content. Its PZC is most acidic, and permanent charge at PZC is highest. These properties are beneficial for sorption both through lipophilic partitioning and through ionic adsorption, hence, sediment-3 was expected to have the highest sorption capacity.

Table I. Properties of Sediments

Sediment	Organic matter content (%)	Clay content (%)	PZC	σ_0 ($\mu C/cm^2$)
Sediment-1	2.09	25.4	4.12	-6.53
Sediment-2	4.31	42.1	3.78	-17.6
Sediment-3	4.46	60.2	3.51	-34.9

PZC is abbreviation of potential of zero charge and σ_0 is permanent charge at zero charge.

Distributions of Butyltins in Natural Sediment

Total concentrations of TBT, DBT and MBT in sediment-1 were 95.8, 39.8 and 116.3 ng Sn/g, respectively. TBT concentration in ambient water was 62.9 ng Sn/L, while DBT and MBT were not detectable (analytical detection limits were 0.5ng/L for DBT and MBT). Hence, sorption constant for TBT was 1.52×10^3 L/Kg. Distributions of butyltins in residue sediment, fulvic acid , and humic acid fractions of sediment-1 were measured (Figure 1). It can be seen that of the three butyltins, with the decrease in the number of butyl groups, butyltin distributions in the fulvic acid fraction and also in the soluble humic matter fraction (fulvic acid + humic acid) increased, while that in the residue sediment decreased. TBT was mainly bound to the residue sediment, while MBT was associated primarily with fulvic acid. The percentage of DBT in residue sediment and fulvic acid was roughly the same. Association of the three butyltins with the humic acid fraction was less important compared to that with the other two fractions. A general conclusion is that percent butyltins present in the soluble humic matter changed in the order of MBT>DBT>TBT agrees with the study of Kuballa (31).

Comparisons of Sorption of Butyltins Species

In order to compare sorption constants of different butyltin species, sorption isotherms and constants of TBT, DBT and MBT on sediment-1 were measured in the laboratory (Table II) using the sediment's ambient natural water (30.8‰ salinity). Sorption isotherms of butyltins followed the Freundlich pattern well, with a correlation coefficient higher than 0.988. Sorption constant of TBT was 576.53 L/Kg, two times less than that from field data. The difference in sorption constants between laboratory and field results using the same sediment and water phase may be caused by the differences in the initial concentrations of TBT and volume ratio of sediment and water phase applied. Furthermore, recent studies on sorption of organic pollutants disclosed that sorption of some organic pollutants on sediment or soil may occur in two phases: fast and slow. Sorption in the slow phase may account for 33-90% of total sorption (*32*). That is to say, sorption occurring in the field has more time to allow slow sorption occur than a sorption experiment conducted in the laboratory. However, whether or not slow sorption occurs for TBT needs further studies.

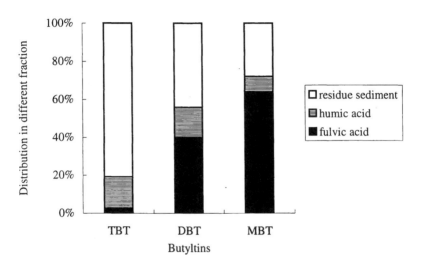

Figure 1. Distribution of butyltins in different fractions of sediment-1

Table II. Sorption Constants of Different Butyltin Specie on Sediment-1

Compound	Freundlich equation	Sorption constant (L/Kg)
TBT	$LnC_s=6.357+0.895LnC_w$	576.53
DBT	$LnC_s=7.328+0.975LnC_w$	1523.3
MBT	$LnC_s=7.935+1.007LnC_w$	2793.9

Sorption constants of the three butyltins changed in the sequence of MBT>DBT>TBT. It has long been noticed that the sorption behaviors of organotin compounds are affected by their molecular structures (*20,26*). However, results are conflicting due to the difference in absorbent, water-phase constitution, and other experimental conditions used in different studies. Randall and Weber reported a sorption affinity sequence of MBT>TBT>DBT on hydrous Fe oxide (26), while a sequence of TBT>DBT>MBT was recently reported by Berg et al. for a sediment-pore water system (17). This indicates that several processes may contribute to control the sorption behavior of butyltins, and sorption constants of butyltins may vary greatly under different situations.

Sorption Constants of TBT on the Three Sediments under Different Conditions

A sorption isotherm of TBT was determined using the three sediments in different water phases (Table III). The original ambient water phases in the natural environment used for sediment –1, -2 and –3 are underlined. In order to investigate the influence of water-phase salinity on the sorption behavior of TBT, sorption isotherms were also measured for sediment-2 and –3 at high salinity (30.8‰). Sorption of TBT in these systems followed the Freundlich equation well, with correlation coefficients being 0.988-1.000 (data not shown).

Table III. Sorption Constants of TBT on Different Sediments

Sediment	Water phase		Sorption constant (L/Kg)	Freundlich equation
	S(‰)	pH		
Sediment-1	30.8	7.57	576.5	$LnCs=6.356+0.895LnCw$
Sediment-2	22.6	7.24	2137.3	$LnCs=7.667+0.923LnCw$
Sediment-2	30.8	7.57	1807.5	$LnCs=7.500+0.925LnCw$
Sediment-3	3.0	6.58	3837.5	$LnCs=8.253+0.902LnCw$
Sediment-3	30.8	7.57	1408.1	$LnCs=7.250+0.904LnCw$

*Original ambient water phases of the sediments are underlined; S means salinity.

Sorption constants of TBT on the three sediments varied greatly. Under the natural aqueous conditions, sorption constants were 576.5, 2137.3 and 3837.5 L/Kg for sediment -1, -2 and -3, respectively. Sediment-3 gave the highest sorption constant. In a recent review paper, Hoch pointed out that cation exchange processes of triorganotin at permanently negative charged surfaces of clays and at deprotonated surface hydroxyl groups sufficiently describe the overall sorption of these compounds to minerals (2). Arnold et al. (18) proposed that sorption of TBT to humic matter is governed by complexation of TBT$^+$ cation by negatively charged ligands of the humic acids. Sediment-3 had the highest humic matter content, clay content and permanent negative charge, and all favorite high sorption of TBT. Overall, TBT was moderately sorbed as indicated by its sorption constants, which means that TBT is likely to be present in the water column as well as in the sediment phase, and therefore it might be available to both pelagic and benthic organisms.

When water-phase salinity was elevated, sorption of TBT on sediment-2 and –3 decreased greatly. For sediment-3, when the water phase was changed from fresh water (salinity of 3.0‰ and pH of 6.58) to salt water (salinity of 30.8‰ and pH of 7.57), the TBT sorption constant decreased by a factor of 2.72. Both changes in salinity and pH are considered to contribute to the sorption reduction. However, based on sorption-pH curves (see Figure 3), 93% of the reduction was attributed to the influence of salinity. This result agrees with the reports that salinity has a negative influence on sorption of TBT when salinity is less than 40‰ (19, 24). A decrease in sorption constant by a factor of 2 over salinity range 0 to 34% was reported by Uger et al. (*19*).

Titration curve of Humic Matter

The primary differential potential titration curve of the extracted humic matter from sediment-1 is shown in Figure 2. It can be seen that there are four active ligand groups in the humic matter. pKa values and contents of each ligand group are listed in Table IV. At the pH below the corresponding pKa, a majority of ligands exist as =ROH, whereas at pH above pKa, negative charge site =RO⁻ is predominant. Elevating pH increases the amount of =RO⁻. It can be seen that at pH >5.73 the first two ligands, which account for 61% of total ligands, exist predominately as =RO⁻ group. These pKa values provide us a good insight into sorption mechanisms of TBT.

Effect of pH on Sorption of Butyltins

Influence of pH on sorption of TBT, DBT and MBT was investigated (Figure 3). It can be seen that the highest sorption occurred at an acidic pH range for all three butyltins (6.0, 5.5 and 5.0 for TBT, DBT and MBT, respectively). A hydrolysis constant of 6.3 was cited for TBTOH (*18*). Using this constant, at pH of 6.0, 60% of the total TBT should be TBT^+ ion. A second peak of sorption

Table IV. pKa and Content of Ligand Groups in Humic Matter

	L_1	L_2	L_3	L_4
pKa	4.26	5.73	8.22	9.97
Content (mol/L)	2.36	1.47	1.79	0.65

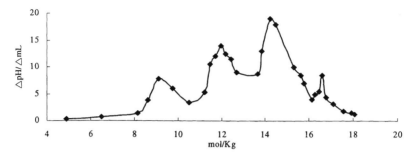

Figure 2. Primary differential potential titration curve of humic matter

occurred after pH 10 for TBT. Under this pH, TBT should exist as neutral molecules. Because this experiment was conducted at a salinity of 30.8‰, TBTCl and TBTOH are expected to be the main species. The two peaks of TBT sorption with pH indicate that sorption of TBT was most likely controlled by both hydrophobic and cationic characteristics. Not only was speciation affected by pH, but also charge properties of mineral and humic matter of the sediment were pH dependent as described above. The highest peak of sorption (6.0) of TBT occurred between pKa of TBTOH (6.3), and PZC (4.12) of the mineral and pKa of the first (4.26) and second ligands (5.73) of the humic matter of the sediment. This indicates that concentration of TBT^+, concentrations of $=XO^-$ groups in the mineral, and $=RO^-$ groups in humic matter of e sediment play important roles on sorption behavior. At this pH range, TBT sorption was mainly controlled by cationic character. At pH above 10, TBT is thought to exist as neutral molecules, with sorption controlled mainly by hydrophobic partitioning into humic matter. This is perhaps why a second peak of sorption occurred after pH 10. Hence, at pH<4, TBT^+ cation is sorbed mainly to the permanent negative mineral charge , and at pH>4, TBT^+ cation is associated with the mineral by cation exchange at the sites of =XOH and with humic matter by complextion at the sites of =ROH. As pH is elevated, neutral molecules, such as TBTCl and TBTOH become dominant, and hydrophobic partitioning gets involved. There are few reports on fine speciation of DBT and MBT, Ionization constants for DBT and MBT are not acquired. It has been reported that at pH<5, main speciation of MBT in sea water is MBT^{3+}, $MBTCl^{2+}$ and $MBTCl_2^+$ (*33*). Because no second peak occurred, we deduce that sorption of DBT and MBT are mainly controlled by cationic character.

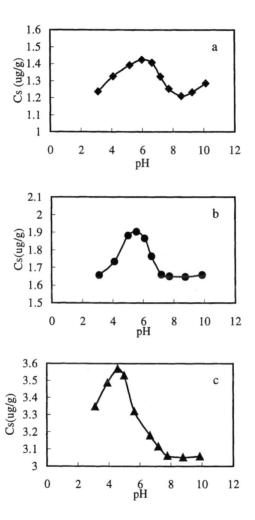

Figure 3. Influence of pH on sorption of TBT (a), DBT (b) and MBT(c)(initial concentration in water phase was 1μg Sn/ml)

IR Spectrum of Humic Matter and TBT

IR spectra of the pure humic matter extracted from sediment-1, pure TBT, and associated products of TBT and humic acid are given in Figure 4. It can be seen that TBT has two characteristic peaks around $3000cm^{-1}$, while no large peaks occur in this area of humic matter. After reacting, the two characteristic peaks of TBT were also observed in the spectrum of the TBT-humic matter sample. After binding to TBT, the spectrum of humic matter changed substantially. The peak at 3422.5 cm^{-1}, which represents the O-H group, decreased significantly. This indicates that association of TBT and humic matter occurred at OH groups. Studies on association of triorganotins with dissolved humic acids also disclosed that sorption of TBT with humic matters was governed by complexation of cations by negatively charged ligands (*i.e.* carboxylate and phenolate groups) of humic acid (*18*).

Conclusions

Sorption constants of different butyltins were quite different, changing in the order of MBT (2.79×10^3 L /Kg) >DBT (1.52×10^3 L/ Kg) >TBT (5.76×10^2 L/ Kg). Their sorption constants were pH-·dependent, with the highest sorption occurring under acidic conditions, being 6.0, 5.5 and 5.0 for TBT, DBT and MBT, respectively. PZC of the mineral was 4.12, and pKa was 4.26 and 5.73 for the first and second ligand groups of humic matter. This indicates that at this pH, butyltins combine with sediment mainly through the corresponding cation with the $=XO^-$ groups in mineral and $=RO^-$ groups in humic matter. At pH<4, binding to the permanent negative charge of mineral may dominate. A second peak was observed for the sorption of TBT after pH 10. When pH was higher that 10, neutral molecules of TBTX (X=OH, Cl etc.) are the main species of TBT, and hydrophobic partitioning of these molecules into humic matter may take control and gave the second peak. Hence, both hydrophobic and cationic character play roles on TBT sorption. Under natural conditions, sorption behavior of TBT is considered to be controlled mainly by cationic character. However, lack of the second sorption peak occurrence for DBT and MBT indicates that sorption of DBT and MBT is most likely controlled mainly by cationic character. Distribution of butyltins in sediments also changes with fractions. Studies on sorption of TBT disclosed that the sorption constant varied with the type of sediment. Sediment-3, which contained the highest content of organic matter, clay and permanent negative charge showed highest sorption capability. Sorption of TBT decreased as salinity increased.

383

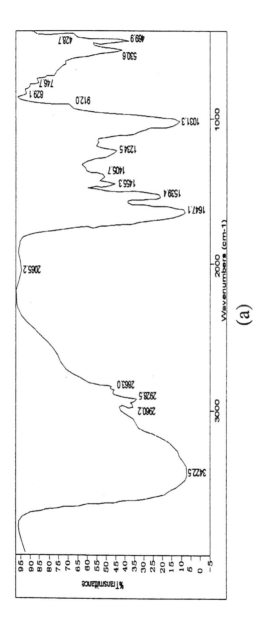

Figure 4. IR spectrum of humic matter (a), TBT (b) and TBT-humic matter (c)

Continued on next page.

384

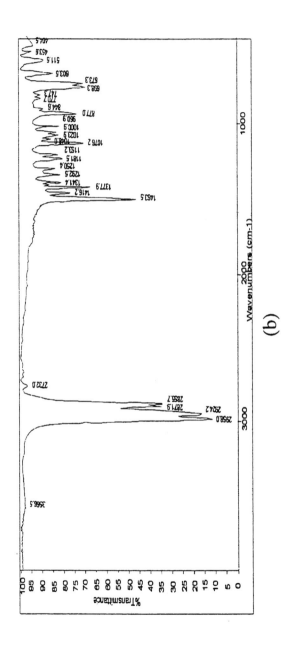

(b)

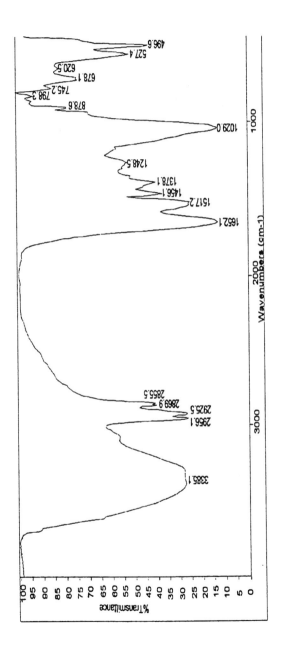

Figure 4. *Continued.*

386

Acknowledgements

This paper was supported by National Foundation of Natural Science, China. No: 29877013

References

1. Craig, P. J. In *Comprehensive Organometallic Chemistry,* Wilkinson, G., Eds.; Pergamon Press: Oxford, 1983; pp 979-1020.
2. Hoch, M. *Appl. Geochem.* **2001**, *16,* 719-743.
3. Bryan, G. W.; Gibbs, P. E. In *Metal Ecotoxicology: Concepts and Applications,* McNewman; McIntosh A.W., Eds; Lewis Publishers: Michigan, 1991; pp 323-362.
4. Yamanoshita, O.; Kurasaki, M.; Saito, T.; Takahasi, K.; Sasaki, H.; Hosokawa, T.; Okabe, M.; Mochida, J.; Iwakuma, T. *Biochem. Biophy. Res. Comm.* **2000,** *272,* 557-562.
5. Jha, A. N.; Hagger, J. A.; Hill, S. J.; Depledge, M. H.*Mar. Environ. Res.* **2000,** *50,* 565-573.
6. Marin, M. G.; Moschino, V.; Cima, F.; Cell, C. *Paracentrotus livius. Mar. Environ.* **2000,** *50,* 231-235.
7. Morcillo, Y.; Porte, C. *Environ. Pollut.* **2000,** *107,* 47-52.
8. Champ, M. A. *Sci. Total Environ.* **2000,** *258,* 21-71.
9. Elgethun, K.; Neumann, C.; Black, P. *Chemosphere* **2000,** *41,* 953-964.
10. Harino, H.; Fukushima, M.; Kawai, S. *Environ. Pollut.* **1999,** *105,* 1-7.
11. Hwang, H. M.; Oh, J. R.; Kahng, S. H.; Lee, K. W. *Mar. Environ. Res.* **1999,** *47,* 61-70.
12. Michel, P.; Averty, B. *Marine Pollut. Bull.* **1999,** *4,* 268-275.
13. Stuer-Lauridsen, F; Dahl, B. *Chemosphere* **1995,** *30,* 831-845.
14. Amouroux, D.; Tessier, E.; Donard, O. F. X. *Environ. Sci. Technol.* **2000,** *34,* 988-995.
15. *Organotin, Environmental Fate and Effects*; Champ, M.A.; Seligman, P. F., Eds.; Chapman & Hall: London, 1996.
16. Surradin, P-M.; Lapaquellerie, Y.; Astruc A.; Latouche, C.; Astruc, M. *Sci. Total Environ.* **1995,** *170,* 59-70.
17. Berg, M.; Arnold, C. G.; Müller, S. R.; Mühlemann, J.; Schwarzenbach, R. P. *Environ. Sci. Technol.* **2001,** *35,* 3151-3157.

18. Arnold, C. G.; Ciani, A.; Müller, S. R.; Amirbahman, A.; Achwarzenbach, R. P. *Environ. Sci. Technol.* **1998**, *32*, 2976-2983.
19. Uger, M.A.; Macintyre, W. G.; Huggett, R. J. *Environ. Toxicol. Chem.* **1988**, *7*, 907-915.
20. Sun, H. W.; Huang, G. L.; Dai, S. G. *Chemosphere* **1996**, *33*, 831-838.
21. Weidenhaupt, A.; Arnold, C.G.; M ü ller, S. R.; Haderlein, S. B.; Schwarzenbach, R. P. *Environ. Sci. Technol.* **1997**, *31*, 2603-2609.
22. Kram, M. L.; Stang, P. M.; Seligman, P. F. *Appl. Organomet. Chem.* **1989**, *3*, 523-536.
23. Dowson, P. H.; Bubb, J. M.; Lester, J. N. *Appl. Organomet. Chem.* **1993**, *7*, 623-633.
24. Langston, W. J.; Pope, N. D. *Mar. Pollut. Bull.* **1995**, *31*, 32-43.
25. Harris, J. R. W.; Cleary, J. J. *Oceans'87 Proceedings of the Fourth International Organotin Symposium,* Institute of Electrical and Electronics Engineering: NY, 1987; pp1370-1374.
26. Randall, L.; Weber, J.H. *Sci. Total Environ.* **1986**, *57*, 191-203.
27. Alzieu, C.; Michel, P.; Tolosa, I.; Bacci, T.; Mee, L. D.; Readerman, J. W. *Mar. Environ. Res.* **1991**, *32*, 7-17.
28. Huang, C. P. In *Adsorption of Inorganics at Solid-liquid Interfaces;* Anderson, M. A.; Rubin, A. J., Eds.; Ann Arbor Science: Ann Arbor, MI, 1981; pp183-217.
29. Tolosa, I.; Bayona, J. M.; Albaiges, J.; Alencastro, L. F. Tarradellas, J. *Fr. J. Analyt. Chem.* **1991**, *339*, 646-653.
30. *Research Methods for Organic Matters in Soil (in Chinese);* Wang, Q. X., Eds.; Agriculture Pub.: Beijing, China, 1984; pp210-217.
31. Kuballa, J.; Desai, M. V. M.; Wilken, R. D. *Appl. Organomet. Chem.* **1995**, *9*, 629-638.
32. Pignatello, J. J.; Xing B. S. Environ. Sci. Technol. **1996**, *30*, 1-11.
33. Hermosin, H.C.; Martin, P.; Cornejo, J. *Environ. Sci. Technol.* **1993**, *27*, 2606-2611.

Chapter 25

Spectroscopic and Modeling Study of Lead Adsorption and Precipitation Reactions on a Mineral Soil

Quy T. Nguyen and Bruce A. Manning

Department of Chemistry and Biochemistry, San Francisco State University, 1600 Holloway Avenue, San Francisco, CA 94132

This study investigated the adsorption reactions of lead (Pb) on a well-characterized soil and examined the Pb adsorption complex using several techniques. Traditional Pb batch adsorption experiments (adsorption isotherms and envelopes), X-ray absorption spectroscopy (XAS), and surface complexation modeling were used. In addition, comparisons were made of the Pb adsorption behavior on soil with Pb adsorption experiments performed on individual soil minerals. The effects of pH on Pb adsorption revealed that the two mechanisms controlling Pb solubility in the presence of soils and soil minerals are adsorption on mineral surfaces and pH-dependent precipitation of a poorly crystalline Pb solid phase (probably $Pb(OH)_2$). The results from EXAFS analysis revealed a possible Pb inner-sphere adsorption mechanism that was used in conjunction with a surface complexation model to quantitatively describe Pb binding to soil.

Sources of lead (Pb) contamination in soils include mining activities, historical use of leaded gasoline, and flue gas waste disposal and incineration. Remediation is required when Pb levels are between 300-500 mg/kg in residential soil and >2000 mg/kg in industrial soil (*1*). The primary mechanisms of Pb attenuation in soil include adsorption by soil mineral surfaces, complexation on organic matter molecules (*2*), and precipitation as insoluble Pb phases such $PbCO_3$ (cerrusite) and $Pb_5(PO_4)_3Cl$ (pyromorphite) (*3*).

Previous work on Pb binding in soil includes spectroscopic studies on minerals and humic/fulvic substances (*2-6*), chelation and extraction methods (*7,8*), and adsorption modeling (*6,9-11*). Despite the importance of precipitation reactions of Pb, very few Pb adsorption studies have carefully considered the pH-dependent effects of $Pb(OH)_2$ precipitation occurring in combination with adsorption. The precipitation of $Pb(OH)_2$ is of interest because this reaction places severe limitations on the solubility of Pb. The objectives of this paper were to assess the adsorption reactions of Pb on a well-characterized soil using several techniques including traditional batch adsorption experiments (adsorption isotherms and envelopes), X-ray absorption spectroscopy (XAS), and surface complexation modeling. In addition to soil, well-characterized minerals were studied including layer silicates (montmorillonite and illite) and synthetic metal hydroxides (FeOOH and $Al(OH)_3$).

Materials and Methods

Adsorbent Materials

Fallbrook sandy loam soil from Fallbrook, CA (Fine-loamy, mixed, thermic, Typic Haploxeralf) was sieved to <2 mm and used without further pretreatment. Soil texture analysis (%sand, silt, and clay) was determined by the hydrometer method (*12*). A sample of the Fallbrook soil clay fraction was separated and subjected to XRD analysis on an oriented slide mount. The specific surface area (BET N_2) of the <2 mm fraction was measured with a Quantichrome Quantisorb Jr. surface area analyzer. Crystalline iron oxyhydroxide (goethite, α-FeOOH) and amorphous aluminum hydroxide (am-$Al(OH)_3$) were synthesized according to previously described methods (*13, 14*). Only very weak Pb binding to am-$Al(OH)_3$ was found in this study, therefore data for Pb reactions with Al hydroxide are not shown. Reference clay minerals (layer silicates) were obtained from the Clay Minerals Society Source Clay Repository (Department of Geology, University of Missouri, Columbia, MO). Samples of sodium montmorillonite (SWy-1) from Crook County, WY and Illite (IMt-2) were sieved to <500 μm. Characterization of the clays by XRD of both random powder and oriented slide mounts indicated that montmorillonite contained a trace impurity of mica, whereas illite was XRD pure.

Pb Adsorption and Precipitation Experiments

In all experiments, a 1000 mg L^{-1} Pb stock solution (as $PbCl_2$ in 0.01 M HNO_3) was used as a source of Pb. Soil adsorption envelopes and isotherms were produced using 2 g soil in 20 mL solution (100 g soil/L) with pH adjusted between 2 and 8 using 0.1 M NaOH and 0.1 M HCl. A solution of 0.1 M NaCl was used to control ionic strength. Adsorption isotherms contained Pb concentration varying from 0.025-2.40 mM Pb whereas adsorption envelopes used a constant 2.40 mM Pb concentration. Adsorption envelopes for layer silicates and metal hydroxides used 50 mg solid in 20 mL (2.5 g/L). Samples were shaken for 24 hours, centrifuged (8,000 rpm, 15 min), followed by pH measurement using an Orion pH meter and a glass Ag/AgCl combination pH electrode. After pH measurement, the supernate solutions were filtered using 0.2 μm-diameter membranes followed by Pb analysis by flame (air-acetylene) atomic absorption spectrometry (217.0 nm). No atmospheric control was used in this study due to recent work suggesting that atmospheric PCO_2 did not effect Pb reactions with α-FeOOH (6). The contribution of organic matter to the Pb adsorption reaction was investigated by oxidizing and removing organic materials from a soil aliquot using warm 3% H_2O_2 followed by rinsing. This step may also remove other soil components such as manganese oxides and possibly sulfides.

The Pb precipitation experiments used 20 mL of Pb solution (2.4 or 4.8 mM Pb) which was titrated with 1 M NaOH solution in 50-mL polycarbonate centrifuge tubes (batch reactions). After titration, the precipitate-solution mixtures were allowed to age 24 h followed by the identical measurement and sample treatment regime given above.

XAS Analysis

Samples were prepared for XAS analysis by reacting solids with 5 mM Pb solution for several hours at pH 4.2. This pH ensured that $Pb(OH)_2$ precipitation was not contributing to the adsorption mechanism on the solids. After reaction with Pb solution, the solids were collected on filter paper and rinsed with deionized water to remove excess unreacted Pb solution. The XAS spectra of all samples were analyzed at the Stanford Synchrotron Radiation Laboratory on beamline 4-2. The Pb L$_{III}$ edge (13035 eV) of all samples was scanned a minimum of six times and the scans were averaged for final data analysis. Data were analyzed and theoretical fits applied using the EXAFSPAK suite of programs (*15*).

Surface Complexation Modeling

Adsorption envelopes were modeled using the constant capacitance model as part of the FITEQL Version 3.1 program (*16*). The following model parameters were used for Pb adsorption on the Fallbrook soil: surface site density ([SOH] = 0.0024 mole sites/L, based on maximum Pb loading from adsorption isotherms), soil surface area (19.4 m^2/g), pH range 2-10, and surface complex species of $SO\text{-}Pb\text{-}OH^0$ and $(SO)_2\text{-}Pb^0$. The reactions and intrinsic equilibrium constant expressions used to describe Pb adsorption are given in Table I. In the Table I equations, SOH is a mineral surface hydroxyl group which can be protonated and deprotonated. Two surface complex types were postulated (monodentate attachment ($SO\text{-}Pb\text{-}OH^0$) and a bidentate attachment ($(SO)_2\text{-}Pb^0$)).

Table I. Definition of intrinsic surface complexation constants

Reaction	log K(int)^a	Equation
$SOH \rightarrow SO^- + H^+$	−10.21	1
$SOH + H^+ \rightarrow SOH_2^+$	2.734	2
$SO^- + PbOH^+ \rightarrow SO\text{-}Pb\text{-}OH^0$	0.825	3
$2SO^- + Pb^{2+} \rightarrow (SO)_2\text{-}Pb^0$	0.050	4

^{a} K(int) = intrinsic surface complexation constant is defined as $K(\text{int}) = K_{eq} \exp(F\Psi_0/RT)$, where K_{eq} is the equilibrium constant, F is the Faraday constant, Ψ_0 is the surface charge generated by the complex, R is the gas constant, and T is temperature (°K).

Results and Discussion

Characterization of the Fallbrook Soil

The Fallbrook soil, which was partly characterized in a previous paper (*17*), had a particle size distribution dominated by the sand fraction (66%) with lesser amounts of silt (20%), and clay (14%) (Table II). A specific surface area of 19.4 m^2 g^{-1} was measured and used as a fixed input parameter for surface complexation modeling. The pH of the soil (1:1 paste) was 6.5. The predominant XRD reflections for the Fallbrook soil clay fraction are shown in Figure 1. The Fallbrook clay particles were composed of predominantly mica and illite with moderate amounts of kaolinite. A predominance of silica was found in the sand fraction XRD (data not shown). We suspect that a considerable amount of surface area in this soil is from the inorganic clay mineral fraction composed of predominantly silicate and aluminosilicate minerals and a low organic C content (<0.5% C w:w).

Table II. Characterization of the Fallbrook soil

Property	Value
pH (1:1)	6.50
Sand, silt, clay (%)	66, 20, 14
Surface area (m^2 g^{-1})	19.4
Al (mmol kg^{-1})	24.7
Fe	115
Mn	3.6
Organic C (%)	0.45

Pb Adsorption and Precipitation

The Pb adsorption characteristics on the Fallbrook soil as a function Pb concentration, pH, and ionic strength are shown in Figures 2 and 3. Adsorption isotherms (Figure 2) at pH 4.2 and 5.7 show the dramatic effect of pH on the adsorption capacity of the Fallbrook soil. The adsorption of Pb increased sharply as pH increased from 2 to 5.5 and increasing the ionic strength of the background electrolyte from 0.0125 to 0.1 M (as NaCl) caused a decrease in Pb adsorption between pH 2.5-5.0 (Figure 3). Ionic strength effects on ion adsorption are evidence for "outer sphere", electrostatic adsorption on charged mineral surfaces (18). Therefore, both outer sphere bonds ($SO^-\cdots PbOH^+$) and "inner sphere", covalent surface linkages ($SO\text{-}Pb\text{-}OH$) are probably contributing to adsorbed Pb. The effects of ionic strength were minor when I = 0.01-0.05 M, however, and were not considered in efforts to model Pb adsorption on soil.

The precipitation of insoluble Pb at increased pH was investigated in parallel experiment where standard solutions of 2.4 and 4.8 mM Pb were pH-adjusted and the removal of dissolved Pb by precipitation of Pb(OH)$_2$ was induced (Figure 4). The presence of another Pb solid phase such as cerussite (PbCO$_3$) was ruled out based on thermodynamic calculations of available

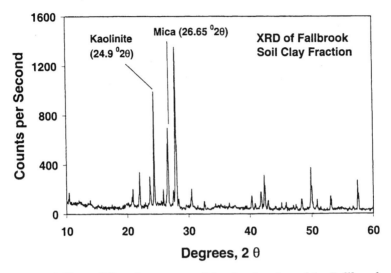

Figure 1. X-ray diffraction pattern of the clay fraction of the Fallbrook soil.

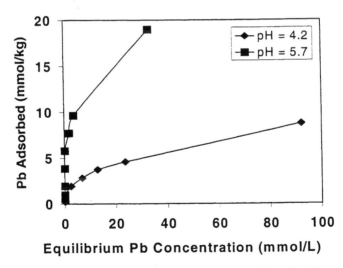

Figure 2. Equilibrium Pb adsorption isotherm on Fallbrook soil at pH 4.2 and 5.7.

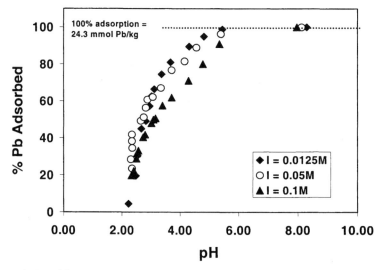

Figure 3. Equilibrium Pb adsorption envelopes on Fallbrook soil at three ionic strengths. Adsorption conditions: 2 g soil in 20 mL of Pb solution.

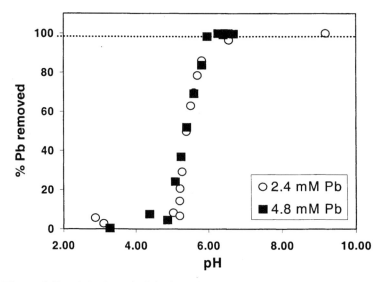

Figure 4. Precipitation of Pb hydroxide (Pb(OH)$_2$) from Pb solution (batch reaction) at two starting concentrations.

dissolved CO_3^{2-} in 20 mL of 0.24 mM Pb solution equilibrated with atmospheric PCO_2. The batch experimental solutions contained less than 1% of the required moles of CO_3^{2-} necessary for complete $PbCO_3$ precipitation in a closed system. These results suggest that at pH greater than 5.5 the Pb removal process included both precipitation of $Pb(OH)_2$ and adsorption on mineral surfaces. The precipitation data from this experiment were compared with the soil adsorption data (Figure 5) and indicated that the mechanism of Pb removal by soil below pH 5.5 was most likely adsorption on reactive soil surfaces such as negatively charged clays and possibly organic matter. The effects of removing organic matter showed that Pb adsorption was independent of the treatment. No significant difference in Pb adsorption behavior between the H_2O_2-treated soil and untreated soil was observed over the pH range 2.5-8.0. This suggested that the Pb adsorption reaction on the Fallbrook soil involved primarily the inorganic mineral fraction and not organic matter. Also, minerals such as Mn oxides which are removed by H_2O_2 treatment did not contribute to sorption.

Individual soil minerals were also investigated as model compounds for Pb adsorption. Due to the importance of negatively charged phyllosilicate clay minerals in adsorption of divalent cations in soil, sodium montmorillonite (SWy-1) was investigated. The effects of both pH and ionic strength on the adsorption of Pb on montmorillonite are shown in Figure 6. Increasing the ionic strength of the background electrolyte from 0.015 to 0.115 M caused a decrease in Pb adsorption between pH 3.0 and 5.5. This behavior suggests that some weak outer-sphere adsorption is occurring at low ionic strength. Interestingly, adsorption of Pb on SWy-1 at high ionic strength was nearly indistinguishable from homogeneous precipitation of $Pb(OH)_2$ from solution suggesting that precipitation may be occurring on the SWy-1 surface. The pH-dependent adsorption of Pb on α-FeOOH is also nearly indistinguishable from $Pb(OH)_2$ precipitation from solution, though a slight shift is evident (Figure 7). The model compounds appear to be not completely representative of the adsorbing mineral phases in the Fallbrook soil below the precipitation cut-off at pH = 5.5.

XAS Analysis

One of the objectives of this study was to use XAS analysis to gain insight into the structure of the Pb adsorption complex in soil. The ultimate goal is to develop accurate quantitative model descriptions of surface reactions using information derived from spectroscopy. The normalized X-ray absorption edge regions of several Pb-treated mineral/soil systems are shown in Figure 8. The first derivative of these spectra (Figure 9) was analyzed and the maximum in the first derivative, corresponding to the X-ray absorption edge energy, was 13056.5 eV for all samples. The increased scatter (decreased signal:noise ratio) in the

396

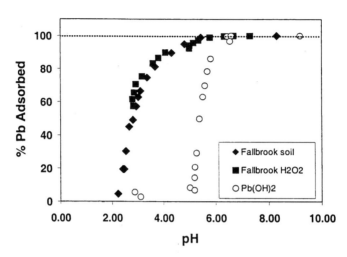

Figure 5. Equilibrium Pb adsorption envelopes on Fallbrook soil, with amd without 3% H₂O₂ treatment. Adsorption conditions: 2 g soil in 20 mL of 2.4 mM Pb solution. (I = 0.0125 M). Precipitation of Pb(OH)₂ from 2.4 mM Pb solution.

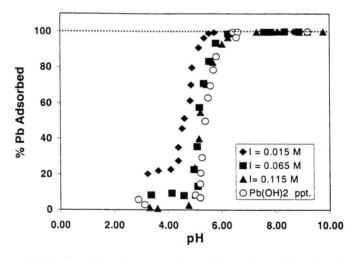

Figure 6. Pb adsorption edges on montmorillonite (SWy-1) at three ionic strengths (as NaCl) and precipitation of Pb(OH)₂ from 2.4 mM Pb solution. Adsorption conditions: 50 mg clay in 20 mL of 2.4 mM Pb solution

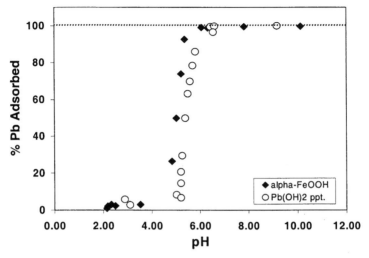

Figure 7. Pb adsorption edge on goethite (α-FeOOH) and precipitation of Pb(OH)₂. Adsorption conditions: 50 mg oxide in 20 mL of 2.4 mM Pb solution.

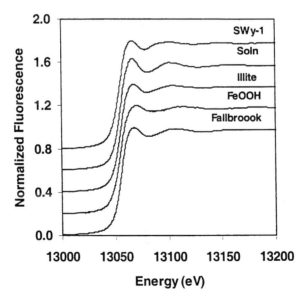

Figure 8. Pb X-ray absorption L III edges of Fallbrook soil and synthetic soil minerals.

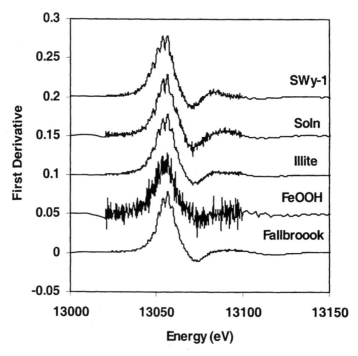

Figure 9. First derivative curves of Pb X-ray absorption L_{III} edges of Fallbrook soil and synthetic soil minerals.

Pb-treated α-FeOOH sample was due to a lower Pb surface coverage compared to the other samples. The Pb-treated illite and montmorillonite samples had similar X-ray absorption edge features to the Fallbrook soil.

The EXAFS data in Figures 10 and 11 correspond to the Pb-Fallbrook soil system at pH 4.5 and a solution of 5.0 mM Pb. The Pb solution data are included as a reference for comparison with the soil data. The Pb solution data shown in Figure 10 were fit with the k^3 weighted $\chi(k)$ EXAFS function ($\chi(k) \times k^3$) comprised of a shell of 6 O atoms at a distance of 2.50 Å. For the Pb-treated Fallbrook soil, two shells of atoms were included in the fit and included a shell of 6 O atoms at a distance of 2.43 Å and a second shell of either Al or Si atoms at a distance of 3.04 Å. The Pb-O interatomic distances are in agreement with previous studies (3, 19). The theoretical fit to the Fallbrook data (dashed line, Figure 10) represents a composite, 2-component sine wave with a strong contribution from the Pb-O shell of atoms plus a smaller, but distinct, contribution from a Pb-Si or Pb-Al second shell of atoms further away from the Pb atom.

Figure 11 shows the Fourier transform of the EXAFS function data for the Pb solution and Pb-treated soil. Though the interatomic distance between Pb and the second shell of atoms was determined to a reasonable degree of certainty, the element identity of the second shell atoms could not be determined at the resolution of this EXAFS spectrum. We were able to satisfactorily fit the EXAFS data using C, Al, or Si as potential backscattering atoms. Verifying the contributions of these light elements (C, Al, and Si) is difficult to resolve by EXAFS, however, without carefully constraining model systems for comparison. Other EXAFS studies of Pb adsorption on natural materials have encountered this problem (2, 19), and a reasonable best guess must be used in the final fitting for heterogeneous materials such as soil. Figure 11 illustrates the theoretical fit using 2 Si atoms as the second shell backscatterer. Soil solid phases such as iron oxide minerals (6, 20), aluminum oxides (4), Mn oxides, and humate materials (2) are potential environmental sorbents of Pb. However, we concluded that neither synthetic amorphous Al oxide or α-FeOOH had affinity for Pb below the pH 5.5 precipitation cutoff. Moreover, because there was no discernable difference between adsorption of Pb on untreated and 3% H_2O_2-treated Fallbrook soil, organic matter was at most a minor contributor to adsorption in this primarily inorganic mineral soil. The Pb-soil EXAFS data is most likely an average of contributions from several functional groups including layer silicates and the reactive functional groups on metal oxides.

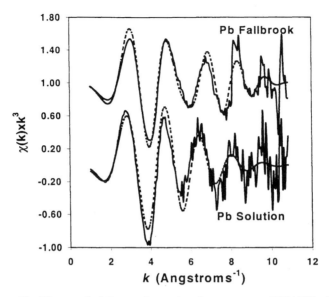

Figure 10. Pb extended X-ray absorption fine structure (EXAFS) plots of Fallbrook soil and Pb solution. Solid lines are experimental data and dotted lines are theoretical fits.

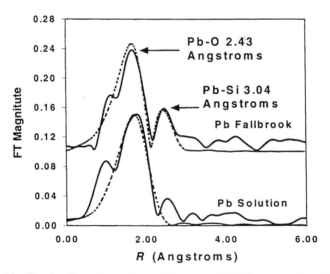

Figure 11. Fourier Transform plots of Pb-treated Fallbrook soil and Pb solution (not phase-corrected). Solid lines are experimental data and dotted lines are theoretical fits.

Surface Complexation Modeling

The Pb adsorption on soil data was described using the constant capacitance surface complexation model and FITEQL software. The constant capacitance model uses the inner-sphere adsorption mechanism as a basis for development of a quantitative equilibrium description of an interfacial reaction. An inner-sphere complex was postulated based on the results of EXAFS data which gave evidence for an atom (probably Al, Fe, or Si) at a distance (3.04 Å) indicative of a covalent linkage with the adsorbing surface.

In this study, two separate assumptions were used to describe the Pb-surface reaction based, in part, on the results of EXAFS analysis. Use of the one K(int) fit (Figure 12) assumed a single surface hydroxyl group (SOH) reacted with the Pb^{2+} ion to form an inner-sphere surface complex (SO-Pb-OH^0) (Eqn. 3). The 2 K(int) fit (Figure 12) used two reactions resulting in a composite fit implying the formation of the inner-sphere monodentate SO-Pb-OH^0 complex plus an additional bidentate (SO)$_2$-Pb^0 complex (Eqns. 3 and 4). The 2 K(int) model fit provided a more accurate description of the Pb adsorption data over the pH range 2.5-8.0 than the 1 K(int) fit. This approach has proven successful in previous efforts to apply surface complexation modeling to Pb adsorption (6). Though this may be, in part, due to increased flexibility in the model due to an additional adjustable parameter, the assumption is valid considering the heterogeneous nature of sorption sites in whole soils.

Conclusions

This work confirmed that the adsorption of Pb on a predominantly inorganic, mineral soil is strongly pH dependent and moderately dependent on ionic strength. The mechanisms of Pb removal from solution in the Fallbrook soil included adsorption on primarily inorganic soil mineral surfaces and probably precipitation of $Pb(OH)_2$ at pH > 5.5. The precipitation of insoluble Pb phases such as $Pb(OH)_2$ and $PbCO_3$ above pH 5.5 are important environmental limitations on Pb solubility and should be considered as an important Pb removal mechanism. It is likely that formation of $Pb(OH)_2$ surface precipitation is enhanced by the presence of mineral surfaces. EXAFS analysis of Pb-treated Fallbrook soil allowed resolution of two atomic shells. These included a Pb-O shell of 6 O atoms at 2.43 Å and a second shell of atoms at a distance of 3.04 Å. This is direct evidence that inner sphere adsorption of Pb is operative in this soil. The adsorption of Pb on soil ultimately involves mixtures of surface species including both inorganic and organic complexes. Modeling the Pb removal mechanism using the constant capacitance surface complexation model was accomplished at pH below 5.5 because complexation on reactive soil surfaces

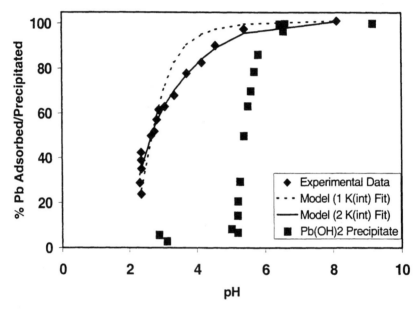

Figure 12. Pb adsorption edge on Fallbrook soil showing the fit results of the constant capacitance model using two model assumptions.

was presumed to be dominant in this region. The most successful model description used a combination of monondentate plus bidentate Pb surface complexes.

References

1. U.S. Environmental Protection Agency. *Soil Screening Guidance: User's Guidance.* USEPA 540/R-96/018. USEPA: Washington, D.C., 1996.
2. Xia, K.; Bleam, W.; Helmke, P. A. *Geochim. Cosmochim. Acta* **1997**, *61*, 2211-2221.
3. Ryan, J. A.; Zhang, P.; Hesterberg, D.; Chou, J.; Sayers, D.E. *Environ. Sci. Technol.* **2001**, *35*, 3798-3803.
4. Bargar, J. R.; Towle, S. N.; Brown, G. E., Jr.; Parks G. A. *Geochim. Cosmochim. Acta* **1996**, *60*, 3541-3547.
5. Cabaniss, S. *Environ. Sci. Technol.* **1992**, *26*, 1133-1138.
6. Villalobos, M.; Trotz, M. A.; Leckie, J. O. *Environ. Sci. Technol.* **2001**, *35*, 3849-3856.
7. Lebourg, A.; Sterckeman, T.; Ciesielski, H.; Proix, N. *J. Environ. Qual.* **1998**, *27*, 584-590.
8. Ma, L. Q.; Choate, A. L.; Rao, G. N. *J. Environ. Qual.* **1997**, *26*, 801-807.
9. Papini, M. P.; Kahie, Y. D.; Troia, B.; Majone, M. *Environ. Sci. Technol.* **1999**, *33*, 4457-4464.
10. Van Benschote, J. E.; Young, W. H.; Matsumoto, M. R.; Reed, B. E. *J. Environ. Qual.* **1998**, *27*, 24-30.
11. McBride, M. B. *Envrironmental Chemistry of Soils.* Oxford University Press: New York, NY, 1994; pp 125-135.
12. Gee, G. W.; Bauder, J. W. In *Methods of Soil Analysis Part 1*, 2nd ed.; Klute, A., Ed.; Agronomy Monograph 9; American Society of Agronomy: Madison, WI., 1986; pp 383-411.
13. McLaughlin, J. R.; Ryden, J. C.; Syers, J. K. *J. Soil Sci.* **1981**, *32*, 365-377.
14. Kyle, J. H.; Posner, A. M.; Quirk, J. P. *J. Soil Sci.* **1975**, *26*, 32-43.
15. George, G. N.; Pickering, I. J. *EXAFSPAK.* Stanford Synchrotron Radiation Laboratory: Stanford, CA, 1993.
16. Herbelin, A. L.; Westall, J. C. *FITEQL Version 3.1.* Rep. 94-01, Oregon State University: Corvallis, OR, 1994.
17. Manning, B. A.; Suarez, D. L. *Soil Sci. Soc. Am. J.* **2000**, *64*, 128-137.
18. Hayes, K. F.; Papelis, C.; Leckie, J. *J. Colloid. Interface Sci.* **1998**, *125*, 717-726.
19. Schneegurt, M. A.; Jain, J. C.; Menicucci, J. A., Jr.; Brown, S. A.; Kemner, K. M.; Garofalo, D. F.; Quallick, M. R.; Neal, C. R.; Kulpa, C.F., Jr. *Environ. Sci. Technol.* **2001**, *35*, 3786-3791.
20. Scheinost, A. C.; Abend, S.; Pandya, K.; Sparks, D.L. *Environ. Sci. Technol.* **2001**, *35*, 1090-1096.

Chapter 26

Heavy Metals in the Coastal Water of Hong Kong

D. C. Wang[1], X. D. Li[1,*], C. X. Wang[1,2], O. W. H. Wai[1], and Y. S. Li[1]

[1]Department of Civil and Structural Engineering, The Hong Kong Polytechnic University, Hung Hom, Kowloon, Hong Kong
[2]State Key Laboratory of Aquatic Environmental Chemistry, Research Center for Eco-Environmental Sciences, CAS, Beijing 100085, China

A chelation method was used to separate labile metal fractions in marine water. A chelating resin (Metalfix® Chelamine®) with immobilized tetraethylenepentamine was used to chelate the divalent metals of Cu, Pb, Zn, Cd, Co and Ni in the water samples. The extracted heavy metals were eluted with 2M HNO_3, deionized water, 0.12M $NH_3 \cdot H_2O$ and deionized water sequentially. The recovery rates of the method for the spiked bivalent metal ions were 96-103% for Zn, 85-91% for Pb, 79-91% for Cd and 93-102% for Ni respectively. The chelation method was used to study the labile heavy metal concentrations in the coastal water of Hong Kong. The metal concentrations in suspended solids, the total dissolved metal concentrations and the labile metal concentrations of the coastal water samples in 10 different locations were analyzed. The results show that labile metals constituted up to 20% of the total dissolved metals in the coastal water of Hong Kong. Among the 10 sampling sites investigated, the sites at Tsuen Wan, Kwun Tong, Tsim Bei Tsui and Luk Keng were more polluted than other sites with higher metal concentrations in suspended solids and dissolved forms. The results reflect the potential contamination sources in the vicinities of these sampling sites.

Hong Kong is situated on the coast of south China, and consists of Hong Kong Island, Kowloon Peninsula, the New Territories, and 235 surrounding islands. With a population of approximately 6.8 million and a limited area of only 1,067 km^2, Hong Kong is one of the most densely populated areas in the world. The discharge of agricultural and urban runoff, industrial effluents and sewage severely affects the quality of the coastal water of Hong Kong (1,2). Amongst the various contaminants being discharged into the coastal waster of Hong Kong, heavy metals pose threats to organisms due to their potential bioaccumulation (3,4).

Metals can exist in various chemical forms in water. It is often assumed that dissolved heavy metals are readily available to phytoplankton and organisms in water. The determination of total dissolved metals and their chemical speciation may provide useful information on metal bioavailability and toxicity in an aquatic environment. Furthermore, many surveys have shown unequivocally that metal toxicity in aquatic systems is controlled by metal labile fractions, and not the total dissolved concentration (5,6,7). The importance of free metal ions and other labile species is evident in the bioavailability of heavy metals to phytoplankton and organisms in water (8,9).

Blackmore (2) reviewed a number of publications about the metal concentrations in water, sediment and seafood in Hong Kong. The results of these publications show that metal pollution is very serious in the coastal areas of Hong Kong, as shown by the elevated metal concentrations in seawater, sediments and biomonitors. There have been many studies on heavy metals in the coastal water of Hong Kong (1,10-14). All have focused on the total concentrations of heavy metals in the water. Metals can be in different speciation in the water, as free ions (M^{2+}), inorganic and/or organic complexes. It is generally known that the bioavailability of heavy metals is directly related to labile (free ion) metal concentrations and/or activities (6,15,16). So far, little attention has been paid to the labile metal concentrations of heavy metals in the coastal water of Hong Kong. Knowledge of heavy metal forms and speciation is very important for evaluating the ecological effects of metals on biosystems. Different metal species will have different mobility, chemical behaviours and toxicities (17). To determine metal speciation in water, suitable methods are required to measure both the total concentration and the concentration of the different chemical species, particularly the labile concentration. Free metal ion concentrations can be determined by selective electrodes or calculation using chemical equilibrium models such as the Windermere Humic Aqueous Model (WHAM) (18). The selective electrode method is a quick and simple way for determination of "true" free metal ions in water samples. However, the technique is limited by its detection limit and interference from salinity when it is applied for marine water sample measurement. The WHAM method is a useful tool to predict the potential availability and toxicity of metals in aquatic environments, but the model needs a large number of data such as major metal ions, major anions, trace metals, DOM, pH, temperature and stability constants of inorganic and organic complexes before any quantitatively proportion of metal speciation can be estimated (*18*). Therefore, these methods may not be suitable for fast determination of labile metal concentrations in the marine water samples of a large coastal area.

The objective of the present study was to use a method that is capable of separating divalent labile metals in marine water samples. The proposed method was applied to investigate the speciation of heavy metals in the coastal water of Hong Kong. The total dissolved metal concentrations and the metal concentrations in the suspended solids in the coastal water of Hong Kong were also studied.

Materials and Methods

Ten sampling stations along the coast of Hong Kong were selected in this study during November 1999. The sampling sites reflected different land use areas and various hydraulic conditions, including Victoria Harbor, industrial areas and country parks (Figure 1). About 4 L of water were collected at each sampling site and filtered through a 0.45 μm membrane. Both the suspended solids on the filter membrane and the filtered water were used for further analysis.

Metalfix® Chelamine® (Sigma Chem. Co. USA) resin was pre-conditioned by washing with deionised water (DIW), 2M HNO_3, DIW and 2M ammonia solution and was rinsed with DIW again before filling the column. The column was purchased from Sigma Chem. Co., and consisted of a 1.5 cm I.D. × 10 cm length of borosilicate glass tubing constructed with fixed polypropylene ends caps and polyethylene bed supports for the resin in the column. A slurry of Metalfix Chelamine resin was aspirated into the column with a syringe, and rinsed with DIW at 4 ml/min until the pH of the effluent was same as the influent. The other end of the column was then sealed into the column body. The column was then connected with a pump. The marine water sample was pumped through the Metalfix Chelamine column at 1ml/min. After the chelating process, the column was eluted three times with 5 ml 2 M HNO_3, 10 ml DIW, 10 ml 0.12 M $NH_3 \cdot H_2O$ and 10 mL DIW at the rate of 1 ml/min.

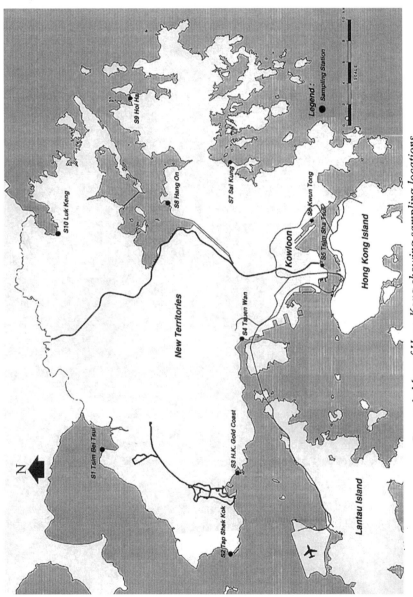

Figure 1. Map of Hong Kong showing sampling locations.

To test the recovery rate of the divalent free metals, the solutions containing 10, 20, 30 and 40 µg/L Cd^{2+}, Ni^{2+}, Pb^{2+} and Zn^{2+} were used as testing water samples. The chelation and rinse procedures were the same as for the marine water samples. The divalent metal ions of the eluted solutions were then analyzed by ICP-MS (Perkin Elmer Elan 6000).

The total dissolved metal concentrations were determined directly after filtration and acidification (pH 2.0) by ICP-MS. For the metal concentrations in suspended particles, the solids on the pre-weighed filter membranes were dried at 105°C for 24 hours. They were weighed, and then digested using concentrated nitric acid and perchloric acid at 150°C for 18 hours (19). The metal concentrations were determined using inductively coupled plasma atomic emission spectrometry (ICP-AES, Perkin-Elmer 3300 DV).

All glassware used in the study was pre-cleaned by soaking in 10% HNO_3 for at least 24 hours, and was rinsed with DIW. All reagents used were of analytical grade or better. Duplicate and blank samples were used throughout the whole analytical process, including the labile fraction (divalent ions) separation process. The precision was below 15% for most metals analysed in this study.

Results and Discussion

Chelation Separation Method for Metal Labile Fractions in Coastal Water Samples

In order to assess the chelation efficiency of the resins for free divalent ions in marine water, an experiment was conducted on spiked water samples with known free divalent metal concentrations. The recovery rates of the divalent ions from these solutions are presented in Table I. The recovery rates ranged from 79% to 102% for Cd^{2+}, Ni^{2+}, Pb^{2+} and Zn^{2+}. The results show that the method was very effective at chelating divalent metal ions in marine waters. It must however be noted that although the recovery rates for the above spiked metals are very good, the metals obtained by this method may include free metal ions, metals bound in dissociable complexes, and a small component of metals in colloids (20,21). Therefore, the fraction measured by this technique is an operationally defined labile fraction extracted by Chelamine with majority metals in free ion forms.

Table I. Recovery rates of spiked bivalent metal ions to 500mL marine
water using Metalfix Chelamine column separation

Ions	Metal Concentration Spiked ($\mu g/L$)				
	0	10	20	30	40
Pb^{2+}	0	8.5(85%)	17.8(89%)	27.2(91%)	33.8(86%)
Zn^{2+}	4.3	14.7(103%)	23.3(96%)	34.8(101%)	43.3(98%)
Cd^{2+}	0	9.1(91%)	17.8(89%)	23.8(79%)	33.1(83%)
Ni^{2+}	0.6	10.2(96%)	19.2(93%)	31.1(102%)	38.1(94%)

Total Dissolved Metal Concentrations

The total dissolved metal concentrations in the coastal water of Hong Kong
at the 10 sampling stations are presented in Figure 2. From the results, there
were no significant spatial variations of the total dissolved metal concentrations
of Pb, Cd, Cu, Ni and Co among the sampling locations. Zn concentration was
elevated at Tsuen Wan (S4) and Luk Keng (S10). The total dissolved metal
concentrations at all the sampling sites were generally lower than the Chronic
Freshwater Criteria for dissolved metals of Cu, Cd, Ni, Zn and Pb recommended
by the US EPA (22).

Table II shows the dissolved metal concentrations of Cd and Pb in the
coastal waters of Hong Kong, as obtained in the present study and from previous
reported results. The total dissolved Cd concentration has generally shown an
increasing trend during the last two decades. The total dissolved Pb
concentration was quite high in the early 1970s in Hong Kong's coastal water.
Recent result (1) and the present study show that Pb concentration in marine
water has decreased in the last 20 years. This may be partially due to the
introduction of unleaded petroleum in Hong Kong in the late 1980s.

Metal Labile Fraction Concentrations

The concentrations of labile metal fractions in the coastal water at the 10
sampling locations in Hong Kong are plotted in Figure 3. There was no
significant variation of the concentrations of Cu and Co among the different
locations. The concentrations of Pb varied greatly among the 10 sampling sites.
Tsuen Wan (S4) had the highest Pb concentration and higher Cd and Ni
concentrations, reflecting the possible influence of industrial activities in the
area.

410

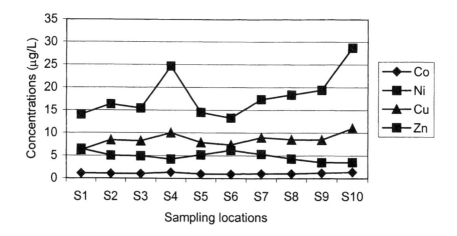

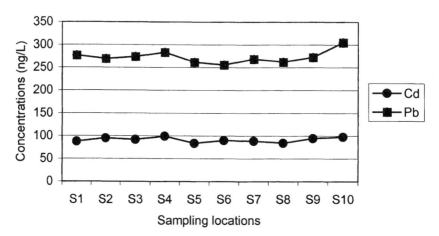

Figure 2. The total dissolved metal concentrations in the coastal water of Hong Kong

Tsim Bei Tsui (S1) had relatively high Ni and Pb concentrations in the marine water. This site is located in Deep Bay, along the border with Shenzhen (see Figure 1). The input of the Shenzhen River discharge and the local sewage outfall could be the cause of the elevated concentrations of labile metal fractions in the water sample. The higher concentrations of labile fractions at the two sampling sites may indicate high bioavailability of metals and greater environmental impacts on the marine ecosystem in the region.

Table II. Comparison of the total dissolved Cd and Pb concentrations in coastal water obtained in this study with previous monitoring results (μg/L)

Sources	Cd	Pb
Present study	0.091 (0.084-0.099)	0.273 (0.256-0.305)
Chan et al., 1974	0.05 (0.01-0.1)	0.66 (0.04-1.41)
Chan, 1995	0.054	0.238 (Victoria Harbor)
	0.35	0.135 (Kwun Tong)

The range and mean concentrations of total dissolved metals and the labile fractions in the marine water at the 10 sampling locations are given in Table III. The percentage of labile metal concentration to total dissolved metal concentration was different for different metals. For Cd, Cu and Pb, the labile fractions accounted for 7-10% of the total dissolved metal concentrations. Labile Ni accounted for around 20% of the total dissolved Ni in the marine water, particularly at the Tsim Bei Tsui site (S1). The lowest ratio was for labile Co, which was only 0.61% of the total dissolved Co. There was also significant correlation between the labile Pb and Cd concentration and the total dissolved Pb and Cd concentrations in these marine water samples (Figure 4). No significant relationship between total dissolved form and labile fraction was found for other metals.

Heavy Metal Concentrations in Suspended Solids

It has been demonstrated that the metal concentrations contained in either suspended solids or sediments are also important in assessing the extent of heavy metal contamination in water systems (*23,24,25*). The suspended solid concentrations in marine water samples are listed in Table IV. According to the results, Luk Keng (S10) and Tsim Bei Tsui (S1) had the highest suspended

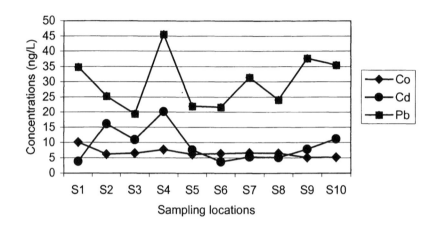

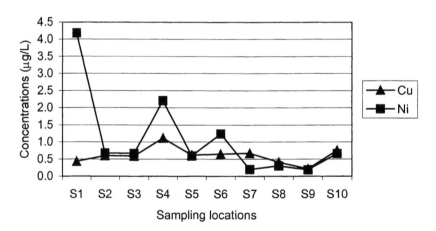

Figure 3. The labile metal concentrations in the coastal water of Hong Kong

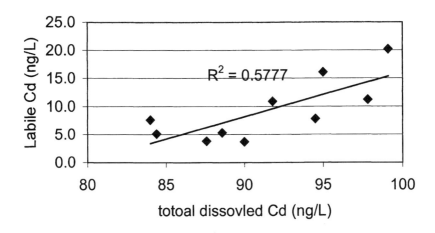

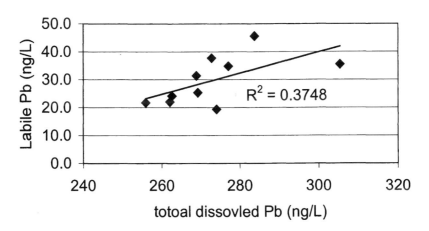

Figure 4. Relationships between labile metals and total dissolved metals in the coastal water of Hong Kong

Table III. The range and mean concentrations of total dissolved metals and labile metals

Metal Dissolved	Total Dissolved (µg/L)	Labile Metals (µg/L)	Labile/Total (%)
Cd	0.091 (0.0839-0.0991)	0.0092 (0.0037-0.0202)	9.84
Pb	0.237 (0.256-0.305)	0.0278 (0.0022-0.0456)	10.2
Co	1.11 (0.929-1.39)	0.0066 (0.0051-0.0101)	0.61
Ni	4.88 (3.53-6.48)	1.091 (0.188-4.19)	20.9
Cu	8.51 (6.23-11.0)	0.604 (0.216-1.11)	7.05

solids in water. These two sites are located in relatively enclosed inlets and are close to mangrove-rich wetland systems.

The metal concentrations in suspended solids of the marine water samples are shown in Figure 5. Cu, Pb and Zn had similar concentration trends among the 10 sampling sites. The concentrations of Cd and Co in suspended solids did not show any marked variations among different sites. Tsim Bei Tsui (S1) had the highest Cu, Zn and Pb concentrations in suspended solids. A relatively higher Pb concentration in suspended solids was also found in Luk Keng (S10). The Pb and Zn concentrations in suspended solids were also high in Kwong Tung (S6). These three locations all had a higher content of suspended solids in water.

Metal concentrations in suspended solids can be closely related to metal contents in surface sediments since metals can be transferred to sediment through the deposition of suspended particles, and metals in sediments can be remobilised back into the water column through re-suspension (26,27). To compare the metal concentrations in suspended solids with the sediments from previous studies, the metal concentrations (mg/kg) in suspended solids were calculated according to the suspended solid concentration in the water samples. The results are presented in Table V. In comparison with the metal concentrations in the sediments of the Pearl River estuary (regional background values) (28), the concentrations of Cu and Zn in suspended solids were higher in Hong Kong, but the concentrations of Pb and Co were lower. This may reflect the local contamination patterns in Hong Kong. The proposed interim sediment quality values (ISQV) for Hong Kong are also listed in Table V for comparison (29). The concentrations of Cd and Cu in suspended solids were higher than the ISQV-low (background values), reflecting the same pattern of elevated values of these two elements in the marine water of Hong Kong.

Table IV. Suspended solid contents at different sampling sites

Sampling Sites	Suspended Solids (mg/L)
Tsim Bei Tsui	15.8
Tap Shek Kok	11.6
HK Gold Coast	6.20
Tsuen Wan	4.00
Tsim Sha Tsui	7.00
Kwun Tong	9.70
Sai Kung	3.80
Hang On	3.60
Hoi Ha	4.90
Luk Keng	17.1

Table V. Comparison of metal concentrations in suspended solids of coastal water in Hong Kong, surface sediments of the Pearl River Estuary, and the proposed ISQV (mg/kg)

Metal	Suspended Solid	Surface Sediment (0-5cm)*	ISQV (low-high)**
Cd	2.31 (0.97-6.64)	--	1.5-9.6
Co	3.97 (1.74-6.46)	16.3 (5.75-18.8)	--
Pb	27.7 (1.42-55.0)	59.5 (27.0-72.0)	75-218
Cu	159 (5.92-540)	40.9 (7.05-63.0)	65-270
Zn	142 (13.9-447)	115 (32.3-210)	200-410

Note: * from Li et al., 2000 (*28*)

    ** interim sediment quality values (ISQV) proposed for HK (*29*)

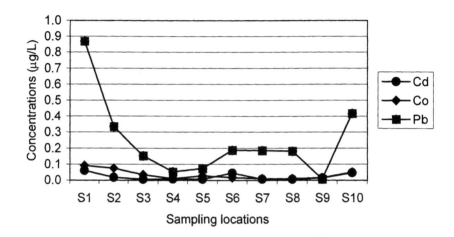

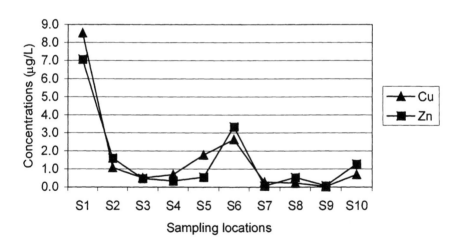

Figure 5. The metal concentrations in suspended solids in the coastal water of Hong Kong

The total metal concentrations of Cu, Pb, Zn and Cd in suspended solids in Victoria Harbor (S5) and Kwun Tong (S6) are presented in Table VI. In comparison with a previous study conducted in 1983 (2), the Pb concentration in suspended solids decreased at both locations. This result showed a similar trend to the total dissolved Pb in marine water (see Table II), reflecting the consequence of the use of unleaded petroleum since the late 1980s. However, the Zn concentration in suspended solids showed an increasing trend at both locations. The Cu concentration in suspended solids increased significantly at Victoria Harbor, but decreased at Kwun Tong following the relocation of the print circuit board industry in the 1980s.

Table VI. Metal concentrations in suspended solids in the coastal water of Hong Kong (ng/L)

Locations		Cd	Cu	Pb	Zn
Present study:	Victoria Harbor	7.0	1780	74	257
	Kwun Tong	45.0	2630	187	3350
Chan, 1983:	Victoria Harbor	8.2	79	132	134
	Kwun Tong	12.4	12800	260	1050

The total concentrations of heavy metals in marine water should include total dissolved metal concentrations and metal concentrations in suspended solids. Therefore, the two industrial sites (Kwun Tong (S6) and Tsuen Wan (S4)) and two shallow water sites (Tsim Bei Tsui (S1) in Deep Bay and Luk Keng (S10) in the Starling Inlet) were more polluted than other sites with higher metal concentrations in suspended solids and dissolved form. The high metal concentrations in the marine water samples reflect the influence of nearby industrial discharge and sewage outfalls.

Conclusions

The separation method for labile metal fractions in marine water samples based on the chelation mechanism has potential applications in metal speciation studies. The method was used to study heavy metal concentrations and speciation in Hong Kong's coastal water. The total dissolved metal concentrations, labile metal concentrations and metal concentrations in suspended solids in marine water varied from location to location, reflecting the variety of pollution sources from industrial and sewage discharges at individual sampling sites. The labile metal fraction constituted up to 20% of total dissolved metals in the coastal water of Hong Kong, particularly for Ni. Pb in suspended solids formed an important part of the total Pb in marine water. In general, Pb concentration in the marine water of Hong Kong has decreased in the last 20 years, while Zn, Cu and Cd have shown an increasing trend in some locations.

Acknowledgements

The project was funded by the Hong Polytechnic University (Project No. G-YY16) and the Research Grants Council of Hong Kong SAR Government (Project No. PolyU 5057/99E). We would like to thank Dr. Yong Cai and the reviewers for constructive comments and suggestions, which greatly improved the quality of the manuscript.

Literature Cited

1. Chan, K M. *Marine Pollution Bulletin* **1995**, *31*, 277-280.
2. Blackmore, G. *The Science of the Total Environment* **1998**, *214*, 21-48
3. Cheung, Y.H.; Wong, M.H. *Environmental Management* **1992**, *16*, 743-751.
4. Cheung, Y.H.; Wong, M.H. *Environmental Management* 1992, *16*, 753-761.
5. Verweij, W.; Glazewski, R.; De Hann H. *Chemical Speciation and Bioavailability*, **1992**, *4*, 43-51.
6. Campbell, P. In *Metal Speciation and Bioavailability in Aquatic Systems*, John Wiley and Sons, Chichester, UK, 1995. 45-102.
7. Allen, H.E.; Batley, G.E. *Human and Ecological Risk Assessment,***1997**, *3*, 397-413.
8. Bell, P.F.; Chaney, R.L.; Angle, J. S. *Plant and Soil* **1991**, *130*, 51-62.
9. Suave, S.; Cook, N.; Hendershot, W.H.; McBride, M.B. *Environmental Pollution* **1996**, *94*, 154-157.
10. Chan, J.P.; Cheung, M.T.; Li, F.P. *Marine Pollution Bulletin* **1974**, *5*, 171-174.
11. Wong, M.H.; Chan, K.C.; Choy, C.K. *Environmental Research* **1978**, *15*, 342-356.
12. Wong, M. H.; Chan, K.Y.; Kwan, S.H.; Mo, C.F. *Marine Pollution Bulletin* **1979**, *10*, 56-59.
13. Wong, M.H.; Kwok, T.T.; Ho, K.C. *Hydrobiology Bulletin* **1982**, *16*, 223-230.

14. Chan, H. M. In The Marine Flora and Fauna of Hong Kong and Southern China III, Proceedings of the Fourth International Marine Biological Workshop. Hong Kong University Press, 1992. pp. 621-628.
15. Sparks, D.L. Soil Science Society of America Journal **1983**, *48*, 514-518.
16. McBride, M. B. *Environmental Chemistry of Soils*. Oxford University Press, 1994.
17. Bryan, G. W. *Phil Trans R Soc Lond* **1979**, *B286*, 483-505.
18. Tipping, E. *Computers and Geosciences* **1994**, *20*, 973-1023.
19. Li, X.D.; Shen, Z.G.; Wai, W.H.O.; Li, Y.S. *Marine Pollution Bulletin* **2001**, *42*, 215-223.
20. Davison, W.; Zhang, H. *Nature* **1994**, *367*, 546-548.
21. Zhang, H.; Davison, W. *Anal. Chem.* **2000**, *72*, 4447-4457.
22. USEPA (United States Environmental Protection Agency) *Water Quality Standards-Revision of Metals Criteria*. Fed. Reg. 60, 22229-22237, 1995.
23. Hiromitsu, S.; Yotaka, K.; Kazuo, S. *Water Research* **1986**, *20*, 559-567.
24. Fuller, C.C.; Davis, J.A.; Cain, D.J.; Lamothe, P.J.; Fries, T.L.; Fernandez, G.; Vargas, J.A.; Murillo, M.M. *Water Research* **1990**, *24*, 805-812.
25. Juan, G.M.; Sosa, A.M. *Water, Air, and Soil Pollution* **1994**, *77*,141-150.
26. Hart, B.T. *Hydrobiologia* **1982**, *91*, 299-313.
27. Mance, G. *Pollution Threat of Heavy Metals in Aquatic Environments*. Elservier, New York, 1987.
28. Li, X.D.; Wai, W.H.O.; Li, Y.S.; Coles, B.J.; Ramsey, M.H.; Thornton, I. *Applied Geochemistry* **2000**, *15*, 567-581.
29. Chapman, P. M.; Allard, P. J.; Vigers, G. A. *Marine Pollution Bulletin* **1999**, *38*,161-169.

Indexes

Author Index

Subject Index